AI-Powered Developments in Medical Robotics

Medical Robotics and Computer Assisted Surgery

AI-Powered Developments in Medical Robotics

Data-Driven Techniques for Enhanced Surgical Efficiency

Edited by

THOMAS HEINRICH MUSIOLIK
Woxsen University, Hyderabad, Telangana, India; University of Europe for Applied Sciences, Potsdam, Germany

HEMACHANDRAN KANNAN
Department of Artificial Intelligence and Business Analytics, Woxsen University, Hyderabad, Telangana, India

RAUL VILLAMARIN RODRIGUEZ
Woxsen University, Hyderabad, Telangana, India

MANJEET REGE
Department of Software Engineering and Data Science, University of St. Thomas, St. Paul, MN, United States

Series Editors

TAMÁS HAIDEGGER

SUBHENDU KUMAR PANI

ELSEVIER

ACADEMIC PRESS
An imprint of Elsevier

Academic Press is an imprint of Elsevier
125 London Wall, London EC2Y 5AS, United Kingdom
50 Hampshire Street, 5th Floor, Cambridge, MA 02139, United States

ISBN: 978-0-443-27783-2

For Information on all Academic Press publications
visit our website at https://www.elsevier.com/books-and-journals

Publisher: Mara Conner
Acquisitions Editor: Sonnini R. Yura
Editorial Project Manager: Siddharth Khattri
Production Project Manager: Fizza Fathima
Cover Designer: Arumugam Kothandan

Typeset by MPS Limited, Chennai, India

Contents

Part V Challenges and future directions

List of contributors

Al Ahmed
Harbourside Institute of Australia, Pyrmont, NSW, Australia

Kande Archana
Department of CSE-Data Science, CMR Engineering College, Hyderabad, Telangana, India

Anto Lourdu Xavier Raj Arockia Selvarathinam
Department of Data Science and Analytics, Grand Valley State University, Allendale, MI, United States

Sowmya Bachigari
Woxsen University, Hyderabad, Telangana, India

Ojashwini Dubey
Woxsen University, Hyderabad, Telangana, India

Palak Dwivedi
AI Research Centre, Woxsen University, Hyderabad, Telangana, India

Zakia Hammouch
Department of Mathematics and Applied Mathematics, School of Science and Technology, Sefako Makgatho Health Sciences University, Ga-Rankuwa, South Africa; Department of Sciences, Ecole Normale Supérieure Moulay Ismail University of Meknès, Meknès, Morocco

Lydia D. Isaac
Department of IT, Sona College of Technology, Salem, Tamil Nadu, India

Pushpa J.
Jain (deemed-to-be) University, School of CS & IT, Bengaluru, Karnataka, India

Madhav Janumula
AI Research Centre, Hyderabad, Telangana, India

Shyam Krishan Joshi
Analytics Department, AI Research Centre, Woxsen University, Hyderabad, Telangana, India

Nargis Khaliq
Department of Computer Science and Applications, Sharda University, Greater Noida, Uttar Pradesh, India

Ravi Teja Kothuru
Shiley-Marcos School of Engineering, Applied Artificial Intelligence, University of San Diego, San Diego, CA, United States

Dharani Kumar
School of Business, Woxsen University, Hyderabad, Telangana, India

Sanju Mannumadam Venugopal
Institute of Electrical and Electronics Engineers, Downey, CA, United States

Geetha Manoharan
School of Business, SR University, Warangal, Telangana, India

Laxmaiah Mettu
Department of CSE-Data Science, CMR Engineering College, Hyderabad, Telangana, India

Seshadri Mohan
Systems Engineering Department, University of Arkansas at Little Rock, Little Rock, AR, United States

Arun N.
Vidyavardhaka College of Engineering, Mysore, Karnataka, India

Daria A. Nefedova
Heller School for Social Policy and Management, Brandeis University, Waltham, MA, United States

M. Nivedha
School of Creative & Cultural Business, Robert Gordon University, Aberdeen, Scotland, United Kingdom

M.R. Padmapriya
Department of CSE (Data Science), Dayananda Sagar College of Engineering, Bangalore, Karnataka, India

Anil Pise
Sr. Data Scientist, X-idian, Johannesburg, Gauteng, South Africa; Technical Partner, Cumulus Solutions, Johannesburg, South Africa; Senior Consultant, Deloitte SA, Johannesburg, South Africa

V. Kamakshi Prasad
Department of CSE, Jawaharlal Nehru Technological University, Hyderabad, Telangana, India

A. Srinivasula Reddy
Department of EEE, CMR Engineering College, Hyderabad, Telangana, India

Manjeet Rege
Department of Software Engineering and Data Science, University of St. Thomas, St. Paul, MN, United States

Deepa S.
Department of Computer Science, Christ University, Bangalore, Karnataka, India

David Samuel S. Azariya
Department of Information Technology, Sona College of Technology, Salem, Tamil Nadu, India

Neetha S.S.
Department of Computer Science, Christ University, Bangalore, Karnataka, India

Swati Sah
School of Computing Science and Engineering, Sharda University, Greater Noida, Uttar Pradesh, India

Sachin Sharma
Amity School of Engineering and Technology, Amity University Punjab, Mohali, Punjab, India

Vishal Kumar Sharma
AI Research Centre, Woxsen University, Hyderabad, Telangana, India

Donald S. Shepard
Heller School for Social Policy and Management, Brandeis University, Waltham, MA, United States

Nisha Soms
Department of Computer Science and Engineering, KPR Institute of Engineering and Technology, Coimbatore, Tamil Nadu, India

K. Suneetha
Jain (deemed-to-be) University, School of CS & IT, Bengaluru, Karnataka, India

Mohanraj V.
Department of IT, Sona College of Technology, Salem, Tamil Nadu, India

Shiji Valsan
Global Health, Boston Children's Hospital, Interventional Radiology, Boston, MA, United States

PART I

Fundamentals and innovations in medical robotics

CHAPTER 1

Fundamentals of medical robotics

Ojashwini Dubey[1], Shyam Krishan Joshi[2], Madhav Janumula[3], and Zakia Hammouch[4,5]
[1]Woxsen University, Hyderabad, Telangana, India,
[2]Analytics Department, AI Research Centre, Woxsen University, Hyderabad, Telangana, India,
[3]AI Research Centre, Hyderabad, Telangana, India,
[4]Department of Mathematics and Applied Mathematics, School of Science and Technology, Sefako Makgatho Health Sciences University, Ga-Rankuwa, South Africa,
[5]Department of Sciences, Ecole Normale Supérieure Moulay Ismail University of Meknès, Meknès, Morocco

1.1 Introduction

By allowing unprecedented degrees of accuracy, control, and minimally invasive operations. The integration of robotics into the medical sector marks a transformative change in healthcare. Combining mechanical engineering, control systems, computer science, and biomedical technology, medical robotics creates intelligent devices that either aid or automate certain tasks in healthcare environments. By increasing surgical accuracy, lowering recuperation periods, and hence improving patient outcomes, this fusion is transforming established procedures (Dario et al., 2004; Okamura, 2009). Medical robotic systems fall generally into four categories: hospital service robots, telepresence and diagnostic robots, surgical robots, and rehabilitation robots. Among these, surgical robots have been the most well-known and have caused a paradigm change in surgical operation technique. Far beyond the natural limits of the human hand, these devices give surgeons exceptional dexterity, motion scaling, tremor filtration, and high-density 3D visualization (Landman et al., 2002). Developed by Intuitive Surgical and approved by the United States, the da Vinci Surgical System is among the most important turning points in the industry. Food and Drug Administration (FDA) 2000: Supporting urologic, gynecologic, cardiothoracic, and general laparoscopic surgeries with amazing success, it is still the most often used robotic surgery system worldwide (Satava, 2002). Its parts consist of a vision system, a patient-side cart with robotic arms, and a surgeon console, all combined to execute minimally invasive procedures with improved ergonomics and control. Other benchmark surgical robots include

- For orthopedic surgeries, including complete hip and knee arthroplasty, the robotic arm-assisted MAKOplasty System (Stryker).
- Designed for neurosurgical and orthopedic procedures, particularly spine and brain surgery, ROSA (Zimmer Biomet).

AI-Powered Developments in Medical Robotics
DOI: https://doi.org/10.1016/B978-0-443-27783-2.00007-7

- Designed as a frameless stereotactic radiosurgery device with submillimeter accuracy, CyberKnife (Accuray) treats tumors.
- Next-generation modular robotic system Versius (CMR Surgical) provides general and laparoscopic surgery (Versius Surgical Robotic System, n.d.) adaptability.
- Applied in robotic hair restoration using AI-driven follicular unit extraction, ARTAS iX (Restoration Robotics).
- Emerging platform Hugo (Medtronic) with open architecture meant to rival da Vinci, allows digital integration and cloud analytics by means of digital integration.

Medical robotics is also very important in microsurgery, where robots like the Symani Surgical System by MMI can do supermicrosurgical operations, including lymphatic vessel anastomosis and nerve repair. The Preceyes Surgical System has been successful in sensitive retinal operations for ophthalmic uses. Medical robotic systems' use has been enhanced even further by the fast development of enabling technologies such as machine learning (ML), haptics, real-time imaging, and tactile sensors. Coupled with the rising need for precision, safety, and minimally invasive technologies, the medical robots industry is expected to develop greatly and reach over USD 25 billion by 2028 (Marketsandmarkets, 2028). Focusing on surgical robotics and benchmark systems, this study offers a thorough analysis of the principles, key technologies, and classifications of medical robots, thusaddressing their technical, clinical, and regulatory settings.

1.2 Historical background and evolution

Robotic surgery was originally meant to be safer, more accurate, and less invasive, especially in complex and dangerous surgeries. By the early 80s, all the major revolutions in mechanical engineering, control systems, computers, and medical imaging together changed the scenario of robotic assistance in medicine completely. Medical robots, which were once limited to research, have now become a must-have tool in clinics all over the world, actively used in theaters, rehab, and diagnostics. Robotics has seen impressive changes in capabilities over the last 40 years due to the development of sensors, real-time imaging, and computer power, along with artificial intelligence (AI). These changes have resulted in the creation of safer surgical interventions, improved therapeutic outcomes, less surgeon fatigue, and more consistent procedures. The development of medical robots can be divided into stages that show their technological and clinical maturation.

1.2.1 Early developments (1980–90s)

In 1985 robotic surgery saw its very first clinical implementation in the history of mankind; to be more precise, the PUMA 560 robot was utilized in computed tomography (CT)-guided stereotactic brain biopsy operations. This momentous step forward was the demonstration of the computer-controlled precision in neurosurgery, where the impossibly small, submillimeter accuracy is a must. This event, which is referred to as the "birth of robot-assisted surgery" by many, in fact, validated robotics as a trustworthy partner for performing delicate medical interventions (Davies, 1992). Nonetheless, the Imperial College London team used the FORCE robot in the late 1980s to carry out prostate surgery. For transurethral resection of the prostate, the FORCE robot showed how robots could take over the tedious and geometrically confined surgical movements with utmost precision, thus reducing human error as well as variability. This was the transition from industrial robot adaptation to medical robot design based on specific tasks. At the same time, the ROBODOC system, which was a joint venture of International Business Machines (IBM) and Integrated Surgical Systems, became the first active robot that was approved for orthopedic surgery. ROBODOC was mainly used in total hip replacement, and it did so by allowing very precise milling of the bone depending on the CT data taken before the operation, thus guaranteeing proper fitting of the implant and stability of the prosthesis for a long duration. The 1990s witnessed the introduction of robotic assistance in the form of minimally invasive endoscopic procedures. Computer Motion Inc. had launched the AESOP (Automated Endoscopic System for Optimal Positioning), which was the first FDA-approved voice-controlled robotic endoscope holder. By providing stable and accurate camera control, AESOP improved visualization, reduced the requirement for surgical assistants, and increased surgeons' comfort, thus paving the way for more sophisticated robotic surgical systems (Davies et al., 1991).

1.2.2 Rising surgical robotics (2000s)

The beginning of the new century was a momentous period in the field of medical robotics, as the FDA-approved the da Vinci Surgical System for application in 2000. This surgical system, created by Intuitive Surgical, the da Vinci, was the first commercially successful, multiarm robotic surgical platform, which was specifically meant for minimally invasive surgeries. The system was a hybrid of the master–slave model, with the surgeons controlling surgical precision through a console that operated the articulated robotic arms fitted with surgical instruments and a high-definition, stereoscopic 3D vision system. The major innovations, such as scaling of motion, eliminating tremor, and increasing dexterity, made a way for the surgeons to perform complex operations, which earlier would not have been possible with conventional laparoscopy (Cinquin, 1998). Consequently, da Vinci quickly became the choice for urology, gynecology, cardiothoracic, and general surgeries, primarily in prostatectomy and hysterectomy operations. At that time, Computer Motion's ZEUS was another surgical robot competing with da Vinci, focusing on telesurgery and minimally invasive surgeries. Moreover, Accuray's CyberKnife system transformed the field of stereotactic radiosurgery by providing robotic arms with real-time imaging to monitor and adjust the position of the tumor during the procedure, thus making it possible to treat without surgery those tumors previously considered inoperable . The year 2003 saw the fusion of the two companies cooperating in a merger, Intuitive Surgical and Computer Motion, which allowed them to share their technical capabilities and build a cutting-edge robotic system for the operation that combined the good features of both da Vinci and ZEUS, therefore reaping the benefits of fast innovation and a stronghold in the medical robotics market.

1.2.3 Advanced integration and expansion—2010s—present

From the mid-2000s onward, medical robotics enjoyed a phase of advanced integration in which they combined many technologies such as real-time imaging, surgical navigation, force feedback (haptics), and ML algorithms. The technologies developed alongside decision-making during the operation, situational awareness, and procedural automation (Marescaux et al., 2001) to a greater extent. The use of guided-missile robots became common practice in interventional radiology, cardiology, and neurosurgery as they provided an excellent targeting accuracy of less than a millimeter for catheter placement, biopsies, and ablation therapies. The increase in the use of robotics in the field of rehabilitation medicine, at the same time, has brought a significant change in neuro-rehabilitation. The use of Lokomat and Armeo (Hocoma) devices has allowed for intensive, repeatable, and patient-specific movement therapy, thus contributing significantly to the recovery of patients' motor functions after stroke and spinal cord injuries (Lum, 2009). The areas where the use of robots would still increase were telemedicine and remote patient care. For instance, the iRobot and InTouch Health-developed RP-VITA system facilitated real-time monitoring of the patients by clinicians, conducting consultations, and even assisting in emergencies—the functions that turned out to be very helpful during the pandemics and in isolated or underprivileged areas. A more recent trend that has emerged is the compactness, modularity, and digital interconnectivity of robotic systems. The CMR Surgical-developed Versius, Medtronic's Hugo RAS System, and MMI's Symani all focus on the priority of portability, scalability, and cost efficiency. In extremely delicate procedures like ophthalmology and microsurgery, robots like Preceyes have made it possible to perform retinal surgeries that demand even micron-level precision, which goes beyond the limits of human physiology.

1.2.4 Major medical robotics milestones

At present, medical robots are not only playing the role of assistants but also being incorporated more and more as main parts in the surgical workflows, rehabilitation programs, and hospital operations. The combination of AI, advanced sensors, high-speed 5G, cloud computing, and miniaturization is leading to increased autonomy, adaptability, and personalization of medical robots. Medical robotics is going to be the game changer in the diagnosis, treatment planning, surgical execution, and patient outcomes, making the medical industry very different in the future, and hence, global healthcare delivery is going to be shaped accordingly (Table 1.1).

1.3 Medical robotics: core technologies

A medical robotic system's clinical efficacy, safety, and utility depend on the way several technical fields are integrated. These comprise AI, mechanical design and kinematics, control theory, real-time imaging and navigation, and haptic feedback systems, and every one of these elements helps the robot to be precise, flexible, and operate in delicate and dynamic healthcare surroundings.

1.3.1 Mechanical planning and kinematics

Any medical robotic system's fundamental layer is the mechanical architecture. Serial manipulators, parallel systems, and continuum robots are just a few of the structural configurations that help to classify these robots. Usually found in surgical systems such as the da Vinci platform (Riener, 2005), serial-link arms offer a broad workspace and versatility. While stiffer and more compact, parallel mechanisms are used in high-precision applications such as orthopedic or neurosurgical robots. Offering access to areas including the gastrointestinal tract, blood arteries, or the spinal cord, continuum and snake-like robots allow movements inside highly limited anatomical routes (Guthart and Salisbury, 2000). Advanced medical robots are made with bioinspired degrees of

Table 1.1 Key milestones.

Year	Milestone
1985	First robot-assisted brain biopsy using PUMA 560 (Davies, 1992)
1988	Development of PROBOT for prostate surgery (Satava, 2002)
1992	Use of ROBODOC in orthopedic procedures (Davies et al., 1991)
1994	FDA approval of AESOP robotic camera holder (Cinquin, 1998)
2000	FDA approval of da Vinci Surgical System
2001	First telesurgery (Lindbergh Operation) across the Atlantic
2003	Merger of Intuitive Surgical and Computer Motion (Lum, 2009)
2005–15	Expansion of AI-driven, image-guided, and rehabilitation robots (Riener, 2005)
2015–Present	Commercial introduction of VERSIUS, Hugo, Symani, and Preceyes ()

freedom that mimic human arm or hand motions, therefore enabling minimally intrusive operation. Important for safe clinical use include load-bearing capability, sterilizability, material biocompatibility, and miniaturization, which are all design problems.

1.3.2 Systems of control

Achieving the intended accuracy, stability, and responsiveness in dynamic surgical and rehabilitative settings depends on control systems in medical robotics. Usually using closed-loop control topologies dependent on continuous sensory feedback, these systems modify actuator commands in real-time (Shoham, 2007). Integrating dynamic models of the robot and surroundings, model-based control systems maximize movement planning and energy economy. Applications involving human–robot interaction also depend on force and impedance control systems since too strong contact forces must be avoided, as in microsurgical operations or robotic rehabilitation. Robotic-assisted orthopedic operations can use hybrid position-force control in which exact trajectory tracking must coexist with compliant behavior to accommodate for anatomical heterogeneity (Siciliano and Khatib, 2016).

Often needing adaptive or intelligent control strategies, robust control design must also solve the problems of latency, system nonlinearities, and interpatient variability.

1.3.3 Imaging and navigating

Commonly integrating a range of preoperative and intraoperative imaging modalities, including magnetic resonance imaging (MRI), CT, and ultrasound, medical robotic systems give visual guidance and improve spatial awareness (De Luca and Lucibello, 2000). By use of 3D reconstruction and surgical planning, these pictures enable the surgeon to visually replicate and practice operations. Advanced technologies register the patient's anatomy in real-time using fiducial markers or electromagnetic tracking on navigational platforms. These systems dynamically change for intraoperative motion or deformation and match the surgical equipment with the preoperative strategy. Moreover, high-precision interventions in neurosurgery and interventional radiology have been made possible by robotic systems linked with real-time imaging, including MRI-compatible robots, thereby reducing radiation exposure (Ortega, 2013). Fluoroscopic guiding is not necessary for these operations.

1.3.4 Haptics and tactile sensation

One of the main restrictions of teleoperation, the loss of tactile sensation, is addressed by including haptic feedback into medical robotic systems. Haptic interfaces provide kinesthetic feedback (force, torque, and stiffness) and tactile feedback (texture and pressure distribution) to the operator (Riener et al., 2005), hence recreating the feeling of touch. This skill is especially important in delicate surgical operations when the surgeon's actions, such as differentiating between soft and hard tissues or spotting puncture events, may be informed by the tactile reaction from tissues. Furthermore, in surgical simulators, haptics is essential since it improves realism and helps new users to acquire skills (Stramigioli, 2001).

Aiming at duplicating human-like touch sensation with high spatial resolution and low latency, emerging technologies in this field include soft robotic sensors, piezoelectric tactile arrays, and electroactive polymers.

1.3.5 Machine learning and artificial intelligence

Medical robotic systems are progressively including AI and ML to support autonomous functionality, improve decision-making, and increase sensory capacity. AI applications find use in:

- For surgical scene awareness, instrument tracking, tissue categorization (Behal et al., 2001), and computer vision.
- Path planning and motion prediction let the robot forecast and react to dynamic environments (Peters and Cleary, 2008).
- Anomaly detection, such as early recognition of system failures or surgical complications (Yaniv, 2011).
- Outcome prediction and decision assistance, whereby patient data analysis by ML models supports intraoperative decision-making (Kuchenbecker and Niemeyer, 2008).

Particularly, convolutional neural networks and recurrent neural networks, deep learning frameworks, have shown amazing accuracy in identifying surgical phases, anatomic recognition, and guiding autonomous movements. These technologies improve repeatability and lower cognitive strain on doctors, therefore opening the path for semiautonomous and completely autonomous robotic surgery (Kaufmann et al., 2012) (Fig. 1.1).

1.4 Classifications of medical robots

Medical robots can be systematically divided and classified according to their main function, clinical application, environment of operation, and level of autonomy. This kind of classification helps to understand the robotic systems that are targeting particular issues in the healthcare provision. Broadly speaking, there are five types of medical robots, which are surgical robots, rehabilitation robots, diagnostic robots, telepresence robots, and hospital service robots. In every single category, a combination of

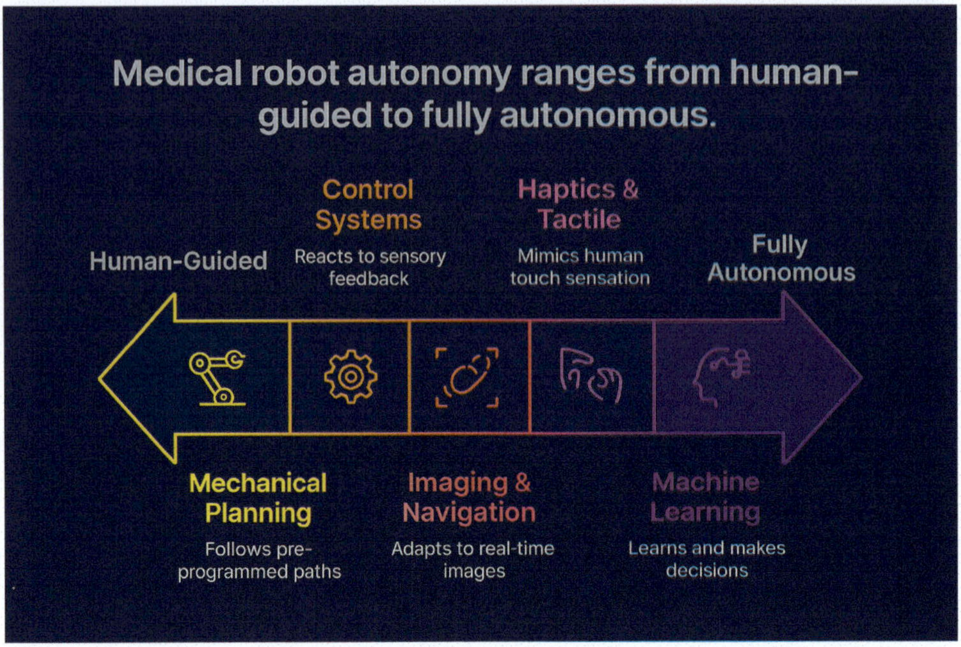

Figure 1.1 Medical robot autonomy ranges from human-guided to fully autonomous.

mechanical design, sensing, control algorithms, and software intelligence is deployed to enhance clinical efficiency, surgical precision, patient safety, and care quality.

1.4.1 Robot surgery

Surgical robots are crafted with the intention of providing surgeons with the help needed to perform difficult operations at a level of precision, dexterity, and invasiveness that is almost nonexistent. Usually, these systems come with robotic arms, high-definition endoscopic imaging systems, and consoles controlled by the surgeon that work in master–slave fashion. Most surgeons prefer not to use these robots with the help of an automatic system, but instead they use the robot to translate their movements into scaled, tremor-free motions inside the patient's body. Through the da Vinci Surgical System, widely known in urology, gynecology, cardiology, and general surgery, the da Vinci Surgical System remains the most popular platform. Its articulated instruments provide the surgeon with multiple degrees of freedom to precisely manipulate the anatomy even in very small spaces. ROSA is one of the other specialized platforms that mainly focuses on neurosurgery and uses an image-guided technique for implanting. Another is MAKO, which is for orthopedic surgeries where the processes of imaging, alignment, and preparation are all done through the robot's interfaces . The pros of robotic surgery include, among others, the patients experiencing much less blood loss during surgery, smaller cuts, less pain after surgery, shorter hospital stays, and quicker recoveries. Notably, these factors are to be considered in surgeries that need high-precision or in areas where the anatomy is very hard to reach and even smaller than normal.

1.4.2 Rehabilitation robots

The invention of rehabilitation robots was for the benefit of patients suffering from neurological and muscular impairments, who were thereby helped to regain their motor functions through repetitive, controlled, and goal-oriented movements. They are the main tools used for recovery of motor function after a stroke, spinal cord injury rehabilitation, orthopedics, and gait training. The robotic rehabilitation systems are divided into two main categories, namely, end effector-based and exoskeleton-based devices. The upper-limb exoskeleton, ArmeoSpring, is one of the systems that by providing adjustable gravity compensation and real-time feedback, supports arm and hand rehabilitation. Similarly, the robotic gait orthosis, Lokomat, allows for intensive and reproducible walking therapy by the synchronization of lower-limb motion with body-weight support. Rehabilitation robots can be programmed to work in different modes, such as passive, active-assistive, or fully active, depending on the patient's capability and therapy goals. These technological advancements will not only enhance the patients' participation in the process but will also standardize the protocols of rehabilitation through objective performance metrics, adaptive training intensity, and continuous therapy delivery.

1.4.3 Robot diagnostics

Improvement of the accuracy, consistency, and efficiency of the diagnostic procedures is the main purpose of the diagnostic robots. They can be utilized for endoscopic navigation, biopsy, imaging, and laboratory automation. The systems make the operators less dependent and improve the precision of the diagnosis, especially in the case of complex as well as minimally invasive procedures. Robotic endoscopy platforms like EndoBot systems allow for semiautonomous or even fully autonomous navigation of the internal organs during colonoscopy and bronchoscopy, with the capturing of high-resolution images for disease to be detected very early. In interventional diagnostics, robotic biopsy systems bring in the real-time imaging data, like CT or MRI, to guide the placement of the needle with very high spatial accuracy, thus minimizing tissue damage and procedural errors (Riener et al., 2005). Diagnostic robots are indispensable in the fields of oncology, gastroenterology, and interventional radiology, where an early and accurate diagnosis has a very significant effect on the prognosis of the treatment.

1.4.4 Virtual reality robots

Telepresence robots are the ones that facilitate remote interaction between healthcare professionals and patients, thus supporting telemedicine, virtual consultations, and remote patient monitoring. These robots, which are equipped with cameras, microphones, displays, sensors, and mobility systems, give the doctors the opportunity to move around in the healthcare facility and communicate with the patients live. Noteworthy instances of this technology include the RP-VITA robot, which was jointly developed by InTouch Health and iRobot and has become the most popular robot in the Intensive Care Units (ICUs) for remote physician rounds and emergency consultations. Besides, VGo's telepresence robots allow continuous audiovisual communication between the patient and the doctor, thus helping the patient be remotely evaluated In the case of telepresence robots, it can be said that their impact has been the greatest in rural healthcare delivery, disaster response, and pandemic situations, where they serve as a bridge between the patients and the doctors and thus overcome the geographical barriers, lower the risk of infection, and provide the specialists to the underserved population.

1.4.5 Robots for hospital services

The nonclinical and yet vital support tasks of hospital service robots have increased the operational efficiency and at the same time reduced the burden of healthcare workers. The logistics, sanitation, and material handling are the main areas where these robots work in a healthcare environment. Among the various robots that are used in hospitals, Timed Up and Go test (TUG), an autonomous mobile robot, has the ability to smoothly and noiselessly hobble through the hospital corridors, bringing along medications, laboratory samples, linens, food trays, and medical supplies. In the war against bacteria and viruses, robots such as Xenex and UVD Robots, which use UV-C light, are cleaning operating rooms, patient wards, and isolating units without human intervention, thus drastically cutting down the rate of pathogen transmission (Table 1.2).

1.5 Challenges and limitations

Although the contribution of medical robotic systems has made great progress and shows great clinical effectiveness, various obstacles and even limits still prevent their general acceptance and best use. These constraints have technical, financial, legal, and ethical aspects as well. Ensuring more general incorporation of robotic technologies into mainstream medicine depends on addressing these problems.

Table 1.2 Classifications of medical robots.

Category	Description	Examples/Applications
Surgical robots	Assist surgeons in performing complex and minimally invasive procedures with precision and control.	*da Vinci Surgical System*, *ROSA*, *MAKO*—used in urology, neurosurgery, and orthopedics.
Rehabilitation robots	Aid recovery for patients with motor impairments and support repetitive movement training.	*ArmeoSpring*, *Lokomat*—used in stroke and spinal cord rehabilitation.
Diagnostic robots	Automate imaging and biopsy processes to improve accuracy and reduce human error.	*EndoBot* and robotic biopsy systems—used in oncology and gastroenterology.
Telepresence robots	Enable remote medical consultations and virtual patient monitoring.	*RP-VITA*, *VGo*—used in ICUs and rural healthcare.
Hospital service robots	Handle logistics and sanitation tasks to improve hospital efficiency.	*TUG*, *Xenex*, *UVD Robots*—deliver supplies and disinfect rooms.

1.5.1 High cost and financial barriers

High initial acquisition cost is the foremost obstacle in the path of medical robotic systems' acceptance. Premium robotic surgery platforms like the da Vinci Surgical System are approximately USD 2 million in price, and this is besides the periodic expenses for annual maintenance contracts, disposable instruments, software upgrades, and specialized operating room infrastructure. The accumulated recurring costs significantly hike the total cost of ownership. Overseas hospitals, especially small- and medium-sized ones in resource-limited or developing areas, usually face prohibitive financial demands that restrict patients from accessing robotic-assisted procedures. In addition, the economic justification of robotic systems is still a topic of dispute, as the cost–benefit analyses yield mixed results when comparing the increased procedural costs against potential long-term benefits such as reduced complication rates and shorter hospital stays.

1.5.2 Technical complexity and training challenges

Medical robotic systems consist of highly sophisticated hardware and software that can be difficult for surgeons, operating room personnel, biomedical engineers, and technicians to learn. The special training programs, which include simulator-based education, certification courses, and supervised clinical practice, are necessary to guarantee the effective and safe use of these systems. The intricate installation, calibration, and troubleshooting processes make the reliance on manufacturer support greater and thus may result in increased downtime and operational delays. Throughout the early adoption phase, user-related errors may happen, which, if not properly handled through structured training and competency assessment, can compromise procedural efficiency as well as patient safety.

1.5.3 Regulatory and compliance challenges

Provisions to ensure patient safety, reliability, and therapeutic efficiency have a tremendous influence on the regulatory path to medical robots. In the European Union, attaining the CE mark and meeting the Medical Device Regulation requirements is compulsory, whereas in the United States, FDA clearance is needed prior to being used in clinical settings. Medical robotics that have the multidisciplinary character of combining mechanical engineering, electronics, computer science, AI, and clinical aspects create a challenge in getting through the regulations. Robots may get continuous updates in the software, have their algorithms refined, and have their hardware upgraded, which can count as a reason for the repeat of validation and regulatory review, thus making the duration and costs of the project longer.

1.5.4 Ethical, legal, and privacy issues

The increasing deployment of assistive and intelligent robotic technologies brings about intricate legal and moral issues, especially in relation to patient autonomy, informed consent, accountability, and data privacy. The use of AI algorithms in clinical decision-making may result in the lack of transparency or explanation, which, in turn, may make it difficult for the patients to comprehend the treatment recommendations and, thus throw the principles of informed consent into question. In case a technical glitch occurs, the software fails, or the patient suffers an adverse outcome, it is going to be very difficult to determine who is legally responsible for the situation among the surgeon, the healthcare institution, the device manufacturer, or the software developer. This is particularly true in robotic telesurgery. What is more, handling and storing of sensitive patient data through networked robotic systems must be in accordance with data protection laws such as HIPAA and GDPR in order to ensure patient confidentiality.

1.5.5 Connectivity and latency challenges

The application of telesurgery and telepresence robotics in terms of safety and effectiveness relies heavily on the promptness of the technology used and robust communication networks. In high-risk surgical procedures, delay caused by data transmission may lead to the surgeon's commands and the robot's movements not being perfectly synchronized, thus risking the required precision. Nevertheless, the new technologies like 5G, edge computing, and sophisticated compression algorithms do give off good signs as far as the reduction of communication lags and even the real-time transmission of high-resolution medical imaging and video are concerned. Yet, the problem of providing consistent, secure, and reliable connectivity across the different regions remains a major challenge. Medical robots, which are capable of changing the world, still have a long way to go before they gain universal acceptance because of the challenges that they are facing. Cost, regulation, complexity, ethical issues, and connectivity issues are the hardest and most rewarding problems all in one when it comes to the acceptance of robotics in medicine. To overcome these barriers and promote equal access to robot-assisted medical care, intensive research, new policy development, and tech innovations will be required.

1.6 Conclusion

Medical robotics stands at the intersection of these domains and is increasingly considered one of the most transformational and exhilarating frontiers of modern medicine. The emergent application of robotic systems into clinical practices changes the ways we approach diagnosis of disease, surgical invasion, rehabilitation, and a patient's continuum of care, which is not only about technological change but also a paradigm shift. Medical robotics is growing in relevance as healthcare systems worldwide grapple with increasing costs, aging populations, the scarcity of medical personnel, and the expectations for precision and efficiency. It can enhance outcomes, access, and quality of care for patients. This effort was designed to disassemble and unify the core components that make up the medical robotics ecosystem. From mechanical structures and actuation systems to control architectures and intelligent planning and reasoning algorithms, the discourse in the previous sections provides a comprehensive outline of the core technologies that safeguard safe and effective robotic processes within healthcare contexts. The multidisciplinary nature of medical robotics, involving mechanical engineering, biomedical imaging, computer vision, AI, and clinical medicine, requires a systems-level understanding, which this effort attempts to facilitate. This review provides an important contribution with its systematic analysis of the technological underpinnings of robotic-assisted medicine. There is a balancing act between the mechanical design of robots for surgical procedures, rehabilitation, and diagnostics, but also the human factors, ergonomics, sterilization requirements, and biocompatibility involved. The control systems in medical robots need to be flexible, robust and have a measured feedback response while finding a balance between real-time control and safety. Imaging integration is another major component, which enables precision of navigation and decision-making, whether the task is searching for a tumor or a minimally invasive procedure. Furthermore, the introduction of AI and ML continues to grow and will enable robots to learn from the data available, adapt to different patient anatomy, and even predict challenges before they arise. This review is of potential interest to several audiences, including researchers, academic institutions, practitioners, the government, as well as manufacturers of medical devices and venture capitalists. For researchers, the review provides information about current developments, identifies technology gaps, and is helpful in encouraging cross-disciplinary collaboration. For practitioners, the review provides information about the scope and limits of robotic systems and how the technology may fit into the treatments that they provide for their patients. The review indicated that for startups and inventors, there will be opportunities for commercialization and development. Researchers, academic institutions, practitioners, but also legislators, developers of medical devices, and venture capitalists, should find this review relevant. The review illustrates current developments, identifies some technology gaps, and promotes multidisciplinary collaboration for academics. For practitioners, the review illustrates the capabilities and limitations of robotic systems, including how such technology might integrate into patient treatments. The review indicates that there are increasing opportunities for commercialization and development for the startup or inventor.

When we look to the future, it is clear that the area of medical robotics will experience many concurrent developments. One of the most significant trends is the ongoing convergence of robotics with advancements in AI and data science. Intelligent robotic systems will support their perceptions of the environment, be able to reason about clinical circumstances, and make intelligent, responsive actions to dynamic patient situations outside of preprogrammed instruction. Robotic surgery, as an example, is bringing together vision-based learning, haptic feedback, and real-time analytics to develop semiautonomous or assisted surgical platforms, and this is particularly relevant to this trend. Miniaturization and portability represent another, even more explosive area of growth. The development of devices, from capsule endoscopy robots to microrobots that will traverse the bloodstream, and the quest for less invasive and more targeted methods of treatment, are giving rise to a new generation of tiny, multifaceted devices. These devices are positing a promise of less traumatic, more accurate, and more available healthcare in resource-limited environments. Similarly, wearable robotic systems (e.g., prosthetic limbs and exoskeletons) are enabling individuals with physical disabilities to regain movement and independence. Connectivity is another transformational driver. The arrival of the Internet of Medical Things, 5G, and edge computing means that the circuitry can now be embedded in wider, cloud-connected ecosystems. Consequently, despite the constraints imposed by local environments, patients are able to have real-time monitoring of their medical condition, remote diagnosis, and possibly telesurgery, providing the opportunity to deliver excellent quality of care across spatial divides, closing the health equity gap, and democratizing access to specialty healthcare in remote and marginalized areas. Of course, whilst great potential also brings new responsibilities and challenges. Considerations of data privacy, algorithmic bias, human-machine relations, and accountability for machine failure pose questions to be solved if the consequences of the ethical implications of medical robotics are to be minimized. Regulatory regimes will need to adapt at the same pace and complexity as the technologies to ensure the safety of patients as well as the sustained innovation and viability of systems. Additionally, the large costs associated with developing, delivering, and servicing robotic systems raise questions about equity, access to technology developments, and sustainability, especially in low- and middle-income countries. Medical robotics will be successful in the next decades, mostly depending on human-centric design. Robotic systems have to be built to increase human capabilities, that is, to improve accuracy, lower tiredness, and enable informed decision-making, not to replace medical experts. How easily these systems are embraced in clinical practice will depend on good training, easy-to-use interfaces, and simple controls. Engineers, doctors, ethicists, and patients working together will be crucial to make sure robotic solutions solve practical problems and enhance patient experiences without sacrificing confidence or autonomy. Finally, this paper offers a thorough and forward-looking picture of medical robots, thus establishing it as a pillar of the upcoming generation of healthcare delivery. It establishes the groundwork for ongoing invention, research, and application by charting the present technological scene and projecting future trends. The ideas offered confirm the important part robotics plays in increasing clinical accuracy, customizing treatment, and finally raising patient outcomes. Stakeholders from academia, business, healthcare, and government must collaborate as we stand on the brink of this robotic revolution in medicine to fully utilize this technology. Medical robotics helps to fulfill a vision of a smarter, safer, and more inclusive healthcare system by means of consistent investment in research and development, careful policymaking, and a dedication to fair access.

References

Behal AK, Dawson DM, De Queiroz MS. Adaptive control of uncertain nonaffine systems: application to robotic systems. IEEE Transactions on Automatic Control. 2001;46:152–7.

Cinquin C. Computer-assisted knee surgery. Clinical Orthopaedics and Related Research 1998:67–76.

Dario, Carrozza MC, Benvenuto A. Medical robotics. IEEE Robotics & Automation Magazine 2004;11:44–56.

Davies B. A robotic system for cementless total hip replacement surgery. Clinical Orthopaedics and Related Research 1992;285:76–82.

Davies BL, Hibberd RD, Ng WS, Timoney AG, Wickham JEA. The development of a surgeon robot for prostatectomies. Proceedings of the Institution of Mechanical Engineers, Part H: Journal of Engineering in Medicine 1991;205(1):35–8. Available from https://doi.org/10.1243/PIME_PROC_1991_205_259_02.

De Luca, Lucibello. A general algorithm for dynamic feedback linearization of robots with elastic joints. IEEE Transactions on Robotics and Automation 2000;16:442–7.

Guthart GS, Salisbury KJ. Intuitive telesurgery system: overview and application. Proceedings - IEEE International Conference on Robotics and Automation 2000;1:618–21.

Kaufmann T, Vecchi MC, Ruther P, Paul O. A computationally efficient numerical model of the offset of CMOS-integrated vertical Hall devices. Sensors and Actuators A: Physical. 2012;178:1–9. Available from https://doi.org/10.1016/j.sna.2012.01.024.

Kuchenbecker K, Niemeyer G. Haptic feedback: enhanced reality for virtual environments. IEEE Transactions on Haptics. 2008;1:20–30.

Landman J, Figenshau RS, Bhayani S, Kibel A, Pettaras JG, Fleshman J, et al. Dual-organ ablative surgery using a hand-assisted laparoscopic technique. A report of four cases. Surgical Endoscopy 2002;16(1):215–6. Available from https://doi.org/10.1007/s004640080174.

Lum MA. Robotic surgery in urology: evolution and applications. The Urologic Clinics of North America 2009;36:493–506.

Marketsandmarkets, Medical Robotics Market - Global Forecast to 2028.

Okamura AM. Haptic feedback in robot-assisted minimally invasive surgery. Current Opinion in Urology 2009;19(1):102–7. Available from https://doi.org/10.1097/MOU.0b013e32831a478c.

Ortega R. Passivity-based control of Euler–Lagrange systems: mechanical, electrical and electromechanical applications. Springer; 2013. 10.1007/978-1-4471-3603-3.

Peters T, Cleary K. Image-guided interventions: technology and applications. Canada: Springer US; 2008. p. 1–557. Available from http://doi.org/10.1007/978-0-387-73858-1, http://www.springerlink.com/openurl.asp?genre=book&isbn=978-0-387–73856-7.

Riener R. Rehabilitation robots: Lokomat and Armeo. European Journal of Physical and Rehabilitation Medicine 2005;41:15–25.

Riener R, Lünenburger, Jezernik S. Rehabilitation robots for the treatment of sensorimotor deficits: a neurophysiological perspective. Journal of Rehabilitation Research and Development 2005;42:509–22.

Satava RM. Surgical robotics: the early chronicles. Surgical Laparoscopy, Endoscopy & Percutaneous Techniques 2002;12(1):6–16. Available from https://doi.org/10.1097/00129689-200202000-00002.

Shoham M. SpineAssist: a miniature robotic system for spinal surgery: design and development. The International Journal of Medical Robotics + Computer Assisted Surgery: MRCAS 2007;3:292–9. Available from https://doi.org/10.1002/rcs.174.

Siciliano B, Khatib O. Springer handbook of robotics. Italy: Springer International Publishing; 2016. p. 1–2227. Available from http://doi.org/10.1007/978-3-319–32552-1, http://doi.org/10.1007/978-3-319-32552-1.

Stramigioli S. Modeling and IPC control of interactive mechanical systems: a coordinate-free approach. Springer; 2001. 10.1007/978-1-4471-0395-0.

Versius Surgical Robotic System; n.d.

Yaniv ZR. A practical framework for MRI-guided robotic surgery. IEEE Transactions on Bio-medical Engineering 2011;58:834–41.

CHAPTER 2

Artificial intelligence and data-driven innovations in medicine

Anto Lourdu Xavier Raj Arockia Selvarathinam[1], and David Samuel S. Azariya[2]
[1]Department of Data Science and Analytics, Grand Valley State University, Allendale, MI, United States,
[2]Department of Information Technology, Sona College of Technology, Salem, Tamil Nadu, India

2.1 Introduction

2.1.1 Overview of artificial intelligence's impact on modern medicine

By including artificial intelligence (AI) in healthcare, we now have better ways to diagnose and treat patients and improve how care is organized. Basically, AI systems depend on algorithms that look through medical data to discover patterns that standard analysis might miss (Liu et al., 2022; Alowais et al., 2023).

Mathematically, AI models in healthcare are built upon supervised learning, where a function $f: X \rightarrow Y$ is trained using labeled datasets $\{(x_i, y_i)\}_{i=1}^{n}$,

where:

- x_i represents a medical input (e.g., image, biomarker vector),
- y_i is the corresponding diagnosis or classification (e.g., disease/no disease),
- f is the learned model that predicts patient outcomes.

This approach has powerful applications, like:

- Disease classification using MRI or CT scans,
- Predicting readmission rates from EHRs (Electronic Health Records),
- Risk stratification for chronic conditions.

2.1.2 Rise of data-driven healthcare solutions

As EHRs, wearables, and high-throughput biomedical technologies have increased, healthcare has become filled with data (Gedikci Ondogan et al., 2023; Sinha, 2024; Tandel et al., 2024). Being able to draw valuable information from massive and varied data has driven more decisions to be made with data.

These datasets are shown in Table 2.1 and are processed using models such as

- Logistic regression for binary outcomes:

$$P(y = 1 \mid x) = \frac{1}{1 + e^{-(\beta_0 + \beta_1 x_1 + \cdots + \beta_n x_n)}}$$

AI-Powered Developments in Medical Robotics
DOI: https://doi.org/10.1016/B978-0-443-27783-2.00008-9

Table 2.1 Types of healthcare data.

Data source	Format	Example use case
EHR	Structured/ Unstructured	Diagnosis history, medications, and lab reports
Medical imaging	Pixel data (DICOM)	Tumor detection in radiology
Genomic sequences	Text/Numeric vectors	Variant detection in cancer genomics
Wearable sensors	Time-series data	Heart rate, sleep, and oxygen saturation

EHR, electronic health records.

- Convolutional neural networks (CNNs) for image-based classification,
- Recurrent neural networks (RNNs) or transformers for sequential health data.

2.1.3 Importance of artificial intelligence in precision medicine, diagnostics, and treatment

Precision medicine is about matching treatment to how people differ in their genes, environments, and lifestyles. With AI, omics data is analyzed, and links are made between genetic markers and how responses to therapies change.

Use case example: Let x be a patient's genomic feature vector, and $T(x)$ be a treatment recommendation function. Then:

$$T(x) = \underset{t \in T}{\operatorname{argmin}} E[U(t, x)]$$

where:

- $\mathcal{T}$ is the set of possible treatments,
- $U(t, x)$ is the expected utility or effectiveness of treatment t for the patient profile x.

AI is also improving:

- Diagnostics by providing real-time image classification (e.g., diabetic retinopathy, skin lesions),
- Treatment planning via predictive analytics (e.g., anticipating chemotherapy toxicity).

The first section of this paper sets the scene for understanding how AI is implemented in healthcare. Today's medical ecosystem gathers enormous volumes of both structured and unstructured data (Cremin et al., 2022). Because of AI technologies, both machine learning (ML) and deep learning, medical professionals now have the help they need to make faster, more precise, and more individualized decisions (Khalifa and Albadawy, 2024; Haleem et al., 2023).

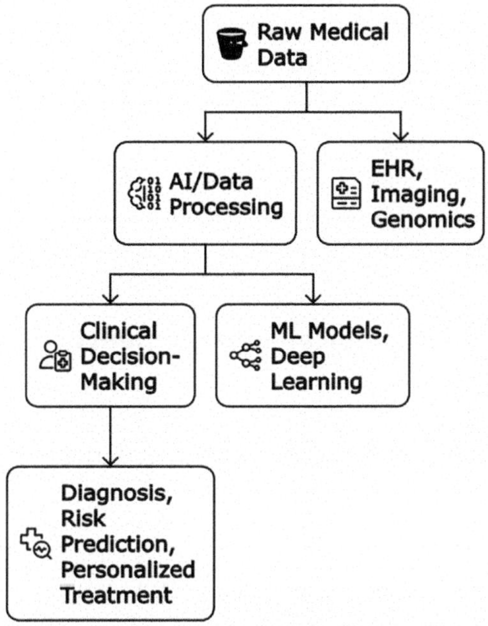

Figure 2.1 AI in the medical data pipeline. *AI*, artificial intelligence.

The basic math behind algorithms helps us understand how to identify patterns in data using regression for risk assessment and neural networks for help in diagnosing diseases from imaging. Since data is from a range of source genes, sensors, and other sources, AI systems should be designed so they can work well with all types of data, and the AI medical data pipeline shown in Fig. 2.1.

Today, health practitioners are moving from using the same treatments for all toward precision medicine that takes individual data into account. With AI, practices can now offer therapies that fit each patient's needs, causing fewer harmful reactions and leading to better patient outcomes. While technology shows great promise, scaling it up means dealing with issues including connecting different types of data, understanding algorithm decisions, and maintaining ethical rules. We look at these matters in the chapters that follow.

2.2 Foundations of artificial intelligence in medicine

The main technologies in AI are causing big changes in how medical data is handled, understood, and applied in healthcare (Gupta and Kumar, 2023). Here, you'll learn about the main AI procedures in healthcare, see the power of big data in helping these advancements, and understand how dealing with various, sometimes confidential health information remains a challenge (Khalid et al., 2023; Hong et al., 2018).

2.2.1 Key artificial intelligence technologies in medicine

Different medical domains, such as diagnostics, prognosis, treatment optimization, and operation management, are supported by several key technologies used in AI in medicine, as shown in Table 2.2.

The data from these technologies is commonly used with others in sophisticated AI platforms that offer full support for different types of input (e.g., testing + computerized images + vital signs).

2.2.2 Role of big data in medical advancements

AI in healthcare is greatly helped by access to big data. The medical sector collects thousands of pieces of big data.

The 5 Vs are a marker for big data: Volume, Variety, Velocity, Veracity, and Value.

Table 2.2 Key AI technologies and their applications in healthcare.

AI technology	Description	Typical applications in medicine
Machine learning (ML)	Algorithms that learn patterns from data to make predictions or decisions	Risk prediction, treatment recommendation, and patient triage
Deep learning (DL)	A subset of ML using multilayered neural networks for complex data patterns	Radiology (X-rays, MRIs), pathology, and ECG interpretation
Natural language processing (NLP)	Algorithms that process and analyze human language (text/speech)	Extraction from clinical notes, chatbot support, and summarization of EHRs
Computer vision (CV)	AI methods that interpret visual data like images or videos	Skin lesion classification, ophthalmology screening, and surgical robotics
Reinforcement learning (RL)	Learning optimal actions through reward-based feedback loops	Dynamic treatment planning, ICU ventilator settings, and robotic control
Generative AI (e.g., Generative Adversarial Networks (GANs), Variational Autoencoders (VAEs))	Models that learn to generate synthetic data based on real distributions	Data augmentation, synthetic medical images, and missing data imputation

EHR, electronic health records.

2.2.2.1 Characteristics of big data in healthcare

- Volume: Petabytes of data generated daily from EHRs, imaging devices, wearables, and omics technologies.
- Variety: Includes structured (lab results), semistructured (EHRs), and unstructured (free-text notes, images).
- Velocity: Real-time data streaming from continuous monitoring systems and sensors.
- Veracity: Ensuring reliability and accuracy amidst noisy, missing, or inconsistent data.
- Value: Extracting actionable insights for diagnostics, treatment planning, and population health.

Mathematically, big data allows AI models to approximate complex functions:

$$f(x) \approx y \text{ where } x \in \mathbb{R}^n \text{ is a high} - \text{dimensional patient vector}$$

Larger, diverse datasets help models generalize better, reducing the risk of overfitting and increasing clinical utility.

This analysis allows doctors to develop ongoing and customized care for people with chronic diseases, and the sample datasets are shown in Table 2.3.

2.2.3 Challenges in healthcare data

Despite its promise, using data in medicine presents several challenges that must be addressed to ensure reliable and ethical AI solutions.

2.2.3.1 Privacy and security

All medical information is private and is regulated by rules like the HIPAA (in the USA) and the GDPR (in the EU). A breach can result in fines from the government and the

Table 2.3 Sample dataset volume (Illustrative example).

Data type	Average size per patient	Source	Application
EHR (clinical notes)	~1–2 MB	Hospitals, Clinics	Risk stratification, NLP extraction
Radiology images (DICOM)	~100 MB per scan	Imaging centers	Cancer detection, segmentation
Genomic sequence data	~200 GB	Biobanks, Labs	Mutation analysis, drug targeting
Wearable data streams	~10–50 KB/min	Smartwatches, Biosensors	Anomaly detection, vitals monitoring

EHR, electronic health records; *NLP*, natural language processing.

loss of people's confidence. For data privacy reasons, secure computation methods like federated learning and differential privacy are frequently used.

Mathematical note—differential privacy: Let $\mathscr{M}(D)$ be an algorithm that outputs on the dataset D. It satisfies ϵ-differential privacy for all datasets D_1, D_2 differing in one record:

$$\Pr[\mathscr{M}(D_1) \in S] \leq e^{\epsilon} \cdot \Pr[\mathscr{M}(D_2) \in S]$$

This ensures minimal disclosure of any individual's data.

2.2.3.2 Standardization and interoperability

Data heterogeneity across healthcare institutions leads to issues in merging and harmonizing datasets. For instance:

- Varying terminology (ICD vs SNOMED)
- Differences in imaging resolutions or lab test units. Standard ontologies (e.g., FHIR, HL7) are being promoted to improve interoperability.

2.2.3.3 Data quality and bias

Incomplete, imbalanced, or biased datasets can produce misleading AI predictions.

- Missing data: Common in real-world clinical settings
- Class imbalance: Rare conditions are underrepresented
- Demographic bias: Models may underperform on minority groups

This calls for preprocessing, data augmentation, and bias mitigation strategies to ensure fairness and accuracy.

The foundations of AI in medicine lie in a synergistic combination of powerful algorithms and rich, diverse datasets. As this section reveals, various AI technologies from classical ML to deep learning and generative models are each suited to different forms of healthcare data and decision tasks.

However, the effectiveness of these technologies is contingent upon the availability of high-quality, standardized, and secure data. The evolution of big data infrastructure in healthcare, driven by IoT devices, genomics, digital records, and mobile health, offers the raw material for intelligent healthcare systems (Haleem et al., 2023). But alongside this opportunity lies a host of technical and ethical challenges, especially regarding privacy protection, integration of disparate data sources, and model fairness.

To make the most of AI in medicine, clinicians, data scientists, and policymakers should work together to solve these issues (Polevikov, 2023). Doing this will make sure AI is used for providing personalized and data-driven health services.

2.3 Artificial intelligence-powered diagnostics and decision support

Making use of AI in diagnosis is one of the biggest advancements in modern medicine (Xu et al., 2023). With AI-guided tools, physicians interpret difficult data more efficiently, make fewer diagnostic errors, and react sooner when needed. Three vital areas where AI is significant, as explained in this section, are medical imaging, finding diseases early, and providing assistance in making clinical decisions (Khalifa and Albadawy, 2024; Chauhan et al., 2024).

2.3.1 Artificial intelligence in medical imaging

Deep learning, which is a type of AI, has greatly improved how medical images are read. CNNs are widely applied to image analysis, with their results as shown in Table 2.4, and nearly matching or beating those provided by people in certain areas.

Example (Radiology): Given a chest X-ray image I, an AI model $f(I) \rightarrow \{0,1\}$ may output the probability of a condition like pneumonia or tuberculosis.

These models are trained on massive datasets such as:

- NIH Chest X-ray14
- HAM10000 (skin disease images)
- The Cancer Genome Atlas (TCGA)

2.3.2 Early disease detection

AI is helping spot chronic and dangerous diseases earlier by finding patterns too small or hidden for humans or traditional data analysis to detect.

Application examples:

- Cancer:
- AI can detect precancerous lesions in Pap smears, polyps in colonoscopy videos, and nodules in lung CTs.
- Models like DeepSurv predict survival probabilities using patient-specific covariates.
- Cardiovascular diseases:

Table 2.4 Key domains of application.

Domain	Imaging modality	AI capabilities
Radiology	X-ray, CT, MRI	Tumor segmentation, fracture detection, and pneumonia screening
Pathology	Histopathological slides	Automated cell counting, cancer grading, anomaly detection
Dermatology	Clinical photographs	Melanoma classification, lesion risk scoring

- Wearable sensors + AI monitor electrocardiogram (ECG) data to detect arrhythmias (e.g., atrial fibrillation).
- Signal processing + Convolutional Neural Networks (CNN) can classify heart sounds into normal or pathological patterns.
- Neurological disorders:
- MRI + AI helps diagnose Alzheimer's disease by tracking changes in brain volume.
- Electroencephalography (EEG) signals analyzed using long short-term memory (LSTM) networks help predict seizure onset.

Mathematical insight: For early detection, a binary classification function is trained:

$$f(x) = \begin{cases} 1 & \text{if disease detected (positive)} \\ 0 & \text{if no disease (negative)} \end{cases}$$

where x represents patient features such as lab values, image vectors, or biosignals.

2.3.3 Artificial intelligence-driven clinical decision support systems

Clinical decision support systems (CDSS) are software tools that help healthcare providers by recommending care based on clear evidence from each patient's data (Sutton et al., 2020). The examples are shown in Table 2.5.

These systems process data from EHRs, labs, genomics, and wearable devices to suggest:

- Diagnostic differentials
- Treatment plans
- Risk scores

Example—sepsis prediction model: Given vitals, lab tests, and patient history $x \in \mathbb{R}^n$, the model outputs:

$$P_{\text{sepsis}} = \sigma(Wx + b)$$

Table 2.5 Types of CDSS.

Type	Functionality	Example
Rule-based CDSS	Follows IF-THEN logic using medical guidelines	Alerts for drug interactions
Machine learning-based CDSS	Learns from historical data to predict outcomes or suggest actions	ICU risk scoring, sepsis prediction
Context-aware CDSS	Adjusts recommendations based on dynamic clinical environments	Real-time triage systems in the ER

CDSS, Clinical decision support systems.

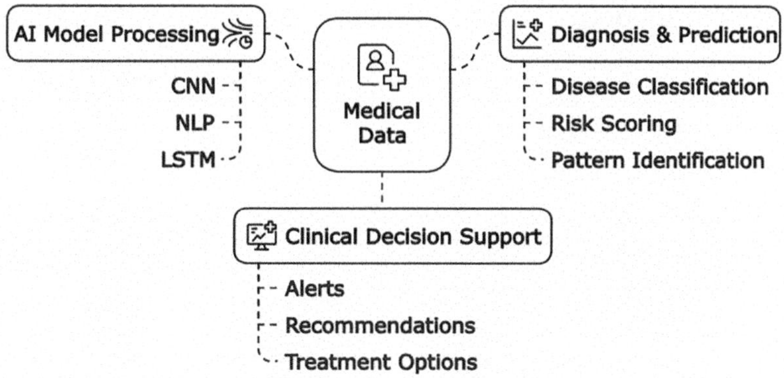

Figure 2.2 AI in diagnostics and decision support. *AI*, artificial intelligence.

where σ is the sigmoid activation function. Alerts are raised when the predicted probability exceeds a critical threshold (Fig. 2.2).

AI-based diagnostics are shown in Fig. 2.2 and are now essential team members in doctors' day-to-day jobs. Using AI helps reduce errors, start treatments earlier, and improve results for people in oncology, cardiology, and neurology (Williams et al., 2024). AI in medical imaging is successful because it detects small details that are hard for most medical practitioners to notice.

Identifying problems early, thanks to sensors and AI designed for time-series analysis, is very promising for people with chronic illnesses and those at risk of hospital stays. Leveraging ongoing data from patients, AI can make real-time predictions that help lead to prompt interventions in risky hospital settings such as ICUs (O'Reilly et al., 2024).

CDSSs are moving medical decisions closer to using insights from evidence for each patient (Karthik et al., 2025). With the help of AI, CDSS moves from fixed rule-based systems to intelligent advisors that merge various types of input. The use of these tools helps clinicians with insights that support, rather than replace, human intuition.

Nonetheless, these applications must be proven useful and understandable and uphold ethical standards to maintain users' trust and not let anyone depend too heavily on them. Collaboration among doctors and AI will play a bigger role in diagnostics than if only AI were used.

2.4 Personalized medicine and drug discovery

Because of AI, we are moving from standard medical treatments to medicines that are personalized for every patient (Mendhi et al., 2025). Such transformation happens when genomics, continuous health data, and advanced AI models are brought together. AI is also greatly transforming how drugs are found and how older drugs may be used for new uses.

2.4.1 Artificial intelligence in genomics and personalized treatment plans

Genomics means exploring DNA data to determine how someone might react to different diseases and medicines. Genomics data is so large and complex that using AI models helps us analyze it more effectively.

Example: personalized cancer therapy

Let $G_i \in \mathbb{R}^n$ represent the gene expression profile of patient i. A deep learning model $f(G_i) \rightarrow \text{Drug}_j$ maps genomic features to the most effective therapeutic compound. These models are trained on databases such as:

- TCGA (The Cancer Genome Atlas)
- GEO (Gene Expression Omnibus)

AI also supports pharmacogenomics, identifying how a patient's genetic profile affects drug response, critical for avoiding adverse drug reactions (ADRs) and maximizing efficacy. The application in genomics is listed in Table 2.6.

2.4.2 Predictive analytics in disease progression

Predictive analytics uses historical and real-time patient data to forecast the course of a disease. AI models, especially RNNs and transformers, are effective at modeling temporal patterns in patient data.

Let x_t be the feature vector (e.g., vitals and labs) at time t. Predictive models aim to learn:

$$y_{t+k} = f(x_t, x_{t-1}, ., x_0)$$

where y_{t+k} is a future health outcome (e.g., organ failure, remission).

2.4.2.1 Examples of predictive applications

- Diabetes: Forecasting glycemic control based on lifestyle and glucose monitoring data.

Table 2.6 AI applications in genomics.

AI technique	Application	Benefits
CNNs/RNNs	Detecting mutations, structural variations	Faster, more accurate interpretation of sequences
Random forests	SNP (single-nucleotide polymorphism) analysis	Feature selection and disease association
Autoencoders	Dimensionality reduction for high-dimensional gene data	Improves model generalizability

CNN, convolutional neural networks; *RNN*, recurrent neural networks.

- Heart failure: Predicting decompensation events days in advance using wearables and EHR data.
- Cancer: Modeling tumor progression using imaging + genomics over time.

Such forecasting allows clinicians to intervene early, personalize monitoring, and adjust therapy dynamically based on predicted disease trajectory and predict disease progression using AI models shown in Table 2.7.

2.4.3 Artificial intelligence-driven drug discovery and repurposing

Traditional drug discovery takes 10–15 years and billions of dollars. AI significantly accelerates this timeline by automating multiple steps in the process (Fu and Chen, 2025). The disease progression is shown in Table 2.8.

2.4.3.1 Key stages where artificial intelligence contributes

1. Target identification: Discovering biological molecules (proteins, genes) involved in a disease.
2. Lead compound generation: Designing molecules likely to bind the target.
3. Molecular property prediction: Predicting ADMET properties=Absorption, Distribution, Metabolism, Excretion, and Toxicity.
4. Drug repurposing: Identifying new uses for approved drugs.

Table 2.7 AI models for predicting disease progression.

Model	Use case	Strengths
LSTM/GRU	Time-series patient records	Captures long-term dependencies
Transformer models	Multivariate longitudinal data	Scales well, parallelizable, and interpretable
Survival analysis + Deep learning	Event prediction (e.g., mortality risk)	Combines statistical survival models with DL flexibility

Table 2.8 AI in drug discovery.

AI tool/Model	Function	Example system
Graph neural networks	Molecule property prediction	DeepChem, Chemprop
Generative models (GANs, VAEs)	De novo molecule generation	REINVENT, MoleculeNet
Knowledge graphs	Drug-target-disease interaction mining	BenevolentAI, IBM Watson
Transfer learning	Predicting efficacy across diseases	MatchMaker, MolMapNet

Given a molecular structure M, represented as a graph $G(V, E)$, a graph neural network computes embeddings h_v for each atom $v \in V$ and predicts properties:

$$\hat{y} = f_{\text{GNN}}(G)$$

AI systems such as AlphaFold have transformed structural biology by predicting protein folding with near-experimental accuracy, essential for rational drug design.

Real-world impact

- Baricitinib, an arthritis drug, was AI-repurposed during COVID-19 as a potential treatment.
- Insilico Medicine used AI to bring a novel fibrosis drug to clinical trials in under 18 months.

The use of AI in biomedicine has made it possible to give patients personalized medical treatment (Taherdoost and Ghofrani, 2024). Before, it was hard to understand genomic insights, but now they can be interpreted by AI tools very quickly, allowing accurate treatments to be used instead of blundering with medicines. These models allow clinicians to guess risks and make appropriate changes before complications occur. The model, as shown in Fig. 2.3.

AI also changes the drug discovery process by bringing together years of experiments into a few months of ML. With AI, more people can be involved in drug development because expenses are reduced, timelines speed up, and new ways are found for existing medicines. This helps enormously during health crises and when dealing with rare illnesses.

However, challenges remain:

- Ensuring model interpretability for regulators and clinicians.
- Avoiding algorithmic bias due to unbalanced training data.
- Navigating ethical issues around genetic data use and commercial drug development.

With continued cross-disciplinary collaboration and responsible innovation, AI promises to bring a new era of fast, accurate, and equitable healthcare innovations.

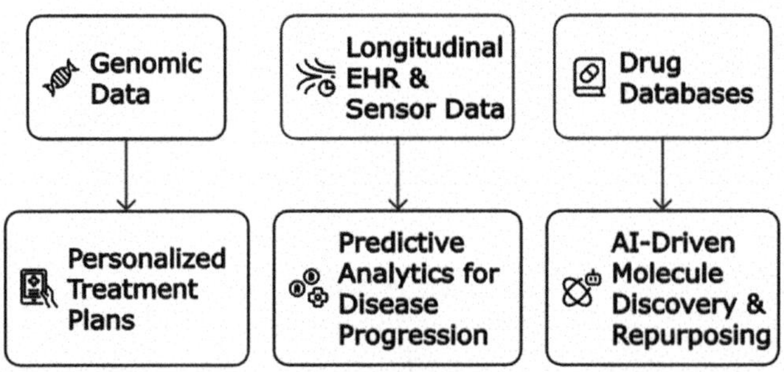

Figure 2.3 AI in personalized medicine and drug discovery. *AI*, artificial intelligence.

2.5 Artificial intelligence in patient monitoring and remote healthcare

With the move to round-the-clock, patient-focused care, technology like AI helps to oversee, examine, and respond to health data as it happens. AI is now able to support patients by using wearables, telemedicine, and remote sensors (Chaturvedi et al., 2025). Three central elements are analyzed in this section: keeping health information current, using digital assistants for care, and predicting risks for early action.

2.5.1 Wearable technology and real-time health tracking

Wearables such as smartwatches, fitness bands, ECG patches, and biosensors continuously collect vital health data, including:

- Heart rate (HR)
- Blood oxygen saturation (SpO_2)
- Body temperature
- Blood glucose
- Respiration rate
- Sleep quality

These devices generate high-frequency time-series data, which AI models analyze for trends, anomalies, and critical health alerts, and the sample wearables are shown in Table 2.9.

2.5.2 Artificial intelligence in telemedicine and virtual healthcare assistants

Telemedicine leverages video consultations and remote diagnostics, but AI enhances its capabilities through:

- Chatbots (e.g., Babylon Health, Ada) that screen symptoms and recommend action.

Table 2.9 Sample wearables and their capabilities.

Device	Sensors	AI capability	Use case
Apple Watch	ECG, SpO_2, accelerometer	Irregular rhythm detection (AFib)	Cardiac monitoring
Fitbit Charge	HR, sleep, skin temp	Sleep stage classification	Chronic stress, recovery
Dexcom G6 (CGM)	Glucose monitoring	Predict hypoglycemia	Diabetes management
BioButton/ WHOOP	HR, RR, skin temp	Early infection detection	Remote patient monitoring

- Speech-to-text NLP tools that automate documentation during calls.
- Image analysis (e.g., skin lesions, eye fundus) during video visits.

Virtual healthcare assistants powered by AI can:

- Schedule appointments.
- Monitor medication adherence.
- Alert caregivers to concerning symptoms.
- Provide mental health support using NLP chat interfaces.

These systems improve access and reduce the burden on human clinicians, especially in rural and underserved areas.

2.5.3 Predictive analytics for patient deterioration detection

AI models review both past and current health records to determine if there is a danger of getting worse so that doctors can step in early. Subtle changes in physiological signals are identified with the aid of deep learning, time-series models, and statistical models.

Let's consider the following sample dataset collected from wearable devices over 24 hours, shown in Table 2.10.

The AI predicts (Fig. 2.4). As a person's heart rate improves and oxygen levels lessen, the risk score will get worse. If the risk hits a certain threshold (0.85), experts in healthcare receive emails as advice to intervene.

The chart demonstrates how vital signs degrade gradually, and AI translates this trend into a nonlinear rise in deterioration risk score. Key takeaways:

- From hour 0 to 10: Early warning signs appear (moderate HR/SpO_2 changes).

Table 2.10 Sample data.

Time (hrs)	Heart rate (bpm)	SpO_2 (%)	Deterioration risk score
0	78	98	0.10
2	80	97	0.12
4	82	97	0.15
6	85	96	0.18
8	90	95	0.25
10	95	94	0.35
12	100	92	0.50
14	110	91	0.65
16	120	89	0.78
18	125	87	0.85
20	130	85	0.92
22	140	82	0.97

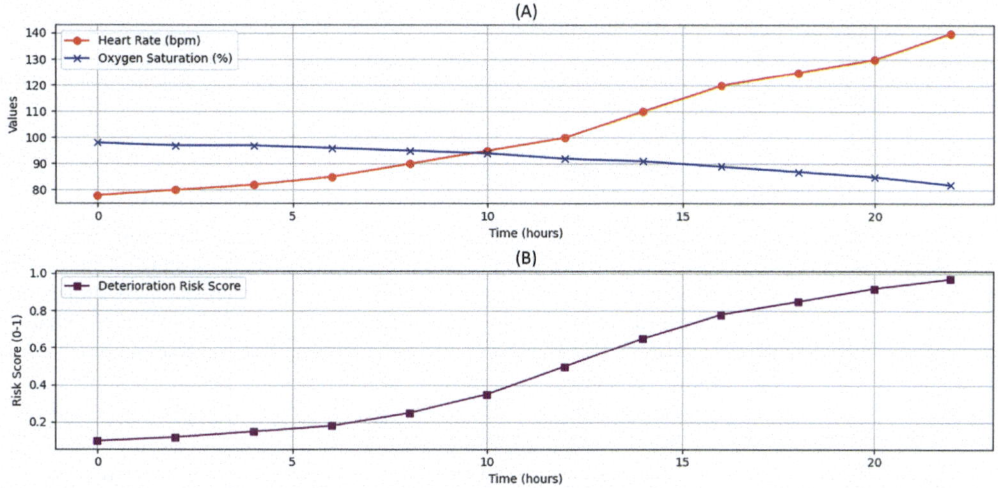

Figure 2.4 (A) Vital signs over time and (B) vital signs over time.

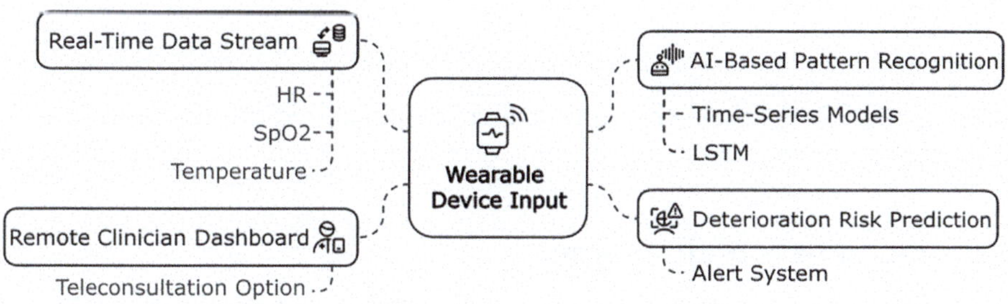

Figure 2.5 AI in remote health monitoring. *AI*, artificial intelligence.

- After hour 12: Risk escalates rapidly, indicating an urgent intervention window.
- By hour 22: The patient is at high risk (>0.95), requiring immediate attention.

AI enables continuous, nonintrusive risk prediction, empowering clinicians to intervene before a crisis, reducing ICU admissions and readmissions.

Here we can see in Fig. 2.5 that data flow, ML, and telemedicine play key parts in providing immediate, accurate, and scalable support to patients in remote healthcare. In light of the shift to these patient-focused models, AI pipelines like these will form the core of preventive and ongoing medical care.

2.6 Robotics and artificial intelligence in surgery

Now, robotic surgeries using AI offer clearer vision, more precise procedures, extra automation, and the ability to be tailored for patients (Moglia et al., 2021). Surgery is undergoing change, thanks to AI-assisted robots, as these allow procedures to be

Table 2.11 Comparison of major robotic surgical systems with AI integration.

Surgical system	AI features	Clinical application	Advantages
da Vinci Surgical System	Motion scaling, tremor filtration, 3D visualization	Prostatectomy, hysterectomy, hernia	Precision, smaller incisions, faster recovery
Mako SmartRobotics	Preoperative CT-based planning with AI	Total knee and hip replacement	Personalized bone cutting, improved implant alignment
ROSA Brain/Spine	AI-enhanced navigation using MRI data	Neurosurgery, spinal fusion	Greater accuracy in electrode/spinal placement
Monarch Platform (Auris)	Computer vision + AI navigation for bronchoscopy	Lung cancer biopsy and diagnostics	Reaches small airways, improving early detection
Versius (CMR Surgical)	AI-driven modular robotic arms	General and colorectal surgeries	Compact design, enhanced ergonomics

performed in a small incision, accurately and based on data and the comparison as shown in Table 2.11.

2.6.1 Evolution of robotic surgery

This type of surgery has changed from a unique study area to being regularly used in many medical fields such as urology, orthopedics, neurology, and cardiology. Because of ML and computer vision, robots can now work independently and keep learning from the information they gather.

2.6.2 Role of artificial intelligence in enhancing surgical capabilities

AI enhances robotic surgery through:

- Preoperative planning: AI analyzes medical imaging to create 3D maps and simulations.
- Intraoperative guidance: Real-time feedback from AI ensures safer navigation and cutting.
- Tissue differentiation: ML models help distinguish between healthy and abnormal tissue.
- Postoperative assessment: AI reviews surgical videos and patient vitals for outcome prediction.

2.6.3 Workflow: how artificial intelligence-powered robotic surgery works

Here's a simplified breakdown of the AI-augmented surgical workflow shown in Fig. 2.6.

AI enables real-time feedback loops between surgical actions and sensor data, increasing both safety and efficiency.

2.6.4 Benefits of robotics and artificial intelligence in surgery

Combining AI and robotics in surgery has changed modern medicine, especially for precision, personalized care, and safe outcomes (Habuza et al., 2021). Thanks to AI helping in imaging, instant analysis, and predictions, robotic surgery is now used widely by various medical professionals, and comparative analysis is shown in Table 2.12.

Through preoperative planning in 3D, Storz AI guides doctors in real-time, identifies different types of tissue during the process, and evaluates results after surgery. Intelligent automation plays a vital role in various da Vinci, Mako, ROSA, and

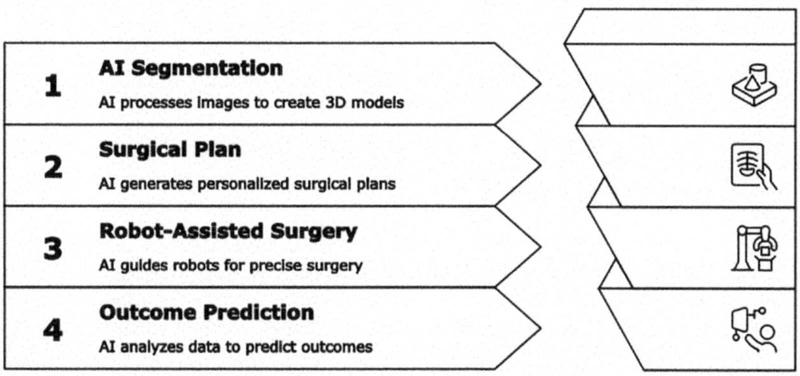

Figure 2.6 AI-augmented surgical workflow. *AI*, artificial intelligence.

Table 2.12 Comparative analysis of traditional versus AI-powered robotic surgery.

Aspect	Traditional surgery	AI-powered robotic surgery
Invasiveness	Large incisions, more trauma	Minimally invasive, faster healing
Precision	Surgeon skill-dependent	Submillimeter accuracy
Surgeon fatigue	High, especially in complex procedures	Reduced via robotic assistance
Learning curve	Steep, especially for new procedures	Shortened through AI training simulations
Data utilization	Mostly manual	Automated analysis of past procedures

Monarch surgeries because each tool is custom-built for different applications and patients.

Data from imaging and modeling is used continuously to help surgeons in the surgical process and make better outcomes possible. Unlike traditional operations, robotic surgeries driven by AI lead to less invasiveness, better accuracy, tired surgeons, and a faster time to learn through AI-based practice and critique.

Overall, using AI in robotic surgery marks a shift to depending more on data and accuracy than in earlier interventions. The expansion of their intelligence and ability to think for themselves is expected to reduce the cost of high-quality surgical treatments, improve patients' recovery, and change the part surgeons play in medical innovation. Close work between experts from multiple fields and thorough testing in healthcare environments will help us reach the full potential here.

2.7 Ethical, legal, and regulatory considerations

AI's increasing role in medicine leads to key ethical, legal, and regulatory issues since it is used for diagnostics, decision-making, patient monitoring, and surgery (Mennella et al., 2024). These issues are essential for AI to be deployed safely and responsibly in healthcare. In this section, I look into the main challenges, what they mean, and the approaches being set up to solve them.

2.7.1 Ethical considerations in artificial intelligence-driven medicine

2.7.1.1 Autonomy and informed consent

AI tools often operate as "black boxes," making decisions that are difficult to interpret or explain. This can undermine the patient's ability to provide informed consent:

- How can a patient agree to an AI-driven treatment if they cannot understand the rationale?
- Who is accountable for explaining the AI's recommendation: the algorithm developer or the clinician?

Solution: Incorporate Explainable AI (XAI) techniques to provide interpretable models and include patients in the decision-making process.

2.7.1.2 Bias, fairness, and health equity

AI models trained on biased or nonrepresentative datasets can exacerbate existing healthcare disparities, and common types are shown in Table 2.13:

- A model trained in urban data may underperform in rural populations.
- Underrepresentation of certain racial, ethnic, or gender groups can lead to diagnostic inaccuracies.

Table 2.13 Common types of bias in AI healthcare systems.

Bias type	Impact	Example
Data bias	Skewed predictions based on nondiverse samples	Lower cancer detection rates in dark skin tones
Algorithmic bias	Model architecture amplifies certain trends	Overdiagnosis in specific age groups
Labeling bias	Inconsistent annotations by clinicians	Confused model learning

Mitigation strategies: Data audits, bias detection tools, fairness-aware ML models, and ethical review boards.

2.7.1.3 Privacy and patient autonomy

AI in healthcare processes massive volumes of sensitive personal data (EHRs, genomic data, and wearable metrics). Misuse or unauthorized access can breach patient confidentiality.

- Use differential privacy, homomorphic encryption, and federated learning to protect identity.
- Ensure that patients are aware of how their data is used and can opt out where necessary.

2.7.2 Legal implications of artificial intelligence in medicine

2.7.2.1 Accountability and liability

In an AI-assisted decision:

- If a misdiagnosis occurs, who is liable: the AI developer, the hospital, or the physician?
- Current legal systems lack clarity on shared responsibility in hybrid human–AI decisions.

Updated medical malpractice laws that factor in AI recommendations and degrees of autonomy.

2.7.2.2 Intellectual property and data ownership

- Who owns the AI model trained on patient data: hospitals, developers, or the public?
- If AI generates a new treatment protocol or drug suggestion, can it be patented?

Emerging view: Patients and communities should retain rights over the data that trains AI models.

2.7.3 Regulatory frameworks and compliance

Global agencies are developing AI regulations specific to healthcare, as shown in Table 2.14:

These frameworks promote transparency, traceability, auditability, and continuous monitoring of AI tools.

2.7.3.1 Emerging ethical concepts for artificial intelligence era

- Human-in-the-loop: Ensures AI decisions are always reviewed and approved by a qualified human clinician.
- Algorithmic accountability: Mandates documentation of training data, model versions, and decision boundaries.
- Right to explanation: Patients should be able to demand an explanation for any AI-driven medical recommendation.

2.7.3.2 Case study: IBM Watson for oncology

IBM Watson, once touted as a revolutionary AI system for cancer care, struggled due to:

- Training on limited datasets.
- Overreliance on U.S.-based treatment guidelines.
- Inconsistent clinical adoption.

This raised serious concerns about validation, reliability, and clinician trust, reinforcing the need for rigorous regulatory oversight.

Table 2.14 Key regulatory bodies and guidelines for AI in healthcare.

Agency	Country/Region	Key regulation or guideline
FDA (U.S. Food & Drug Admin)	United States	AI/ML-Based Software as a Medical Device (SaMD) Action Plan
EMA (European Medicines Agency)	European Union	Guideline on computerized systems and electronic data in clinical trials
CDSCO	India	Draft guidance on medical device software, including AI/ML
MHRA	United Kingdom	Software and AI as a Medical Device Change Program
WHO	Global	Ethics and governance of AI for health report (2021)

ML, machine learning.

2.8 Future trends and innovations in artificial intelligence-driven medicine

The convergence of AI, big data, robotics, and biotechnology is paving the way for a new era of intelligent, proactive, and personalized healthcare (Rashid and Kausik, 2024). While current AI applications are transforming diagnostics, treatment planning, and monitoring, the future promises even more disruptive innovations. This section highlights the key trends, technologies, and shifts expected to shape the next generation of AI in medicine.

2.8.1 Hyperpersonalized medicine

Advancements in multiomics (genomics, proteomics, and metabolomics) and individual health trajectories are enabling the shift toward hyper-personalized treatment plans.

Example: AI models can integrate genomic mutations, lifestyle data, microbiome profiles, and environmental exposures to deliver individualized therapy plans for complex diseases like cancer and autoimmune disorders. The evolution of personalized is shown in Table 2.15.

2.8.2 Federated learning for privacy-preserving collaboration

One of the biggest challenges in healthcare AI is access to large, diverse, and secure datasets. Federated learning enables training AI models across decentralized hospitals or regions without moving raw data.

Example: A cancer prediction model can be trained across hospitals in different countries without sharing patient data, while maintaining both performance and compliance with privacy laws (e.g., GDPR, HIPAA).

Benefits:

- Improved generalization
- Enhanced data security
- Cross-institutional collaboration

Table 2.15 Evolution from ersonalized to hyperersonalized medicine.

Feature	Current personalized medicine	Future hyperpersonalized medicine
Data sources	Genomics, clinical records	Genomics + lifestyle + environment + behavior
AI models	Risk scoring, treatment guidance	Dynamic adaptation based on real-time feedback
Application level	Population subgroups	Individualized profiles

2.8.3 Artificial intelligence-driven preventive and proactive healthcare

With real-time data from wearables, EHRs, and digital biomarkers, the future of AI in healthcare is shifting from reactive care to proactive prevention.

Predictive AI models:

- Forecast disease onset before symptoms appear
- Recommend lifestyle adjustments
- Trigger alerts for anomalies

Data snapshot: According to the World Health Organisation, up to 80% of premature heart disease and diabetes could be prevented with early intervention. AI will be central in identifying early risk indicators.

2.8.4 Digital twins for healthcare simulation

A digital twin is a real-time digital replica of a patient, created using AI, simulation, and sensor data, the sample elements listed in Table 2.16, and these twins can simulate:

- Drug responses
- Disease progression
- Surgery outcomes

Future impact: A digital twin of a heart patient can simulate how different medications or lifestyle changes may affect their health, allowing physicians to test treatments before applying them.

2.8.5 Explainable and trustworthy artificial intelligence

To promote adoption and ensure patient safety, the future of AI in healthcare must prioritize explainability and transparency. Techniques like SHAP (SHapley Additive exPlanations) and LIME (Local Interpretable Model-agnostic Explanations) will become standard.

Need: Regulatory bodies and clinicians require justification for AI-driven decisions, especially in life-critical settings such as intensive care unit (ICU) management and cancer treatment.

Table 2.16 Core elements of AI-driven precision healthcare systems.

Element	Description
Input data	EHR, wearables, genomics, lab results
AI role	Learning system behavior from patient-specific data
Application	Precision simulations, outcome prediction

EHR, electronic health record.

2.8.6 Artificial intelligence in mental and behavioral health

AI is expected to play a pivotal role in mental health, an area often underdiagnosed and undertreated. Future systems will:

- Analyze speech patterns, facial expressions, and typing behavior
- Track mood changes using wearables
- Deliver real-time interventions and therapy via virtual agents

Data: Studies show that AI-powered mental health chatbots can reduce depression and anxiety symptoms by up to 30% over eight weeks.

2.8.7 Integration of quantum computing with artificial intelligence for drug discovery

As quantum computing matures, it will revolutionize AI's ability to simulate molecular interactions, protein folding, and drug responses at unprecedented speeds. The comparison listed in Table 2.17.

2.8.8 Autonomous artificial intelligence agents in healthcare

In the future, autonomous agents (like healthcare robots or virtual clinicians) powered by AI will be able to:

- Manage hospital logistics
- Handle triage and patient flow
- Monitor ICU patients in real-time
- Assist elderly patients with medication, alerts, and companionship

These agents will use reinforcement learning to improve continuously based on real-world interactions.

2.8.9 Predictive healthcare infrastructure

Hospitals and health systems will evolve into smart environments, where predictive AI manages and the trends shown in Table 2.18:

Table 2.17 Comparison between classical AI and quantum AI in drug discovery.

Aspect	Classical AI	Quantum AI
Drug search space	Millions of combinations	Billions processed simultaneously
Simulation time	Weeks or months	Minutes or hours
Application stage	Clinical trials, posthoc simulations	Early-stage design and molecule selection

Table 2.18 AI-driven future roadmap.

Trend	Impact area	Expected timeline
Hyperpersonalized treatment	Oncology, Rare diseases	Ongoing
Federated AI collaboration	Diagnostics, Research	2025–30
Preventive digital health	Chronic diseases	2025–35
AI-powered digital twins	Precision medicine	Emerging (by 2030)
Explainable AI frameworks	Regulatory and clinical trust	In development
Quantum AI for drug discovery	Pharmacogenomics	Experimental (2030+)
Autonomous healthcare agents	Elderly care, ICU, home monitoring	2030+

- Resource allocation (ICU beds, staff shifts)
- Medical supply chains
- Emergency response systems

An AI system at a smart hospital in South Korea was able to reduce ER waiting time by 27% by dynamically adjusting resources based on real-time predictions.

The following decade will welcome a big change in healthcare, driven by the expanding intelligence, widespread use, and individualized approach possible with AI. For it to achieve its best results, stakeholders should concentrate on the following:

- Interdisciplinary collaboration
- Robust regulation and ethics
- Open and inclusive data ecosystems
- Patient empowerment and education

With these pillars in place, AI will not just assist medicine; it will redefine it.

2.9 Conclusion

Modern medicine is being transformed by the introduction of AI. AI helps make faster, more accurate, and better-supported decisions in diagnostics, drug research, remote monitoring, and personal treatment than people could have imagined. This chapter explains that the combination of AI tools and detailed healthcare information is doing more than just making current approaches better; it is leading to predictive, preventive, and more precise care.

Thanks to AI's strong ability to work with huge data, it can discover patterns missed by human observation. Radiologists use AI for examining images, genetics uses it to find mutations, and wearable devices use it to track patients, all of which support clinicians. With AI applied to CDSS, the prediction of disease risks and instant diagnostics is making both outcomes and patient care safer and more accurate. However, as AI technologies become more pervasive, the importance of addressing ethical, legal, and regulatory concerns cannot be overstated. Ensuring transparency, explainability, fairness, and accountability is critical for trust and long-term success. Similarly, issues like data privacy, bias mitigation, and responsible AI governance must be proactively managed through collaborative frameworks involving technologists, healthcare professionals, policymakers, and patients.

We are on the cusp of an exciting future, and new inventions such as digital twins, federated learning, and AI-based drug production will take the lead. However, AI's full medical benefits will be achieved if we give equal consideration to technology, ethics, equal data availability, teamwork among different fields, and broad innovation.

In short, AI and new data-based methods are helping medicine shift to a new era, where healthcare is both reactive and capable of being predicted, personalized, and precise. Thinking carefully about this change will help build a healthcare future that is more advanced, practical, and friendly to people.

References

Alowais SA, Alghamdi SS, Alsuhebany N, Alqahtani T, Alshaya AI, Almohareb SN, et al. Revolutionizing healthcare: the role of artificial intelligence in clinical practice. BMC Medical Education 2023;23(1). Available from https://doi.org/10.1186/s12909-023-04698-z, https://bmcmededuc.biomedcentral.com/.

Chaturvedi U, Chauhan SB, Singh I. The impact of artificial intelligence on remote healthcare: enhancing patient engagement, connectivity, and overcoming challenges. Intelligent Pharmacy 2025. Available from https://doi.org/10.1016/j.ipha.2024.12.003.

Chauhan AS, Singh R, Priyadarshi N, Twala B, Suthar S, Swami S. Unleashing the power of advanced technologies for revolutionary medical imaging: pioneering the healthcare frontier with artificial intelligence. Discover Artificial Intelligence 2024;4(1). Available from https://doi.org/10.1007/s44163-024-00161-0, https://link.springer.com/journal/44163.

Cremin CJ, Dash S, Huang X. Big data: historic advances and emerging trends in biomedical research. Current Research in Biotechnology 2022;4:138–51. Available from https://doi.org/10.1016/j.crbiot.2022.02.004.

Fu C, Chen Q. The future of pharmaceuticals: artificial intelligence in drug discovery and development. Journal of Pharmaceutical Analysis 2025:101248. Available from https://doi.org/10.1016/j.jpha.2025.101248.

Gedikci Ondogan A, Sargin M, Canoz K. Use of electronic medical records in the digital healthcare system and its role in communication and medical information sharing among healthcare professionals. Informatics in Medicine Unlocked 2023;42:101373. Available from https://doi.org/10.1016/j.imu.2023.101373.

Gupta NS, Kumar P. Perspective of artificial intelligence in healthcare data management: a journey towards precision medicine. Computers in Biology and Medicine 2023;162:107051. Available from https://doi.org/10.1016/j.compbiomed.2023.107051.

Habuza T, Navaz AN, Hashim F, Alnajjar F, Zaki N, Serhani MA, et al. AI applications in robotics, diagnostic image analysis and precision medicine: current limitations, future trends, guidelines on CAD systems for medicine. Informatics in Medicine Unlocked 2021;24:100596. Available from https://doi.org/10.1016/j.imu.2021.100596.

Haleem A, Javaid M, Pratap Singh R, Suman R. Exploring the revolution in healthcare systems through the applications of digital twin technology. Biomedical Technology 2023;4:28–38. Available from https://doi.org/10.1016/j.bmt.2023.02.001.

Hong L, Luo M, Wang R, Lu P, Lu W, Lu L. Big data in health care: applications and challenges. Data and Information Management 2018;2(3):175–97. Available from https://doi.org/10.2478/dim-2018-0014.

Karthik K, Kamath SS, Supreetha R, Katlam A. Content-based medical retrieval systems with evidence-based diagnosis for enhanced clinical decision support. Expert Systems with Applications 2025;272:126678. Available from https://doi.org/10.1016/j.eswa.2025.126678.

Khalid N, Qayyum A, Bilal M, Al-Fuqaha A, Qadir J. Privacy-preserving artificial intelligence in healthcare: techniques and applications. Computers in Biology and Medicine 2023;158:106848. Available from https://doi.org/10.1016/j.compbiomed.2023.106848.

Khalifa M, Albadawy M. AI in diagnostic imaging: revolutionising accuracy and efficiency. Computer Methods and Programs in Biomedicine Update 2024;5:100146. Available from https://doi.org/10.1016/j.cmpbup.2024.100146.

Liu X, Glocker B, McCradden MM, Ghassemi M, Denniston AK, Oakden-Rayner L. The medical algorithmic audit. The Lancet Digital Health 2022;4(5):e384. Available from https://doi.org/10.1016/S2589-7500(22)00003-6, https://www.sciencedirect.com/journal/the-lancet-digital-health.

Mendhi S, Sawarkar K, Shete A, Vinchurkar K, Mali SS, Singh S, et al. Smart healthcare: artificial intelligences impact on drug development and patient care. Intelligent Pharmacy 2025. Available from https://doi.org/10.1016/j.ipha.2025.01.003.

Mennella C, Maniscalco U, De Pietro G, Esposito M. Ethical and regulatory challenges of AI technologies in healthcare: a narrative review. Heliyon 2024;10(4):e26297. Available from https://doi.org/10.1016/j.heliyon.2024.e26297.

Moglia A, Georgiou K, Georgiou E, Satava RM, Cuschieri A. A systematic review on artificial intelligence in robot-assisted surgery. International Journal of Surgery 2021;95:106151. Available from https://doi.org/10.1016/j.ijsu.2021.106151.

O'Reilly D, McGrath J, Martin-Loeches I. Optimizing artificial intelligence in sepsis management: opportunities in the present and looking closely to the future. Journal of Intensive Medicine 2024;4(1):34–45. Available from https://doi.org/10.1016/j.jointm.2023.10.001.

Polevikov S. Advancing AI in healthcare: a comprehensive review of best practices. Clinica Chimica Acta 2023;548:117519. Available from https://doi.org/10.1016/j.cca.2023.117519.

Rashid AB, Kausik MDAK. AI revolutionizing industries worldwide: a comprehensive overview of its diverse applications. Hybrid Advances 2024;7:100277. Available from https://doi.org/10.1016/j.hybadv.2024.100277.

Sinha R. The role and impact of new technologies on healthcare systems. Discover Health Systems 2024;3(1). Available from https://doi.org/10.1007/s44250-024-00163-w.

Sutton RT, Pincock D, Baumgart DC, Sadowski DC, Fedorak RN, Kroeker KI. An overview of clinical decision support systems: benefits, risks, and strategies for success. npj Digital Medicine 2020;3(1). Available from https://doi.org/10.1038/s41746-020-0221-y, https://www.nature.com/npjdigitalmed/.

Taherdoost H, Ghofrani A. AI's role in revolutionizing personalized medicine by reshaping pharmacogenomics and drug therapy. Intelligent Pharmacy 2024;2(5):643–50. Available from https://doi.org/10.1016/j.ipha.2024.08.005.

Tandel V, Kumari A, Tanwar S, Singh A, Sharma R, Yamsani N. Intelligent wearable-assisted digital healthcare industry 5.0. Artificial Intelligence in Medicine 2024;157:103000. Available from https://doi.org/10.1016/j.artmed.2024.103000.

Williams R, Anderson S, Cresswell K, Kannelønning MS, Mozaffar H, Yang X. Domesticating AI in medical diagnosis. Technology in Society 2024;76:102469. Available from https://doi.org/10.1016/j.techsoc.2024.102469.

Xu N, Yang D, Arikawa K, Bai C. Application of artificial intelligence in modern medicine. Clinical eHealth 2023;6:130–7. Available from https://doi.org/10.1016/j.ceh.2023.09.001.

PART II

Regulatory, training, and adoption challenges

CHAPTER 3

Regulatory requirements and clinical approval pathways for artificial intelligence-driven robotics

Sanju Mannumadam Venugopal
Institute of Electrical and Electronics Engineers, Downey, CA, United States

3.1 Introduction

The integration of artificial intelligence (AI) within robotic surgical systems is reshaping and transforming modern surgical practices. The AI driven surgical robotic systems are designed and developed to considerably improve various aspects of surgery, such as the precision of surgery, optimization of intraoperative decision-making, improvement in clinical outcomes, etc. This is achieved through integrating advanced computational techniques and capabilities within the AI-based robotic systems. For example, in minimally invasive procedures such as prostatectomies and gynecologic surgeries, robotically-assisted systems have shown reduced tissue trauma, lower postoperative pain, and shorter recovery times. Therefore the convergence of AI and robotic surgery represents a significant advancement in surgical innovation and the clinical outcomes and the quality of patient care (Iftikhar et al., 2024).

This chapter aims to provide an understanding of the regulatory landscape governing AI-enabled robotic surgical devices. It begins by outlining AI in robotic surgery and the importance of alignment with regulations. This is followed by general medical device regulatory frameworks from various regulatory agencies, such as the United States Food and Drug Administration (US FDA), the EU European Medicines Agency (EMA), the United Kingdom's Medicines and Healthcare Products Regulatory Agency (UK MHRA), Health Canada, and Japan's Pharmaceuticals and Medical Devices Agency (PMDA) and their approval pathways and subsequently focuses on the applicable regulatory requirements and clinical approval pathways for AI-driven robotics surgical systems. Through this review, the chapter can serve as a practical reference for medical device developers, clinical researchers, regulatory professionals, and other stakeholders engaged in the development and evaluation of advanced surgical systems.

AI-Powered Developments in Medical Robotics
DOI: https://doi.org/10.1016/B978-0-443-27783-2.00014-4

3.2 The importance of regulation in artificial intelligence-driven robotic clinical surgery

The pace of evolution of these advanced technologies, such as AI and machine learning (ML) integrated into robotics in medical devices, presents considerable regulatory challenges. Regulatory agencies such as the US FDA, the EMA, the UK MHRA, Health Canada, etc., are required to facilitate innovation in medical devices along with stringent evaluation of these advanced devices before being marketed, which incorporate elements of autonomy and continuous learning, while focusing on their overall goal of public health protection. The current regulations and guidelines by the regulatory agencies were developed for medical devices that were nonautonomous and operated under the supervision of the surgeons (Mennella et al., 2024; Yang et al., 2017). These regulations are often found to lack clear steps or might need significant updates to address the complexity and dynamic behavior of AI-driven surgical robotic systems. Some of the key aspects in the regulations include establishing accountability in the event of adverse outcomes, ensuring transparency in decision-making algorithms, and validating clinical safety and efficacy under real-world clinical conditions. However, developers and manufacturers of the AI-enabled robotic surgical devices are required to have a strategic approach that embeds the expected and potential regulatory considerations, such as clinical efficacy, algorithmic transparency, and patient safety, from regulatory agencies throughout their product development lifecycle to bring their devices to market (Yang et al., 2017; Elendu et al., 2023). This includes aligning the technical design of the devices with applicable regulatory requirements, proactively engaging with regulatory bodies, and identifying necessary evidence for market authorization. Moreover, early collaboration with the regulators, researchers, clinicians, and relevant industry stakeholders can facilitate a shared understanding of these regulatory expectations, which can significantly reduce development inefficiencies while complying with regulations for a safe introduction of advanced medical surgical devices into clinical settings (Mennella et al., 2024; Yang et al., 2017; Elendu et al., 2023).

3.3 Global regulatory frameworks

3.3.1 The United States (Food and Drug Administration)

In the United States, the FDA plays an important role in overseeing the regulation of medical devices through its Center for Devices and Radiological Health. According to the FD&C Act, a medical device is broadly defined as any instrument, apparatus, machine, implant, or in vitro reagent that either appears in the U.S. Pharmacopoeia or National Formulary, is intended for use in the diagnosis or treatment of disease, or is designed to affect the structure or function of the human body (Van Norman, 2016; FDA, How to Determine if Your Product is a Medical Device, n.d.). The FDA's risk-based classification system of medical devices groups them into three categories: Class I, II, and III, which are based on their potential risk to patients. Class I devices are considered low risk and are subjected to general controls. Some examples include elastic bandages, manual stethoscopes, and so on. Class II devices are considered moderate-risk, which require both general and special controls to ensure their safety and effectiveness for patients. Some examples include infusion pumps, surgical drapes, and so on. Class III devices support or sustain human life and carry the highest potential risk if they fail, which requires the most rigorous level of regulatory evaluation (FDA, Classify Your Medical Device, n.d.). Some examples include implantable pacemakers, deep-brain stimulators, and so on. This risk-based classification of these devices determines the type of regulatory pathway required to be followed to obtain FDA clearance for the marketing of the device(s) (Mennella et al., 2024).

In 1993 the FDA (George et al., 2018) approved the first surgical robot, called the Automated Endoscopic System for Optimal Positioning. Since then, advanced surgical robots have been introduced into the US market, providing a rapid pace of innovation and advancements in the field of surgical robotics. Most of the current surgical robots in use require a surgeon's directions to be part of key clinical decisions and to carry out their activities. For example, a robotic system may assist in generating patient-specific operative plans or in performing precision tasks like bone milling, but they are always executed under the surgeon's supervision and direction. Interestingly, considering the aspect of devices' capabilities that required surgeons' supervision and were not made fully autonomous or semiautonomous, the FDA renamed the term "surgical robots," which was used till 2015, to be named robotically-assisted surgical devices (RAS) (Lee et al., 2024). The implementation of ML and AI into the surgical robotic devices is beginning to have an altered view on the notion of surgeon-assisted robotic devices with a potential to have a rise in semiautonomous and autonomous behavior of these devices in critical decision-making and action-taking abilities. This integration can lead to higher accuracy in the precision of the surgery, efficiency, and accessibility, allowing real-time interpretation of surgical field images and enhancing the precision of instrument movements for surgeons (Lee et al., 2024). Advancements in these surgical robotic devices come with the challenge of navigating through the regulatory pathways for these ML and AI-enabled surgical robots, which is crucial for their widespread adoption. As AI and ML continue to be integrated into robotic platforms, developers, manufacturers, and regulators are required to navigate a complex and evolving landscape of regulatory approvals. Understanding how these emerging technologies fit within existing regulatory frameworks is critical to ensuring their safe and effective adoption in clinical practice.

3.3.1.1 Regulatory pathways for medical devices in the United States

As surgical robotic devices evolve with the integration of AI and ML, navigating the regulatory landscape becomes critical. These devices are not only technologically complex but also have concerns about their safety, reliability, and accountability. The US FDA will carefully evaluate these factors before providing its market authorization. For example, before a device can be used widely in clinical practice, it is required to pass through one of several FDA regulatory pathways, each tailored to the level of risk and novelty associated with the device. Before exploring the regulatory requirements for AI/ML-enabled robotic surgical medical devices, it's important to understand the current regulatory pathways for contemporary medical devices, as understanding them is essential for developers and clinicians working with AI-enabled surgical technologies, as this will not only help in addressing the unique challenges these advanced systems present but also provides a structured framework for bringing them safely and efficiently to the clinical settings while ensuring patient safety.

This section outlines the primary FDA regulatory pathways for obtaining marketing authorization, such as the premarket notification via 510(k) clearance, De Novo classification, and premarket approval (PMA). This section also explores the expanded access pathways (Van Norman, 2018), which enable patients with serious or life-threatening conditions to have access to investigational devices when there are no satisfactory alternatives.

3.3.1.1.1 Premarket notification via 510(k)

The premarket notification pathway, commonly known as the 510(k), was established through the Medical Device Amendments to the Federal Food, Drug, and Cosmetic (FD&C) Act on May 28, 1976. It remains one of the most widely used regulatory pathways for bringing low- to moderate-risk medical devices (classified as Class I and II devices) to the U.S. market. This pathway mandates demonstrating substantial equivalence (SE) between a new device, referred to as the "subject device," and an already FDA-cleared, legally marketed device, called the "predicate device." SE can be established with the demonstration of the new device having the same intended use, technological characteristics, design, principles of operation, material, and energy sources as the predicate device or that any differences do not raise new questions of safety or effectiveness (Van Norman, 2016; FDA, The 510k Program: Evaluating Substantial Equivalence in Premarket Notifications 510k, n.d.). This pathway can be considered to be based more on comparative analysis of the devices (subject and predicate devices) rather than an independent demonstration of their safety and effectiveness. For example, a company developing a new laparoscopic instrument might compare it to an existing, approved device with a similar design and energy source, such as electrosurgical functionality. The FDA maintains a public database of all its legally marketed devices, which the developers and manufacturers can utilize to identify suitable predicate devices for comparison (FDA, Establishment Registration & Device Listing, n.d.). This pathway is particularly valuable for introducing a device to the market for the first time that has a suitable predicate device or reintroducing a previously FDA-cleared device that has undergone significant modifications in materials, design, principles of operation, or technological features that could affect its safety or performance.

Though the 510(k) pathway is often considered more streamlined than other approval pathways, it still requires manufacturers to submit a detailed notification to the FDA at least 90 days before the intended commercial launch. If the FDA determines that the subject device is not substantially equivalent (NSE) to the predicate devices, it will be classified as a Class III (high-risk) device, which requires a more rigorous PMA (FDA, The 510k Program: Evaluating Substantial Equivalence in Premarket Notifications 510k, n.d.). To assist manufacturers and developers in preparing these submissions, the FDA has published a detailed guidance document titled "The 510(k) Program: Evaluating Substantial Equivalence in Premarket Notifications 510(k)" (FDA, The 510k Program: Evaluating Substantial Equivalence in Premarket Notifications 510k, n.d.), which outlines the criteria and expectations for demonstrating SE (Fig. 3.1).

3.3.1.1.2 De Novo pathway

The De Novo classification pathway was established under the Food and Drug Administration Modernization Act of 1997. This pathway helps in the introduction of novel medical devices for which no legally marketed predicate device is available to follow the 510(k) submission process. This pathway uses a detailed risk assessment-based approach to classify new devices into Class I or Class II, where information on the device's intended use, potential health risks, expected benefits, and risk mitigation strategies is documented. It is required to be demonstrated how the device can be reasonably assured to be safe and effective through the application of general controls alone (for Class I devices) or a combination of general and special controls (for Class II devices) (Van Norman, 2016; FDA, De Novo Classification Request, n.d.; FDA, De Novo Classification Process Evaluation of Automatic Class III Designation, n.d.). For example, a novel wearable biosensor that tracks a new physiological parameter not previously monitored by any FDA-cleared device may be required to follow a De Novo classification pathway if it offers a clinical application with assurance in safety and effectiveness. If cleared, this device then becomes the first in a new classification category and can become a predicate device for future devices through the 510(k) pathway. The De Novo pathway provides a way for the manufacturers whose devices have NSE when submitted for initial 510(k). Manufacturers can utilize De Novo classification to get premarket authorization rather than having to follow the rigorous PMA pathway.

The De Novo pathway is becoming important for supporting the growth of innovative and technologically sophisticated medical devices by enabling the FDA to update the existing list of advanced predicate devices (FDA, Device Classification Under Section 513(f)(2)(De Novo), n.d.). The De Novo process helps ensure that new

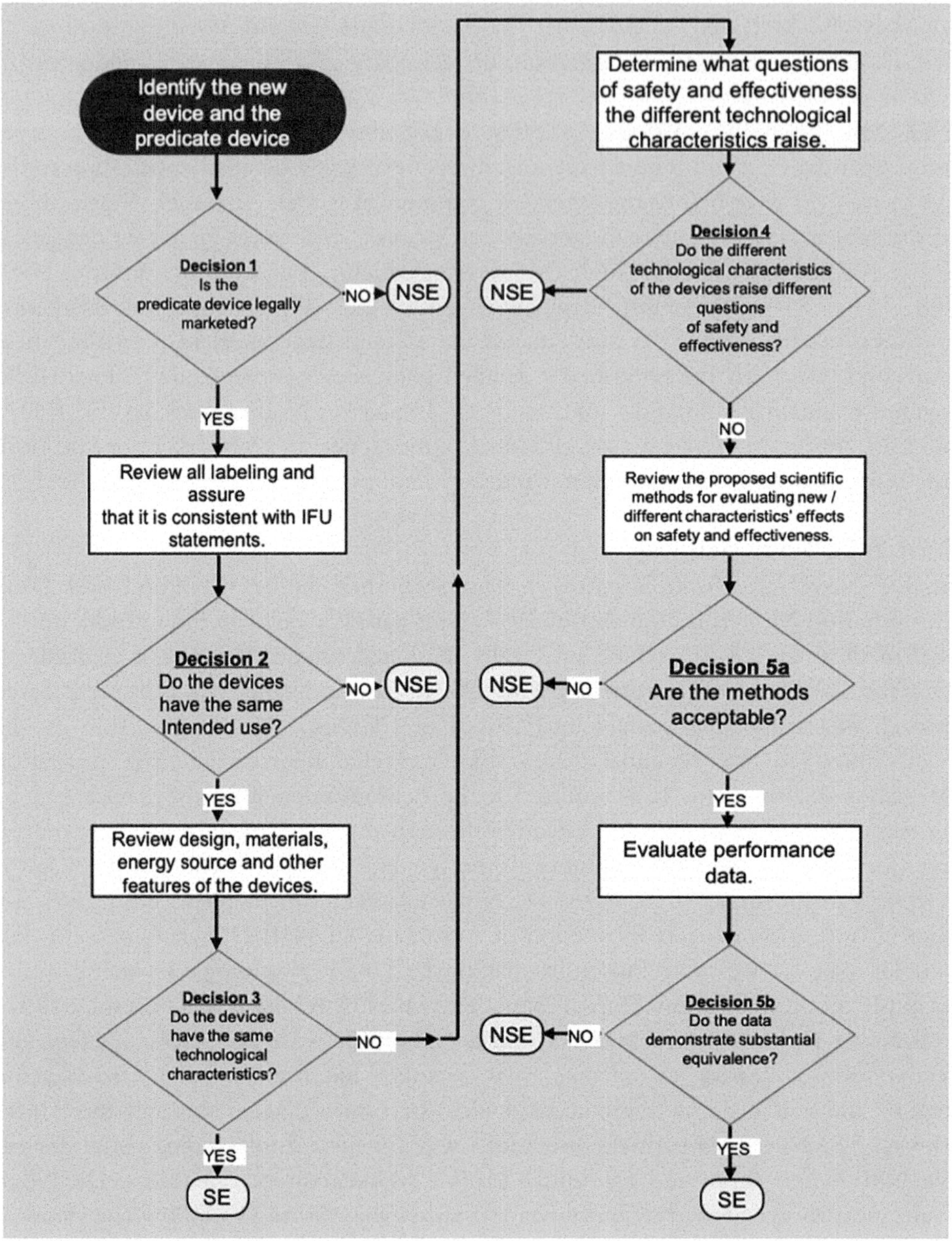

Figure 3.1 510(k) workflow process (The 510(k) Program: Evaluating Substantial Equivalence in Premarket Notifications, 2025).

510(k) submissions are compared to devices that reflect current scientific and clinical standards. However, the De Novo requests are not always approved, where the FDA may decline if the device does not meet the criteria for Class I or II, if additional information is required, or if a benefit-risk analysis made of the device reveals that the device's risks outweigh its potential benefits. In these scenarios, the device must be classified as Class III and must follow the PMA process. For manufacturers and developers working to introduce advanced technologies without clear predicate devices, the De Novo classification offers a viable, less burdensome pathway (FDA, De Novo Classification Process Evaluation of Automatic Class III Designation, n.d.). To assist manufacturers in preparing these De Novo submissions, the FDA has published a detailed guidance document titled "De Novo Classification Process (Evaluation of Automatic Class III Designation)" (FDA, De Novo Classification Process Evaluation of Automatic Class III Designation, n.d.), which outlines the criteria and expectations for providing a detailed assessment (Fig. 3.2).

3.3.1.1.3 Premarket approval

The PMA pathway is the most rigorous regulatory pathway for bringing novel, high-risk medical devices. These devices are classified as Class III, which typically represent entirely new technologies that often have no legally marketed predicate devices to support a 510(k) submission. PMA is primarily intended for devices that support or sustain human life, are used to treat or diagnose life-threatening conditions, or play a crucial role in preventing serious health impairments. Because of the critical nature of these devices, manufacturers are required to provide robust scientific, clinical, and nonclinical evidence to demonstrate the device's safety and effectiveness for its intended use (FDA, Premarket Approval PMA, n.d.; FDA, Premarket Approval Application and Humanitarian Device Exemption Modular Review, n.d.). The PMA's review process is highly structured and demands comprehensive documentation, including detailed descriptions of the device, manufacturing processes, proposed labeling, and full results from clinical trials. The list of medical devices approved via the PMA pathway is available in the Premarket Approval Database (FDA, Premarket Approval PMA Database, n.d.).

There are primarily four types of PMA submission pathways (FDA, PMA Application Methods, n.d.), each designed to accommodate different stages of device development and specific use cases:

1. Traditional PMA: In this standard approach, the manufacturer submits a complete PMA application as a single package. The FDA's review begins only after the full application is received. This method is suitable for devices that have completed clinical studies and may already be approved for use outside the United States (FDA, PMA Application Methods, n.d.).

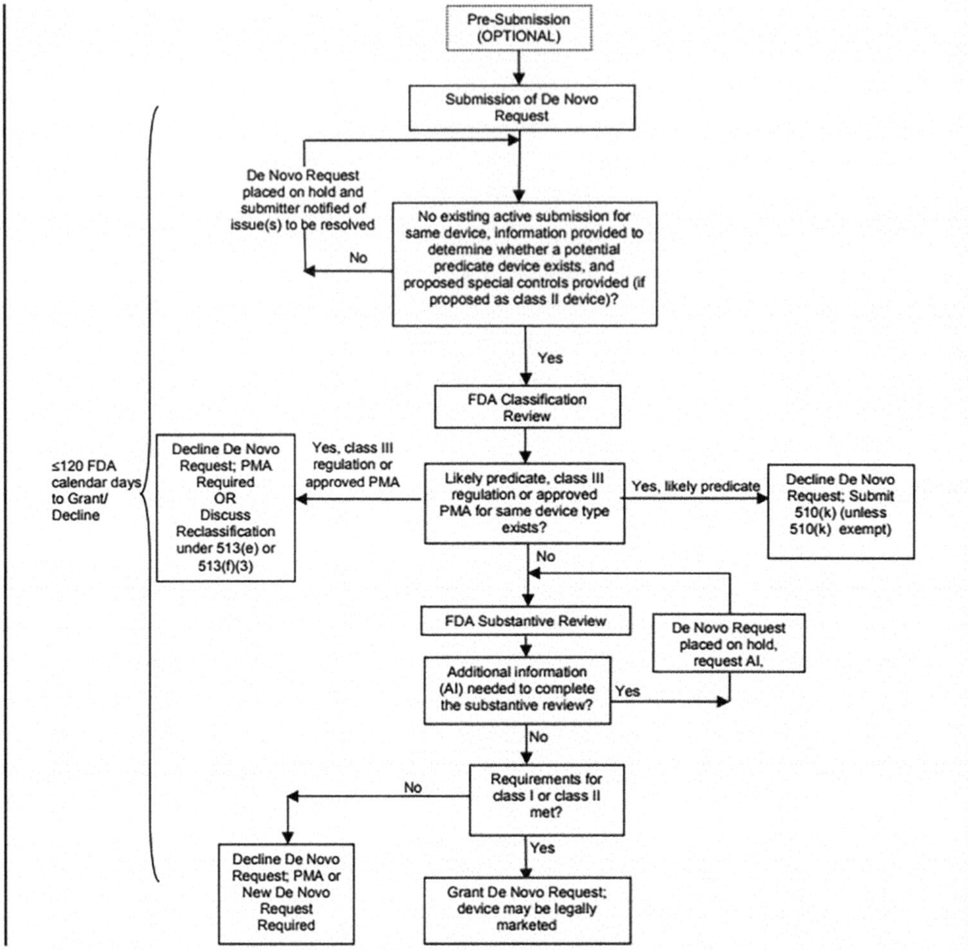

Figure 3.2 De Novo classification workflow process (RAPS, De Novo Classifications: FDA Drafts Guidance on Acceptance Reviews, Finalizes Guidance From, 2014).

2. Modular PMA: The modular approach allows the PMA to be submitted in parts or "modules" over a defined period of time. The process begins with a "shell" outlining the content and structure of the planned modules, which is developed collaboratively between the manufacturer and the FDA. As modules are submitted, they are reviewed individually, allowing the final review to proceed more rapidly and efficiently once the last module is received. This is especially useful for devices in early stages of their development, where not all data may be available at once (FDA, PMA Application Methods, n.d.).
3. Product Development Protocol (PDP): The PDP process involves early agreement between the FDA and the manufacturer on specific development milestones, study

endpoints, and performance criteria. It is particularly suited for devices based on well-established technologies with clearly defined objectives. The PDP ensures transparency and alignment throughout the development process, with regular reporting and milestone tracking to keep the FDA informed (FDA, PMA Application Methods, n.d.).

4. Humanitarian Device Exemption (HDE): The HDE pathway is tailored for devices intended to treat or diagnose conditions affecting rare populations, defined as fewer than 8000 individuals per year in the United States. Unlike traditional PMA, the HDE pathway requires evidence of probable benefit to patients, rather than a full demonstration of effectiveness. Devices approved under HDE are reviewed and monitored by Institutional Review Boards, and they may be sold for profit provided distribution stays within the FDA-established Annual Distribution Number (FDA, PMA Application Methods, n.d.; FDA, Humanitarian Device Exemption, n.d.; Bernad, 2009) (Fig. 3.3).

3.3.1.2 Regulatory pathway for artificial intelligence-driven robotic surgery devices in the United States

As robotic surgical devices continue their advancement with the integration of AI and ML, navigating their regulatory environment has become ever more important since these devices are required to undergo careful evaluation from the FDA before being marketed and widely adopted in clinical settings. Their regulatory pathway has to be governed through a series of structured steps tailored to the device's level of technological sophistication, risk, and intended use. One of the primary challenges in regulating such complex systems is accounting for the varying degrees of autonomy, particularly as AI-based robotic devices integrate AI capabilities to assist or even execute surgical tasks (Yang et al., 2017). To navigate this complexity, the Levels of Autonomy in Surgical Robotics framework is helpful and increasingly being used to classify surgical robots across various levels, from Level 0 (manual operation) to Level 5 (fully autonomous) (Yang et al., 2017; Lee et al., 2024).

3.3.1.3 Classification of surgical robotics autonomy and its regulatory pathway

- Level 0: These are considered to be nonrobotic systems where all surgical tasks are performed manually by the surgeon. There is no robotic or AI assistance involved. These devices do not fall under autonomous regulation and follow conventional device regulatory pathways (Yang et al., 2017; Lee et al., 2024).
- Level 1: There are systems in which the surgeon maintains continuous control over all robotic movements, though the robot may provide passive or active guidance (e.g., tremor filtration, tool tracking). These systems typically lack autonomous decision-making capabilities. The FDA has predominantly approved Level 1 systems

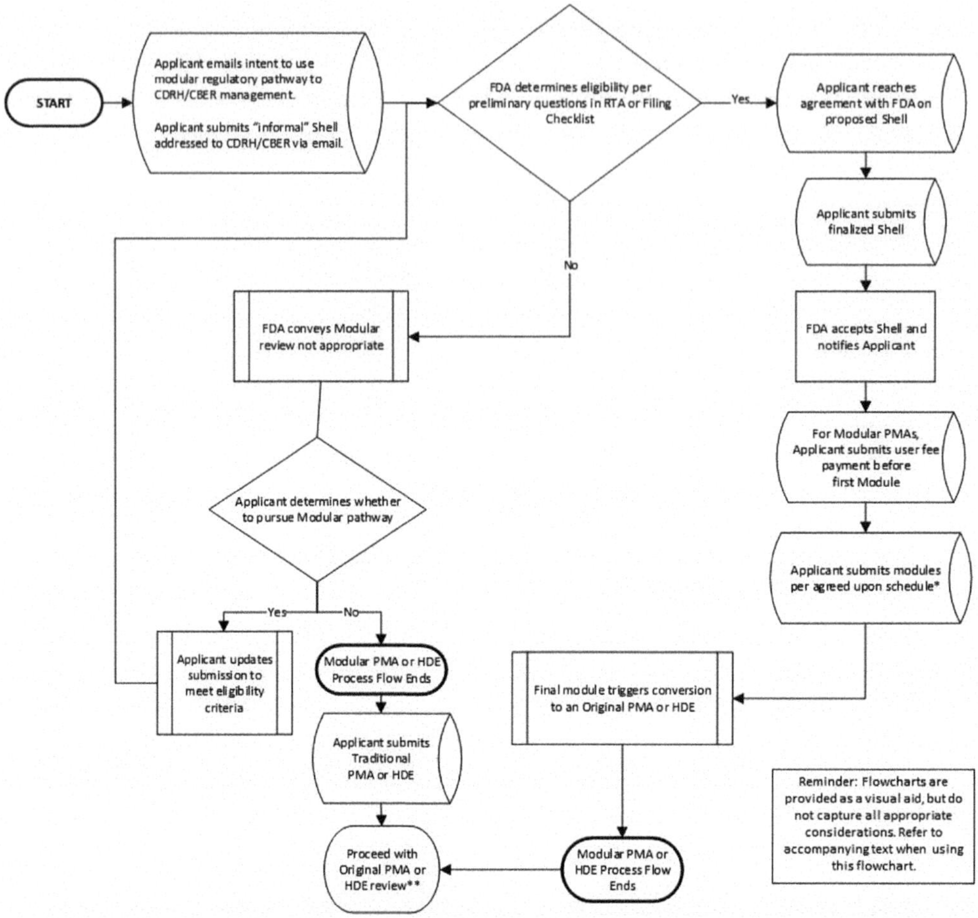

Figure 3.3 Modular PMA or HDE workflow process (FDA, Premarket Approval Application and Humanitarian Device Exemption Modular Review, n.d.). *PMA*, premarket approval; *HDE*, humanitarian device exemption.

through the 510(k) pathway, as the system's hardware characteristics closely resemble existing predicate devices (Yang et al., 2017; Lee et al., 2024).

- Level 2: These are the systems that can perform only the preprogrammed surgical tasks set by the surgeons but would still require the surgeon's approval for initiating these tasks. They support partial automation, such as camera positioning or intraoperative visualization assistance during the surgery. While many Level 2 devices have received clearance via the 510(k) pathway (Aboy et al., 2024; Lee et al., 2024), newer systems with more advanced AI capabilities may require the De Novo pathway if no suitable predicate device exists (Aboy et al., 2024). As of 2023, there were only four devices considered to be level 2 that had been approved by the FDA.

So, new AI surgical robots with level 2 autonomy shall need to have hardware characteristics similar to one of these four predicate devices (as of 2023) (Yang et al., 2017; Lee et al., 2024).

- Level 3: These systems are capable of autonomously executing tasks that have been approved by the surgeon and can continuously update patient-specific surgical strategies, such as real-time intraoperative imaging. As of 2023, these represent the most advanced class of surgical robots that have received FDA clearance, with three such devices approved through the 510(k) pathway. These devices were categorized as Class II. However, the 510(k) review process primarily assessed the hardware components for SE to a predicate device and did not account for the novelty or complexity of the embedded AI algorithms (Aboy et al., 2024). Given that these systems rely heavily on AI/ML algorithms for real-time decision-making, issues such as algorithmic transparency, data privacy, security, and safety become increasingly critical. As a result, future submissions of similar or more advanced devices may require evaluation through the De Novo or PMA pathways to adequately address these considerations (Yang et al., 2017; Lee et al., 2024).
- Level 4: These systems operate semiindependently and are capable of planning and executing critical surgical steps with limited oversight of the surgeon but would still require the surgeon's intervention in case of an emergency or critical situation affecting patient safety or effectiveness of the surgery. As of 2023, there are no level 4 robots approved by the FDA, and there are no predicate systems for these systems. Due to their increased autonomy in performing critical surgical tasks with high risk, there is speculation that these devices may be classified as Class III (high risk) devices, which are expected to be regulated under the PMA pathway. These systems present challenges, including ensuring human oversight, algorithm explainability, and rigorous validation of their autonomous functions in clinical settings (Yang et al., 2017; Lee et al., 2024).
- Level 5: These are fully autonomous systems capable of performing entire surgeries without human intervention. These devices are hypothetical in the current regulatory framework and will follow the PMA pathway due to their high risk potential and the unreasonable risk of illness or injury to the patients undergoing the surgery. Their regulatory approval will require robust clinical evidence, strict oversight protocols, and a well-defined framework for postmarket surveillance (PMS) (Yang et al., 2017; Lee et al., 2024).

3.3.1.4 The Q-Submission process

For developers and manufacturers of AI-enabled surgical robotic systems, early and strategic engagement with the United States. The FDA is essential due to the complexity and novelty of these technologies. The FDA's Q-Submission (Q-Sub) program enables early discussions on critical aspects such as algorithm validation, trial design, data management, periodic review of outputs from AI algorithms, etc. This is especially valuable for ML-based components, which differ from traditional deterministic software (FDA, Requests for Feedback and Meetings for Medical Device Submissions: The Q-Submission Program, n.d.). For high-risk Class III devices, a PMA submission is typically required with robust clinical and nonclinical data under an Investigational Device Exemption. Some of the PMA pathways that are applicable include traditional or modular, based on their developmental stage (FDA, Premarket Approval PMA, n.d.; FDA, Premarket Approval Application and Humanitarian Device Exemption Modular Review, n.d.). The PreSubmission Program (part of Q-Sub) can be a very good option for AI-enabled surgical robots to have the FDA for early interaction to align on evidence expectations, minimizing delays (FDA, Requests for Feedback and Meetings for Medical Device Submissions: The Q-Submission Program, n.d.). Programs such as the Early Payor Feedback Program support alignment between payer and provider bodies, ensuring that clinical endpoints meet both regulatory and economic value criteria (Lee et al., 2024).

3.3.1.5 Change management process

A Predetermined Change Control Plan (PCCP) outlines the expected software and algorithm changes and how they will be safely implemented. PCCP is required as part of the device's marketing submission, such as a 510(k), De Novo, or PMA. Their core details include: (1) a description of the types of expected changes (e.g., model parameter updates, model retraining, feature engineering, etc.), (2) the methods for assessing risks of these changes on the device, and (3) a robust procedure for testing, validating, and documenting the changes prior to their production deployment. PCCPs are particularly required for AI-based robotic devices due to their probabilistic nature in decision-making, which may require critical algorithmic updates in response to new surgical datasets or performance improvements. This approach also aligns with the trend toward lifecycle-based regulatory models for software as a medical device (SaMD), offering a scalable pathway for innovation in intelligent surgical devices. PCCPs bridge the gap between the iterative nature of ML and the static nature of regulatory oversight (FDA, Predetermined Change Control Plans for Machine Learning-Enabled Medical Devices: Guiding Principles, n.d.; FDA, Marketing Submission Recommendations for a Predetermined Change Control Plan for Artificial Intelligence-Enabled Device Software Functions, n.d.).

The FDA's guidance, titled "Marketing Submission Recommendations for a Predetermined Change Control Plan for Artificial Intelligence-Enabled Device Software Functions" (FDA, Marketing Submission Recommendations for a Predetermined Change Control Plan for Artificial Intelligence-Enabled Device

Software Functions, n.d.), provides detailed expectations for PCCPs, including transparency, change protocols, and documentation standards.

3.3.1.6 Postmarketing surveillance under the Food and Drug Administration

PMS is a critical regulatory requirement for medical devices and is especially required for AI-enabled robotic surgical systems, ensuring continued device safety and effectiveness after their market approval. While PMA establishes initial safety under controlled conditions, they often involve trials done with a limited patient population and cannot capture long-term risks or potential failures that might come up in real-world clinical settings. Robotic surgical systems, due to their complexity and evolving AI capabilities, require robust postmarket monitoring mechanisms to detect issues such as software anomalies, incorrect algorithmic actions, or adverse human–robot–AI interactions, etc. The FDA requires manufacturers to implement comprehensive PMS programs, which include adverse event (AE) reporting such as MedWatch, Manufacturer and User Facility Device Experience, periodic safety update reports (PSURs), postmarket clinical follow-up (PMCF) studies, and device registries (FDA, Computer-Assisted Surgical Systems, n.d.). The FDA has also launched initiatives such as the Medical Product Safety Network (MedSun) survey to gather insights from experienced surgeons and is collaborating with professional societies to improve training and education on surgical robotic systems (FDA, Computer-Assisted Surgical Systems, n.d.). The FDA encourages the establishment of patient registries to support real-world evidence collection, particularly for devices used in the prevention of life-threatening diseases and their treatment.

3.3.2 Europe (European Medical Agency)

While the US FDA provides a centralized and standardized regulatory pathway for the approval of medical devices, the European Union (EU) operates under a decentralized but coordinated framework governed by the Medical Device Regulation (MDR) (EU) 2017/745. The MDR, which came into effect in May 2021, replaced the previous Medical Device Directive (MDD 93/42/EEC) and the Active Implantable Medical Device Directive (AIMDD 90/385/EEC). The MDR's objective is to enhance patient safety and ensure that all medical devices approved to be marketed on the EU market meet stringent standards of quality, safety, and clinical performance. Under MDR, medical devices are required to undergo a conformity assessment process to obtain the CE (Conformité Européenne) Mark. The CE Mark is a prerequisite for gaining market access in EU member states. The CE Mark signifies that a device complies with applicable EU legislation and has met the essential safety and performance requirements. Based on the classification of devices, the conformity assessment involves a comprehensive review by a designated notified body (NB). These bodies, which are appointed and monitored by national competent authorities (CAs), are responsible for evaluating technical documentation (TD), clinical data, and the manufacturer's quality management system (QMS) (European Medical Agency, Medical devices, n.d.). Unlike the FDA, which functions as a single centralized regulatory authority, the EU's device approval process involves several bodies, including the CAs, NBs, and Authorized Representatives (Institut, 2023). The role of the Authorized Representative (required for nonEU manufacturers) is to act as a bridge with regulatory bodies and ensure ongoing compliance of the devices. Moreover, as specified in MDR, the CE certification is valid for a maximum of five years, after which re-certification for that device is required. This periodic recertification process becomes very significant for AI-driven robotic surgical systems, where iterative software updates and adaptive learning-based techniques seen in the AI algorithms shall require periodic reassessment of safety and performance (European Medical Agency, Medical devices, n.d.; Nüssler, 2023).

An additional regulatory body, called the Medical Device Coordination Group (MDCG), plays a crucial role in harmonizing interpretations of MDR requirements across European Member States. The MDCG issues guidance documents, including those relevant to emerging technologies such as AI and robotics, to reduce variability in assessment outcomes between NBs and CAs (Medical Device Coordination Group MDCG Working Groups, n.d.). As of June 2024, the EU has designated 50 NBs under MDR and IVDR (In Vitro Diagnostic Device Regulation), with 38 NBs specializing in MDR alone (Study supporting the monitoring of the availability of medical devices on the EU market, n.d.). A provision under MDR allows manufacturers of legacy medical devices an additional time, until the end of 2027 or 2028, depending on risk class, to follow the new MDR regulations, given an application was submitted to a NB by May 2024 (European Medical Agency, Extension of the MDR transitional period and removal of the 'sell off' periods, n.d.).

3.3.2.1 Regulatory pathway for medical devices in Europe

Before CE marking and marketing of medical devices in Europe, classifying these devices, based on their intended use and risks, is required. The classification is done under MDR No. 2017/745 for medical devices or active implantable medical devices, or IVDR No. 2017/746 for IVD devices (European Medical Agency, Medical devices, n.d.). This classification not only provides the applicable regulatory pathway but also affects the required documentation, conformity assessment procedures, and overall approval timelines. Under the MDR 2017/745, the classification is determined using a set of rules. These rules form a risk-based framework where higher-risk devices are subject to more rigorous regulatory scrutiny and compliance obligations (Guidance on Classification Rules for in vitro Diagnostic Medical Devices under Regulation EU 2017/746, 2025).

3.3.2.2 Device classification

Before CE marking, all the medical devices are categorized into one of the four categories based on Medical Device Regulation (EU) 2017/745 (MDR), which are outlined in the classification rules listed in Directive 93/42/EEC Annex IX (Guidance on Classification Rules for in vitro Diagnostic Medical Devices under Regulation EU 2017/746, 2025; European Medical Agency, Guidance on classification of medical devices, n.d.). The categories are

- Class I (including Is & Im)
- Class IIa
- Class IIb
- Class III

Among the above classes, Class III is ranked the highest. The higher the devices are classified, the greater the level of conformity assessment required, evidence of clinical evaluation, and documentation to prove their appropriate classification. Moreover, the classification of a medical device is based on its intended use rather than its specific technical features. Some of the key factors that determine classification basis include the specific medical purpose of the device, continuity in its usage, the duration of the device's interaction with the body, degree of invasiveness (invasive, noninvasive, or surgically invasive), etc. (Institut, 2023; European Medical Agency, Guidance on classification of medical devices, n.d.). Most AI-based robotic surgical devices can be categorized into either Class IIb, Class IIb, or Class III.

3.3.2.2.1 Class I devices

Class I devices represent low to almost no risk to the patient's safety, which require very little regulatory oversight from the EMA, and are typically noninvasive and for transient use.

3.3.2.2.2 Class IIa devices

Class IIa devices represent a moderate-risk level, requiring more oversight than Class I devices but not as stringent as Class IIb or III. These devices are generally surgically invasive and for short-term use.

3.3.2.2.3 Class IIb devices

Class IIb devices represent medium- to high-risk levels, requiring more oversight than Class II devices but not as stringent as Class III. These devices often include devices with a higher risk of harm, such as those intended for long-term use in the body and those used for life-sustaining procedures.

3.3.2.2.4 Class III devices

Class III devices are the highest risk among all the classes, requiring the strongest oversight. These devices are life-supporting, life-sustaining, or present a high degree of invasiveness, complexity, or potential for harm.

3.3.2.3 Conformity assessment process for various classes of devices

The objective of conformity assessment is for medical device manufacturers to evaluate the conformity of their devices with the general safety and performance requirements (GSPR) of the EU MDRs (European Medical Agency, Annex I – General safety and performance requirements, n.d.). Manufacturers are required to involve a NB in the conformity assessment, except for Class I devices (European Medical Agency, Medical devices, n.d.; European Medical Agency, Conformity assessment, n.d.) (Fig. 3.4).

3.3.2.3.1 Conformity assessment pathway for Class I devices

As Class I medical devices represent the lowest risk category, they are subjected to a comparatively less severe conformity assessment process. This regulatory pathway is primarily applicable to standard low-risk devices that do not involve advanced technologies such as AI or robotic capabilities. Manufacturers are responsible for independently demonstrating MDR compliance by preparing and maintaining

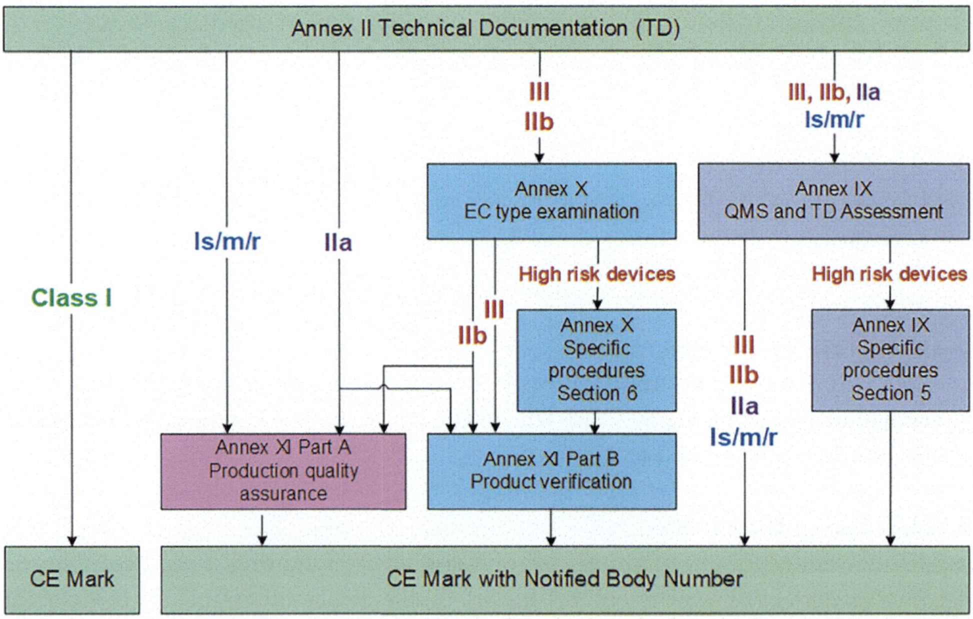

Figure 3.4 Conformity assessment procedures specified by the MDR Regulation (EMA) (Confins, Conformity Assessment Procedures and Premarket Scrutiny under EU MDR, n.d.). *MDR*, Medical Device Regulation; *EMA*, European Medicines Agency.

comprehensive TD. This includes evidence of conformity with GSPR, risk management procedures, clinical evaluations, and appropriate labeling (European Medical Agency, CG 2019-15 rev.1 Guidance notes for manufacturers of Class I Medical Devices, n.d.; European Medical Agency, MDR Article 52, Conformity assessment procedures, n.d.). Once the TD is complete, the manufacturer issues a Declaration of Conformity (DoC) under Annex VII (Module A), signifying that the device meets all applicable regulatory standards. For standard Class I devices that are nonsterile, nonmeasuring, and not reusable surgical instruments, no involvement from a NB is required. However, Class I devices that are sterile (Class Is), have a measuring function (Class Im), or are reusable surgical instruments (Class Ir) require NB review for those specific characteristics of sterility, measurability, and reusability. Additionally, it is important to note that AI-enabled robotic surgical devices are excluded from this class of device (Class I) due to their complexity and inherent risk and are therefore subject to more rigorous conformity assessment pathways under MDR (European Medical Agency, Conformity assessment, n.d.; European Medical Agency, CG 2019-15 rev.1 Guidance notes for manufacturers of Class I Medical Devices, n.d.; European Medical Agency, MDR Article 52, Conformity assessment procedures, n.d.).

3.3.2.3.2 Conformity assessment pathway for Class II devices

Under the MDR, Class IIa and IIb devices are considered medium to medium-high-risk devices, respectively. Their conformity assessment procedures require the involvement of a NB. The complexity and stringency increase from Class IIa to IIb, particularly for implantable devices. Class IIa medical devices are subject to NB oversight to ensure safety, performance, and regulatory compliance. Developers and manufacturers of devices falling in this category may choose from one of the below three conformity assessment pathways, depending on their operational capability and readiness (European Medical Agency, Notified Bodies, n.d.; European Medical Agency, MDR Article 52, Conformity assessment procedures, n.d.).

1. Quality Management System and Technical Documentation Review (Annex IX (Chapters I & III)): This is the most widely followed pathway for Class IIa devices that involves a full audit of the manufacturer's QMS by an NB and a targeted review of TD for at least one representative device within each generic device group. This pathway requires the establishment of robust PMS and PMCF procedures. According to MDCG 2020-6, sufficient clinical evidence is required to demonstrate safety and performance, either through clinical investigation or a well-substantiated equivalence approach (European Medical Agency, Annex IX, Conformity assessment based on a quality management system and on assessment of technical documentation, n.d.).
2. Type Examination and Product Conformity Verification (Annex X + Annex XI (Part A or B)): In cases where the manufacturer does not operate under a fully

certified QMS, the conformity assessment may involve type examination under Annex X, where the NB reviews the design and tests a representative device sample. This is followed by either Annex XI Part A, an audit of the production QMS, or Annex XI Part B, which requires batch or sample verification by the NB prior to the device's market release (European Medical Agency, Annex X, Conformity assessment based on Type examination, n.d.; European Medical Agency, Annex XI, Conformity assessment based on product conformity verification, n.d.).

3. Evaluation of Quality Assurance System (Annex IX (Chapter I)): This pathway requires an end-to-end QMS evaluation, including design, production, and postmarket processes in place. The NB is required to assess at least one representative device per generic device group (European Medical Agency, Annex IX, Conformity assessment based on a quality management system and on assessment of technical documentation, n.d.).

These pathways ensure that AI-enabled Class IIa robotic surgery systems meet stringent clinical and technical standards before CE marking. While distinct from the U.S. FDA's 510(k) or De Novo pathways, the EU system offers flexibility based on manufacturer capability and device complexity, emphasizing lifecycle oversight and evidence-based safety validation.

3.3.2.3.3 Conformity assessment pathway for Class III medical devices

Class III medical devices are subjected to the most stringent conformity assessment procedures. These devices pose the greatest potential impact on patient safety and would require comprehensive oversight by a NB. Manufacturers and developers of these devices are required to demonstrate compliance with GSPRs, provide sufficient clinical evidence, and ensure a robust QMS in place throughout the device's lifecycle (European Medical Agency, Guidance on classification of medical devices, n.d.; European Medical Agency, Annex IX, Conformity assessment based on a quality management system and on assessment of technical documentation, n.d.).

Under MDR, three conformity assessment pathways are available for Class III devices:

1. Conformity Assessment via QMS and Technical Documentation Review (Annex IX): This is the most frequently selected pathway for a Class III device. It requires the implementation of a certified full-scope QMS, typically aligned with ISO 13485, involving design, manufacturing, and PMS activities. The NB performs a detailed audit of the QMS and conducts a rigorous assessment of the TD for every individual device or device group. A critical component of this process is the design dossier review, which evaluates device design, clinical data, risk management, and compliance with applicable standards. Compared to the other lower-risk class devices, the design dossier must be approved by the NB before the CE mark can be

provided. Additionally, ongoing surveillance audits are required to ensure the devices' continued compliance (European Medical Agency, Annex IX, Conformity assessment based on a quality management system and on assessment of technical documentation, n.d.).

2. Conformity Assessment via Type Examination (Annex X) Combined with Production Conformity Verification (Annex XI): This two-part pathway begins with a Type Examination under Annex X (European Medical Agency, Annex X, Conformity assessment based on Type examination, n.d.), where the NB conducts a comprehensive review of the device's design dossier and performs inspections or laboratory testing on a representative sample of the device. This includes verification against harmonized standards and direct evaluation of the device's prototype to confirm alignment with GSPRs (European Medical Agency, Guidance on classification of medical devices, n.d.). If successful, a Type Examination Certificate is issued (European Medical Agency, Annex X, Conformity assessment based on Type examination, n.d.). Following this certification, the manufacturer is required to complete additional steps to ensure that ongoing production matches the examined representative sample of the device, which is carried out via Annex XI (European Medical Agency, Annex XI, Conformity assessment based on product conformity verification, n.d.) as below:
3. Part A—Production Quality Assurance: The NB audits the production process to ensure consistency with the approved device type and ensures QMS is implemented. The NB issues a Production Quality Assurance certificate upon assessing and approving the QMS. Once the NB identifies the manufacturer and obtains the Type Examination Certificate and Production Quality Assurance Certificate, the manufacturer can affix the CE mark to their Class III medical device (European Medical Agency, Annex XI, Conformity assessment based on product conformity verification, n.d.).
4. Part B—Product Verification: The NB performs direct inspection and testing of each individual device or batch through physical examination or statistical sampling. This provides high assurance of conformity with the issuance of a Product Verification certificate. This pathway allows manufacturers of Class III medical devices to ensure compliance through individual product assessments. Once the NB identifies the manufacturer obtains a Type Examination Certificate and Product Verification certificate, the manufacturer can affix the CE mark to their Class III medical device (European Medical Agency, Annex XI, Conformity assessment based on product conformity verification, n.d.).

3.3.2.4 Clinical requirements

Manufacturers are required to conduct a systematic and planned process to collect, evaluate, and analyze clinical data to ensure device safety, performance, and clinical benefits align with its intended purpose. Manufacturers are required to specify and justify the level of clinical evidence necessary to demonstrate conformity relevant to GSPRs. For Class III devices (of high risk), clinical investigations are required unless specific exemptions apply (European Medical Agency, MDR – Article 61 – Clinical evaluation, n.d.). A methodological framework is required to outline the framework for clinical evaluations, such as the Clinical Evaluation Plan and the Clinical Evaluation Report. This ongoing assessment ensures that any new information impacting the device's safety and performance is appropriately addressed, maintaining compliance with the MDR (European Medical Agency, MDR – Article 61 – Clinical evaluation, n.d.; European Medical Agency, Annex XIV - Clinical evaluation and post-market clinical follow-up, n.d.; European Medical Agency, Regulation EU 2017/745: Clinical evidence needed for medical devices previously CE marked under Directives 93/42/EEC or 90/385/EEC, n.d.).

3.3.2.5 Regulatory pathway for artificial intelligence-driven robotic surgery devices in Europe

Under the EU MDR 2017/745, Machine Learning-based Software as a Medical Device (MML-SaMD) is classified according to the potential clinical risk it poses. Rule 11 of Annex VIII (European Medical Agency, Annex VIII, n.d.) specifically governs software classification, focusing on its use in diagnosis, therapy, and physiological monitoring. MML-based devices are hardly classified as Class I unless these devices do not influence diagnostic or therapeutic decisions and pose minimal risk; these may be self-certified by manufacturers. Most MML-SaMD, however, fall under Class IIa or IIb. Class IIa includes AI/ML software that is used to support diagnosis or treatment where the clinical decision does not present a serious risk to the patient or is used for monitoring nonvital physiological processes. Class IIb applies to software supporting decisions that may cause serious deterioration of health or involve surgical intervention and software monitoring vital parameters such as heart rate or respiration when real-time variations could present immediate danger. If the software directly contributes to decisions likely to cause death or irreversible harm, it is classified as Class III, the highest risk category. These classifications determine the level of conformity assessment and whether oversight by an NB is required, ensuring proportionate regulation based on the level of patient risk (European Medical Agency, Annex VIII, n.d.; European Medical Agency, Artificial Intelligence in Medical Devices Questionnaire, n.d.). Additionally, the EU AI Act (2024) focuses on using a risk-based approach for classifying AI-based systems. The risk classification includes Unacceptable Risk, High Risk, and Low or Minimal Risk. It is observed that most of the AI-enabled surgical robots would fall in the category of "High" risk, which would require human oversight, transparency, a risk management system, and dual-human confirmation for certain actions (European

Medical Agency, Article 6: Classification Rules for High-Risk AI Systems, n.d.; European Medical Agency, AI Act, n.d.; European Medical Agency, Artificial Intelligence in Medical Devices Questionnaire, n.d.).

So, it is required for the device manufacturers to classify their devices as per the above-identified criteria as per Rule 11 of Annex VIII and the EU AI Act (2024) and have the appropriate regulatory pathway followed for obtaining the market authorization for their AI-enabled surgical robotic devices.

3.3.2.6 Scientific advice procedures and expert panel consultation

For developers and manufacturers of high-risk, AI-driven robotic surgical devices, early regulatory engagement is a strategic imperative to clarify evidentiary expectations and reduce approval uncertainties. Similar to the FDA's Q-Sub process, scientific advice procedures, and expert panel consultations, particularly under Regulation (EU) 2017/745 for medical devices. The EMA, in collaboration with national CAss and the Medical Device Expert Panels, offers structured presubmission consultation to manufacturers of Class III and certain Class IIb devices that incorporate novel or high-risk features, such as autonomous AI capabilities or adaptive algorithms. These consultations are designed to assess whether the clinical investigation design and proposed evidence generation activities are appropriate for demonstrating compliance with the GSPRs under the MDR (European Medical Agency, High-risk medical devices: consultation procedures and advice, n.d.). The expert panels, composed of independent multidisciplinary specialists, issue nonbinding opinions on the clinical, scientific, and technical aspects of the proposed studies. The expert panel consultation is mandatory for certain high-risk Class III and implantable devices before an NB finalizes its clinical evaluation (European Medical Agency, Medical device expert panels, n.d.). Similar to the FDA's Q-Sub pathway, EMA's expert engagement supports scientific robustness, transparency, and predictability in the clinical and technical validation of AI-based robotic surgery systems in Europe.

3.3.2.7 Change management process

Unlike the FDA's Q-Sub process, which allows manufacturers to obtain nonbinding, premarket feedback on clinical protocols and testing strategies with the PCCP framework, the European Union has no direct equivalent to the PCCP and has no provision for preapproved change protocols. Any significant postapproval change to device design, intended purpose, or manufacturing processes is required to be reviewed and approved individually by an NB before implementation. Manufacturers are required to notify their NB, provide updated risk analyses or clinical data, and receive approval before deployment into production (European Medical Agency, Questions & Answers for applicants, marketing authorisation holders of medicinal products and notified bodies with respect to the implementation of the Regulations on medical devices and in vitro diagnostic medical devices Regulations EU, 2017).

3.3.2.8 Postmarket surveillance under the European Medicines Agency

Manufacturers of AI-driven robotic surgery devices are required to implement robust PMS systems to ensure ongoing safety and clinical performance throughout the product lifecycle. This requirement is especially critical for high-risk implantable and Class III devices, where the complexity of AI algorithms and the surgical context heighten the potential for clinical variability. For Class III and implantable devices, manufacturers are required to generate and submit a PSUR on an annual basis, which is subject to assessment by the relevant NB (MDR Article 86). The PSUR is required to provide an analysis of benefit-risk balance, a summary of results from PMCF activities, sales and usage data, and conclusions from vigilance activities. For Class IIb devices, the PSUR is also required annually, whereas for Class IIa, it is submitted every two years. When premarket approval is based on equivalence or when long-term clinical data is insufficient, particularly for AI systems, manufacturers are required to initiate a PMCF plan. PMCF plans may include targeted clinical investigations or leverage structured data sources (e.g., national registries or digital health platforms) to address unresolved risks or clinical uncertainties. For lower-risk devices such as Class I (including Class Is, Im, and Ir), manufacturers are required to prepare a PostMarket Surveillance Report (MDR Article 85), which summarizes PMS data and is required to be made available to CAs upon request. Such a decentralized surveillance model, enforced through national CAs, and authorized NBs, is critical to ensuring the continued safety, performance, and trust in AI-enabled robotic surgical technologies across the European Union (European Medical Agency, EU postmarket surveillance plans for medical devices, n.d.; European Medical Agency, Market surveillance and vigilance, n.d.).

3.3.3 United Kingdom (Medicines and Healthcare products Regulatory Agency)

In the United Kingdom (UK), the MHRA is the regulatory agency responsible for regulating medical devices, which are governed by the UK Medical Devices Regulations 2002 (UK MDR 2002), SI 2002 No. 618, as amended. To legally market a medical device in the UK, manufacturers are required to register with the MHRA (Regulating medical devices in the UK, n.d.; The Medical Devices Regulations, 2002). Non-UK-based device manufacturers are required to have a UK Responsible Person, who acts on behalf of manufacturers and developers to ensure regulatory compliance, such as completing declarations of conformity, TD, and ensuring UK MDR 2002 meets conformity assessment procedures. The UK Responsible Person is also requried to support MHRA inspections, report AEs, and initiate corrective actions as and when necessary (Regulating medical devices in the UK, n.d.).

3.3.3.1 Regulatory pathway for medical devices in the UK

For getting authorization to market access, manufacturers are required to have the UK Conformity Assessed (UKCA) marking, which confirms compliance with UK MDR 2002. While CE marking was accepted until June 30, 2023, the UK government has extended its acceptance to support transitional arrangements. Devices that require third-party conformity assessment are required to engage a UK-approved body, listed in the UK Market Conformity Assessment Bodies database (Regulating medical devices in the UK, n.d.). In Northern Ireland, manufacturers are required to follow EU rules and affix CE and UKNI markings (as compared to UKCA), with MHRA-designated UK Notified Bodies conducting relevant assessments (Regulating medical devices in the UK, n.d.).

3.3.3.2 Device classification

Similar to CE marking done in the EU, all the medical devices should be classified appropriately before getting the UKCA marking (UK MHRA, Chapter 2: Classification, n.d.). The categories are

- Class I
- Class IIa
- Class IIb
- Class III

The basis for classification of the devices in the MHRA is similar to the EMA as well, where Class III devices are ranked the highest for their risk to patient safety and effectiveness. And the higher classification of the devices results in a greater level of conformity assessment, with increased evidence requirements of clinical evaluation and documentation. In the UK, the appropriate classification is required to be provided by approved bodies to the MHRA. The same factors that were applicable to EMA are applicable in MHRA as well. These include the body system affected, duration of contact with the body, degree of invasiveness, and local versus systemic effects (UK MHRA, Chapter 2: Classification, n.d.).

3.3.3.2.1 Class I devices

Class I devices represent low to no risk to patient safety and require minimal regulatory oversight. These devices are usually noninvasive and intended for use for a very limited time (UK MHRA, Chapter 2: Classification, n.d.).

3.3.3.2.2 Class IIa devices

Class IIa devices are moderate-risk devices, often surgically invasive, used for short-term use (UK MHRA, Chapter 2: Classification, n.d.).

3.3.3.2.3 Class IIb devices

Class IIa devices are medium- to high-risk devices where the devices are commonly used for long-term or life-sustaining functions (UK MHRA, Chapter 2: Classification, n.d.).

3.3.3.2.4 Class III devices

Class III devices are the highest-risk devices with potential for very long-term usage and having life-sustaining functions (UK MHRA, Chapter 2: Classification, n.d.).

3.3.3.3 Conformity assessment process for various classes of devices

Under the UK Medical Devices Regulations 2002 (UK MDR 2002), the conformity assessment process varies based on the classification, with increasing security for higher-risk devices (UK MHRA, UK conformity assessment, n.d.; Medical devices: conformity assessment and the UKCA mark, n.d.).

3.3.3.3.1 Class I devices

Class I devices represent the lowest risk category of nonsterile, nonmeasuring devices. These devices can be self-certified and ensure their compliance with UKCA marking requirements. The regulatory pathway for their conformity assessment involves the preparation of a DoC, TD, and notification provided to the MHRA. However, for Class I devices, which are sterile (Class Is) and measuring devices (Class Im), assessment by a UK-approved body is mandatory. These devices do not qualify for self-certification due to the added risk posed by sterility breaches or measurement errors (UK MHRA, UK conformity assessment, n.d.; Medical devices: conformity assessment and the UKCA mark, n.d.; Guidance on Class I medical devices, n.d.).

3.3.3.3.2 Class IIa devices

Class IIa devices represent low- to moderate-risk devices. These include short-term, surgically invasive devices. Manufacturers are required to have one of the types of assessment, such as a third-party conformity assessment involving examination and testing of each product or homogenous batch of products (Annex IV), an audit of the production quality assurance system (Annex V), an audit of final inspection and testing (Annex VI), and an audit of the full quality assurance system (Annex II). Once approved by the UK-approved bodies from one of the above conformity assessments, the manufacturers can affix the UKCA to their devices (Medical devices: conformity assessment and the UKCA mark, n.d.).

3.3.3.3.3 Class IIb devices

Class IIb devices are categorized as medium- to high-risk, typically involving prolonged internal use or life-sustaining functions. These require comprehensive quality system audits or batch/sample testing. AI-driven robotic surgical systems frequently fall into this category. Two primary conformity assessment options exist for Class IIb devices including:

1. Full Quality System Audit (Annex II): This pathway includes a comprehensive audit of the manufacturer's entire QMS, covering their design controls, production processes, risk mitigation, and PMS. Maintaining a robust in-house QMS system aligned with ISO 13485 will be an ideal scenario for this pathway.
2. Type Examination with Production Control (Annex III + Annex IV or V or VI): The approved body first evaluates a representative sample of the device for compliance. This is followed by one of the three options:

- Annex III—Type examination
- Annex IV—Batch verification through physical inspection or testing
- Annex V—Audit of production quality assurance system
- Annex VI—Audit of final inspection/testing.

Once approved by the UK-approved bodies from one of the above conformity assessments, the manufacturers can affix the UKCA to their devices (Medical devices: conformity assessment and the UKCA mark, n.d.).

3.3.3.3.4 Class III devices

Class III devices present the highest risk and demand the most intensive regulatory scrutiny. They include life-sustaining, life-supporting, or highly invasive systems. Class III devices demand full regulatory scrutiny, including independent design dossier review and end-to-end quality system evaluations. Manufacturers can follow either one of the below two conformity assessments:

1. Full QMS Review and Design Dossier Review (Annex II): A UK-approved body conducts a full-scope audit of the manufacturer's QMS, including postmarket systems, and independently evaluates the design dossier. This includes validation of device specifications, clinical evidence, risk controls, and conformity with applicable standards.
2. Type Examination and Production Control (Annex III + Annex IV or V or VI): Similar to Class IIb, this pathway involves representative sample review, followed by ongoing production oversight.

For all Class III assessments, the manufacturer is required to provide robust clinical data and risk-benefit analysis. Once approved by the UK-approved bodies from one of the above conformity assessments, the manufacturers can affix the UKCA to their devices (Medical devices: conformity assessment and the UKCA mark, n.d.).

3.3.3.4 Regulatory pathway for artificial intelligence-driven robotic surgery devices in the UK

To address the unique challenges of complex software and adaptive algorithmic technologies such as AI/ML, the MHRA launched the Software and AI as a Medical Device (AIaMD) Change Programme (UK MHRA, Software and AI as a Medical Device Change Programme roadmap, n.d.). This Program introduces a lifecycle-based regulatory model specifying and emphasizing transparency, algorithmic accountability, and real-world performance monitoring. This initiative, detailed in a roadmap released in October 2022, supports innovation by considering the evolving nature of AI-based systems (UK MHRA, Software and AI as a Medical Device Change Programme roadmap, n.d.). The regulatory model enables risk-based classification, with AI-enabled robotic systems categorized primarily as Class IIb or Class III devices. Upon successful assessment, the devices receive a UKCA mark, authorizing their devices to be marketed in the United Kingdom. Manufacturers are required to implement robust PMS systems, including real-world data collection and vigilance protocols to monitor for model drift or performance degradation.

Complementing this regulatory framework, the National Health Service (NHS) has provided a structured strategy for robotic-assisted surgery (RAS) implementation through the "Getting It Right First Time" (GIRFT) initiative. This strategy provides standardized protocols for clinical integration, workforce training, and hospital commissioning (National Health Service England, n.d.; National Health Service England, Implementation of robotic assisted surgery RAS in England, n.d.). Emphasis is placed on clinical effectiveness, the accreditation learning curve, infrastructure readiness, and performance benchmarking through data analytics. There are over 140 RAS systems currently in use across NHS trusts (National Health Service England, Implementation of robotic assisted surgery RAS in England, n.d.). Together, the MHRA's adaptive regulatory approach and the NHS's implementation strategy demonstrate a comprehensive pathway for integrating AI-powered robotic systems into surgical practice while ensuring patient safety and clinical efficacy (Aisu et al., 2022).

3.3.3.5 Change management process

To address the dynamic and adaptive nature of AI-enabled surgical systems, the MHRA introduced the AIaMD Change Protocol (2023) (UK MHRA, Software and AI as a Medical Device Change Programme roadmap, n.d.). This framework aligns the FDA's PCCPs best practices to be included in AIaMD, where predefined strategies for managing postmarket modifications in machine learning-enabled medical devices (MLMDs) are detailed. Moreover, the MHRA mandates that all postmarket updates adhere to performance standards defined by the GIRFT program. This framework ensures safety, usability, and clinical value throughout the lifecycle of AI-driven robotic surgery systems while ensuring alignment with NHS implementation priorities (National Health Service England, n.d.).

3.3.3.6 Postmarket surveillance under Medicines and Healthcare Products Regulatory Agency

The new regulation Medical Devices (PMS Requirements) (Amendment) (Great Britain) Regulations 2024 amends the UK Medical Devices Regulations (MDR) 2002, and this new amendment will be effective from June 16, 2025. The amendment requires manufacturers to establish and maintain a comprehensive PMS system to actively monitor the device's performance, safety, and quality throughout the device's product lifecycle. Some of the key requirements in the amended PMS include timely reporting of serious incidents caused by devices, implementation of field safety corrective actions, and submission of PSURs for medium- and high-risk medical devices. These are applicable to AI/ML-enabled surgical robots as well. Manufacturers are required to conduct trend reporting, detect potential safety signals, and update TD accordingly. The regulations apply to general medical devices, in vitro diagnostic devices, and active implantable medical devices, with specific provisions for different device classes (The Medical Devices Post-market Surveillance Requirements Amendment Great Britain Regulations 2024: guidance on implementation, 2024).

3.3.4 Canada (Health Canada)

In Canada, the regulation of medical devices is overseen by Health Canada under the Medical Devices Regulations (SOR/98–282) to ensure the safety, efficacy, and quality of medical device products. The regulations were put into effect under the Food and Drugs Act (Health Canada, Medical devices, n.d.; Health Canada, Medical Devices Regulations SOR/98-282, n.d.; RAPS, The evolution of Canada's medical device regulatory framework, n.d.). These regulations define a risk-based classification system comprising four categories, between Class I and Class IV, with Class I representing the lowest risk and Class IV the highest. This categorization is based on classification rules as per Schedule 1 of the regulation (Health Canada, Medical Devices Regulations SOR/98-282, n.d.). These rules were significantly borrowed from the European Union's Council Directive 93/42/EEC.

3.3.4.1 Regulatory pathway for medical devices in Canada

Health Canada is responsible for both premarket evaluation and postmarket oversight. For medical devices classified as Class II, III, or IV, manufacturers of the device are required to obtain either a medical device license (MDL) or a medical device establishment license (MDEL) through a structured conformity assessment process (Health Canada, Medical Devices Establishment Licence (MDEL) Listing, n.d.). This includes the submission of clinical data, safety and performance evidence, and a risk-benefit analysis aligned with the device's risk details (Health Canada, Medical Devices Regulations SOR/98-282, n.d.). Additionally, manufacturers are required to implement a QMS certified to ISO 13485:2016, assessed by a registrar recognized under the Medical Device Single Audit Program (Health Canada, Medical Device Single Audit Program MDSAP, n.d.). NonCanadian manufacturers willing to market and distribute medical devices in Canada are required to appoint a Canadian Regulatory Representative, similar to the UK Responsible Person. This representative ensures compliance with Canadian regulatory requirements, including maintaining TD, managing AE reporting, facilitating inspections, and coordinating recalls when necessary. Health Canada also requires robust PMS by manufacturers, such as mandatory reporting of adverse incidents under the Mandatory Problem Reporting framework. Except for Class I devices, only the devices that appear in this listing are offered approval to market their devices in Canada (Health Canada, Medical Devices Active Licence Listing MDALL, n.d.). In recent years, Health Canada has been involved in the regulatory modernization efforts to adapt to the evolving technological advancements, such as ML and AI. One of the initiatives from Health Canada includes the development of specific guidance for MLMDs to ensure the safety and effectiveness of using ML-learning-based devices. This guidance introduces PCCP for MLMDs for preauthorization of planned changes to these devices. Additionally, active participation in global harmonization initiatives, such as the International Medical Device Regulators Forum (IMDRF), provide a harmonized structure for medical device regulatory activities (Health Canada, Pre-market guidance for machine learning-enabled medical devices, n.d.).

3.3.4.2 Device classification

Under Canada's Medical Devices Regulations (SOR/98–282), all medical devices are required to be classified before being licensed for market entry (Health Canada, Medical Devices Regulations SOR/98-282, n.d.). Device classification is based on the body system affected, duration of contact with the body, degree of invasiveness, and local versus systemic effects (Health Canada, Guidance Document - Guidance on the Risk-based Classification System for Non-In Vitro Diagnostic Devices non-IVDDs, n.d.). The Health Canada classification rules are close to the EMA's classification rules. The four classes include:

- Class I devices: These represent the lowest risk. Class I devices include noninvasive technologies with minimal patient interaction (Health Canada, Medical Devices Regulations SOR/98-282, n.d.). These devices generally correspond to Class I devices of EMA.

- Class II devices: These moderate-risk devices include diagnostic software and simple monitoring tools (Health Canada, Medical Devices Regulations SOR/98-282, n.d.). These devices generally correspond to Class IIa devices of EMA.
- Class III Devices: Devices in this category pose a higher risk due to their invasive nature or significant clinical function (Health Canada, Medical Devices Regulations SOR/98-282, n.d.). These devices generally correspond to Class IIb devices of EMA.
- Class IV devices: These represent the highest risk, such as life-sustaining systems or AI devices exhibiting a high degree of autonomy (Health Canada, Medical Devices Regulations SOR/98-282, n.d.). These devices generally correspond to Class III devices of EMA.

3.3.4.3 Conformity assessment process for devices

Under the Canadian Medical Devices Regulations (SOR/98–282), medical devices are classified into four risk-based categories: Class I, II, III, and IV (Health Canada, Medical Devices Regulations SOR/98-282, n.d.).

Class I devices: These devices represent the lowest risk category (e.g., surgical scissors and stethoscopes). Manufacturers and importers are required to obtain an MDEL, but no MDL is required. A DoC with basic safety and labeling requirements is sufficient, with no premarket review required by Health Canada. These devices are subjected to basic postmarket controls.

Class II devices: These devices represent moderate-risk devices, such as catheters and contact lenses. Manufacturers are required to obtain an MDL, submit documentation describing device function, intended use, and labeling, and demonstrate compliance with ISO 13485-certified QMS. They require an MDL and QMS certification under ISO 13485, with TD reviewed by Health Canada.

Class III devices: These devices represent substantial risk devices. An MDL is mandatory, along with a detailed application covering device specifications, risk assessments, sterilization methods, and supporting clinical evidence. The QMS certificate (ISO 13485) is a prerequisite. They require rigorous safety, effectiveness, and design documentation. Applications for licensing are required to contain evidence of their clinical effectiveness. Such evidence includes clinical trials, clinical reviews, meta-analyses, and real-world evidence reviews.

Class IV devices: These devices represent the highest risk. Requirements include all Class III documentation plus in-depth clinical data, risk-benefit analysis, and a summary of global usage history and any reported incidents or recalls. Health Canada conducts a comprehensive regulatory and scientific review. They undergo extensive conformity assessment, including clinical evidence and risk-benefit analysis. Similar to Class III devices, applications for licensing must contain evidence of their clinical effectiveness, such as clinical trials, clinical reviews, meta-analyses, and real-world evidence reviews.

Canada's regulatory approach ensures that higher-risk, AI-enabled surgical technologies are thoroughly assessed for safety, effectiveness, and long-term clinical performance.

3.3.4.4 Regulatory pathway for artificial intelligence-driven robotic surgery devices in Canada

For AI- or ML-enabled devices, Health Canada's 2022 "Guidance Document: Premarket Requirements for Machine Learning-enabled Medical Devices (MLMD)," updated in 2025, provides a dedicated regulatory framework that aligns with global standards established by the IMDRF. Class I MLMDs are low-risk devices that do not significantly influence clinical decisions and outcomes. Class II MLMDs that do not pose high risks but are used for medical purposes, such as monitoring or supporting clinical workflows without being critical to patient safety. For Class II MLMDs, clinical validation data should be provided to the regulatory agency upon request, though it is not required to be sent initially. Class III MLMDs are higher-risk devices that support diagnosis or treatment that may cause serious health consequences if errors occur. For these devices, the documentation must include clinical validation data to demonstrate their safety and effectiveness. Class IV MLMDs pose the highest risk, involving devices used in life-threatening conditions or where incorrect output could cause severe harm or lead to death. Applications for Class IV MLMDs must include robust clinical validation studies to demonstrate safety and effectiveness in the intended population (Health Canada, Pre-market guidance for machine learning-enabled medical devices, n.d.).

3.3.4.5 Change management process

As previously seen with the US FDA, PCCP is a regulatory mechanism that has been adopted by Health Canada to manage planned modifications to ML- and AI-based devices (Health Canada, Pre-market guidance for machine learning-enabled medical devices, n.d.; Health Canada, Predetermined change control plans for machine learning-enabled medical devices: Guiding principles, n.d.). This mechanism is specifically required for iterative updates, and adaptive changes to these AI and ML algorithms in devices are expected. This ensures preauthorization of certain modifications to the devices, which ensures timely updates while helping the device remain within its intended use and meet safety and effectiveness standards. Health Canada specifies that, for PCCPs, it is a risk-based approach and should be supported by appropriate evidence to provide a high degree of transparency. All modifications mentioned in a PCCP must ensure that the device continues to operate within its intended use. Changes that alter the medical purpose or intended use of the device require a separate MDL amendment application (Health Canada, Predetermined change control plans for machine learning-enabled medical devices: Guiding principles, n.d.).

3.3.4.6 Scientific advice and expert consultation

For developers of innovative medical technologies, such as AI-driven surgical robotic systems, early regulatory engagement is crucial. Health Canada offers presubmission or preapplication meetings, similar in purpose to the FDA's Q-Sub program, to enable proactive discussion between manufacturers and regulatory authorities before filing a submission. These meetings provide an opportunity to discuss critical elements such as clinical trial strategy, data requirements, regulatory pathways, and the suitability of third-party data. Manufacturers may request a meeting for device submissions, clinical trial applications, priority reviews, or responses to regulatory deficiencies. Requests are required to be submitted at least three months in advance and must include product details, key questions, and proposed dates. These meetings help clarify Health Canada's expectations, identify potential issues early, and allow manufacturers to align submission content with regulatory requirements. Following the meeting, manufacturers are required to submit minutes of the meeting and reference this discussion in their final submission for review and approval. This structured engagement improves submission quality, minimizes review delays, and supports the efficient regulatory evaluation of novel medical technologies in Canada (Health Canada, Guidance Document: The Management of Drug Submissions and Applications, n.d.).

3.3.4.7 Postmarket requirements

PMS is a critical regulatory obligation for medical device manufacturers in Canada. Manufacturers are required to implement PMS systems to continuously monitor device performance once they are on the market. This includes actively collecting and analyzing AE data and reporting incidents to Health Canada in compliance with the Mandatory Problem Reporting framework. Specifically for Class III and IV devices, manufacturers are required to submit PSURs that provide a comprehensive summary of safety-related data, including reported incidents, benefit-risk reassessment, and any corrective or preventive actions taken (Health Canada, Questions and Answers Regarding the Implementation of a Risk-Prioritized Periodic Safety Update Report Regulatory Review Pilot PSUR-RRP at Health Canada, n.d.). In the scenarios where premarket clinical data is limited or based on equivalence, PMCF may also be required. PMCF is particularly required for AI/ML-enabled surgical systems, where long-term safety and performance data may evolve postapproval (Health Canada, Guidance on clinical evidence requirements for medical devices: Additional obligations during post-market phase, n.d.). Additionally, manufacturers are required to manage and report any recalls to Health Canada (Health Canada, Incident reporting for medical devices: Guidance document, n.d.). These postmarket requirements ensure a lifecycle approach to device oversight, enhancing patient safety and facilitating continuous risk management for complex, AI-integrated surgical technologies deployed in real-world clinical settings (Health Canada, Questions and Answers Regarding the Implementation of a Risk-Prioritized Periodic Safety Update Report Regulatory Review Pilot PSUR-RRP at Health Canada, n.d.; Health Canada, Guidance on clinical evidence requirements for medical devices: Additional obligations during post-market phase, n.d.; Health Canada, Incident reporting for medical devices: Guidance document, n.d.).

3.3.5 Japan (Pharmaceuticals and Medical Devices Agency)

Japan's regulation on medical devices is jointly overseen by two key authorities, the Ministry of Health, Labour and Welfare (MHLW) and the PMDA. The overarching regulatory framework for devices in Japan is governed by the Pharmaceuticals and Medical Devices (PMD) Act. The MHLW is responsible for overseeing the overall policies on medical device regulation, decisions on device approval, and final classification of products as medical devices, while the PMDA works as a scientific and technical authority that conducts premarket reviews of the devices, provides consultation for manufacturers, and oversees postmarket safety surveillance activities.

3.3.5.1 Regulatory pathway for medical devices in Japan

Together, the MHLW and PMDA provide an integrated regulatory framework designed to support the implementation of innovative technologies enabling technological advancements while ensuring stringent safety, efficacy, and quality standards in the deployment of these complex technologies (Japan PMDA, Regulations and Approval/Certification of Medical Devices, n.d.). Non-Japanese manufacturers who want to market their devices in Japan are required to have a Marketing Authorization Holder (MAH) or Designated Marketing Authorization Holder (DMAH), similar to the UK Responsible Person in the United Kingdom, for ensuring ongoing compliance, including periodic safety reporting and postmarket safety surveillance activities (Japan PMDA, Regulations and Approval/Certification of Medical Devices, n.d.; Sharma et al., 2023). In the year 2005, a revision in the PMD Act ensured alignment of Japan's regulatory framework with global regulatory standards and international regulatory harmonization, such as the IMDRF (Sharma et al., 2023). Japan also emphasizes the importance of aligning with the regulation of SaMD, to include ML and AI algorithms, with evolving guidance on continuous learning systems and adaptive AI models (Japan PMDA, Report on AI-based Software as a Medical Device SaMD, n.d.).

3.3.5.2 Device classification

In Japan, medical devices are classified into four categories based on varying levels of risk to the patients (Japan PMDA, Regulations and Approval/Certification of Medical Devices, n.d.).

- Class I
- Class II
- Class III
- Class IV

3.3.5.2.1 Class I devices

Class I devices are considered extremely low risk to the patients (Japan PMDA, Regulations and Approval/Certification of Medical Devices, n.d.).

3.3.5.2.2 Class II devices

Class II devices are considered low risk to the patients (Japan PMDA, Regulations and Approval/Certification of Medical Devices, n.d.).

3.3.5.2.3 Class III devices

Class III devices are considered medium risk to the patients (Japan PMDA, Regulations and Approval/Certification of Medical Devices, n.d.).

3.3.5.2.4 Class IV devices

Class IV devices are considered high risk to patients' lives (Japan PMDA, Regulations and Approval/Certification of Medical Devices, n.d.).

3.3.5.3 Regulatory pathway for medical devices in Japan

As mentioned above, all the medical devices are categorized into one of the four risk-based classes (Class I through IV), with classification determined using the Japanese Medical Device Nomenclature (JMDN) database. Based on the classified devices, there are individual regulatory pathways for the classified devices and marketing the devices in Japan. Non-Japanese manufacturers are first required to appoint a DMAH in Japan and complete Foreign Manufacturer Establishment Registration with the PMDA. Also, manufacturers are required to implement a QMS compliant with MHLW Ordinance No. 169, which aligns with ISO 13485:2016 (Japan PMDA, Revision of Japanese Medical Device QMS requirements, n.d.).

Class I devices: These devices represent an extremely low risk to the patients. The regulatory pathway involves providing a notification to the regulatory agency, PMDA, as part of registering their device, where a self-declaration by the MAH on behalf of the manufacturers is sufficient for marketing the device. Review and approval from MHLW or a Registered Certification Body (RCB) is not required (Japan PMDA, Regulations and Approval/Certification of Medical Devices, n.d.).

Class II devices: These devices represent a low risk to patients. The regulatory pathway involves either certification or an approval process, depending on the nature of the device and the reviewing authority. The pathway involves certification by Japan's

RCB if an existing JMDN code is available for these devices. However, if these devices do not have a precedent device or if they requires more clinical evidence to prove their appropriate classification, they are required to undergo a review and approval process by the MHLW (Japan PMDA, Regulations and Approval/Certification of Medical Devices, n.d.).

Class III devices: These devices represent a medium risk to patients. The regulatory pathway may involve a certification process with RCB certification. However, most Class C devices require review and approval from MHLW. Clinical studies are needed, particularly for novel or complex devices coming under Class C, supported by clinical data and detailed TD, for safety and efficacy, with reviews conducted exclusively by the MHLW (Japan PMDA, Regulations and Approval/Certification of Medical Devices, n.d.).

Class IV devices: These devices represent a serious risk to the lives of the patients. The regulatory pathway does not involve RCB certification but is required to undergo a rigorous evaluation and formal approval by the MHLW. Clinical studies are mandatory, often required prior to submission, which involves a comprehensive review of clinical and manufacturing controls to ensure safety and efficacy (Japan PMDA, Regulations and Approval/Certification of Medical Devices, n.d.).

Irrespective of the class of the device, MAH is responsible for ensuring the safety, efficacy, and quality of the device based on the evidence before the device is either notified to PMDA, sent for certification to the RCB, or reviewed and approved by the MHLW. With the AI-driven surgical robotic systems falling under Class III or Class IV due to their complexity and potential for having a critical clinical impact on the health and lives of patients.

3.3.5.4 Regulatory pathway for artificial intelligence-driven robotic surgery devices in Japan

In Japan, SaMD is regulated under the PMD Act, which classifies medical devices into four risk-based categories. SaMDs are subject to this classification, where Class I represents the lowest-risk and Class IV the highest-risk devices. The classification of SaMDs based on risk shall have the regulatory pathway based on their classes. The first ML/DL-based device was approved in December 2018. The PMDA oversees the review and approval processes, ensuring that SaMDs meet stringent safety, efficacy, and quality standards. Additionally, Japan has implemented a two-step approval process for certain SaMDs. The first step grants approval based on analytical performance, allowing the product to enter the market. The second step requires additional clinical evidence to confirm the device's clinical benefits. This phased approach enables earlier access to innovative technologies while ensuring comprehensive evaluation of their clinical utility (Japan PMDA, Report on AI-based Software as a Medical Device SaMD, n.d.).

3.3.5.5 Change management process

To address the unique challenges of AI/ML-enabled SaMDs, Japan has introduced the Post-Approval Change Management Protocol (PACMP). PACMP allows for predefined changes to be made postapproval, facilitating continuous learning and improvement of AI algorithms while maintaining regulatory compliance. This approach ensures that updates enhancing device performance can be implemented efficiently without compromising safety. Japan's regulatory pathway to a structured regulatory pathway for managing anticipated postmarket modifications to medical devices, particularly those incorporating AI technologies, is the PACMP. PACMP facilitates continuous improvement by allowing manufacturers to predefine potential changes, such as adjustments in device size, components, or performance enhancements, and outline corresponding validation protocols. This ensures that modifications aimed at improving diagnostic accuracy or integrating real-world data can be implemented efficiently without compromising safety or efficacy (International Medical Device Regulators Forum IMDRF, Regulatory Update from PMDA – Japan, n.d.).

3.3.5.6 Scientific advice and expert consultation

The PMDA provides consultations to support the development and regulatory approval process of medical devices. These consultations provide scientific and regulatory guidance on trial design, data requirements, and compliance with Japanese regulations. Proactive engagement with PMDA assists the manufacturers in identifying potential issues, aligning development strategies with regulatory expectations, and streamlining the approval process. Consultations can be requested at various stages, including preclinical, preIND, and preapplication phases. For non-Japanese manufacturers, MAH facilitates communication and ensures adherence to local requirements. This proactive approach enhances the likelihood of successful product development and timely market entry in Japan (Japan PMDA, Consultations, n.d.).

3.3.5.7 Postmarket requirements

The PMDA mandates a comprehensive PMS system to ensure the ongoing safety and efficacy of medical devices throughout their lifecycle. Under the PMD Act, manufacturers and MAHs are mandated to report AEs and device malfunctions within 15–30 days, depending on severity. The PMDA follows a multistep process of collecting safety data from MAHs and healthcare providers, analyzing this collected information through expert consultations, and implementing necessary safety measures to record the safety and efficacy details. Some of the measures include device recalls, modifications, labeling updates, or enhanced monitoring (Japan PMDA, Post-marketing Safety Measures of medical devices, n.d.)

References

Aboy M, Crespo C, Stern A. Beyond the 510(k): the regulation of novel moderate-risk medical devices, intellectual property considerations, and innovation incentives in the FDA's De Novo pathway. Digital Medicine 2024;7(1). Available from https://doi.org/10.1038/s41746-024-01021-y.

Aisu N, Miyake M, Takeshita K, Akiyama M, Kawasaki R, Kashiwagi K, et al. Regulatory-approved deep learning/machine learning-based medical devices in Japan as of 2020: a systematic review. PLOS Digital Health 2022;1(1):e0000001. Available from https://doi.org/10.1371/journal.pdig.0000001.

Bernad DM. Humanitarian use device and humanitarian device exemption regulatory programs: pros and cons. Expert Review of Medical Devices 2009;6(2):137–45. Available from https://doi.org/10.1586/17434440.6.2.137Canada, http://www.expert-reviews.com/doi/pdf/10.1586/17434440.6.2.137.

Confins, Conformity assessment procedures and premarket scrutiny under EU MDR; n.d. https://confinis.com/conformity-assessment-procedures-and-premarket-scrutiny-under-eu-mdr/.

Elendu C, Amaechi DC, Elendu TC, Jingwa KA, Okoye OK, John Okah M, et al. Ethical implications of AI and robotics in healthcare: a review. Medicine (United States) 2023;102(50):E36671. Available from https://doi.org/10.1097/MD.0000000000036671, https://journals.lww.com/md-journal/pages/default.aspx.

European Medical Agency, AI Act; n.d. https://digital-strategy.ec.europa.eu/en/policies/regulatory-framework-ai.

European Medical Agency, Annex I – general safety and performance requirements; n.d. https://www.medical-device-regulation.eu/2019/07/23/annex-i-general-safety-and-performance-requirements/.

European Medical Agency, Annex IX, conformity assessment based on a quality management system and on assessment of technical documentation; n.d. https://www.medical-device-regulation.eu/2019/08/14/annex-ix/.

European Medical Agency, Annex VIII; n.d. https://www.medical-device-regulation.eu/2019/08/08/annex-viii/.

European Medical Agency, Annex X, conformity assessment based on type examination; n.d. https://www.medical-device-regulation.eu/2019/08/14/annex-x/.

European Medical Agency, Annex XI, conformity assessment based on product conformity verification; n.d. https://www.medical-device-regulation.eu/2019/08/14/annex-xi/.

European Medical Agency, Annex XIV - clinical evaluation and post-market clinical follow-up; n.d. https://www.medical-device-regulation.eu/2019/08/14/annex-xiv/.

European Medical Agency, Article 6: classification rules for high-risk AI systems; n.d. https://artificialintelligenceact.eu/article/6/.

European Medical Agency, Artificial intelligence in medical devices questionnaire; n.d. https://www.team-nb.org/wp-content/uploads/2024/11/Team-NB-PositionPaper-AI-in-MD-Questionnaire-V1-20241125.pdf.

European Medical Agency, CG 2019-15 rev.1 Guidance notes for manufacturers of Class I Medical Devices; n.d. https://health.ec.europa.eu/document/download/349a2d4c-ea2a-4279-861c-7c063bc077e4_en?filename=md_guidance-manufacturers_en.pdf.

European Medical Agency, Conformity assessment; n.d. https://single-market-economy.ec.europa.eu/single-market/goods/building-blocks/conformity-assessment_en.

European Medical Agency, EU postmarket surveillance plans for medical devices; n.d. https://pmc.ncbi.nlm.nih.gov/articles/PMC6771951/.

European Medical Agency, Extension of the MDR transitional period and removal of the 'sell off' periods; n.d. https://health.ec.europa.eu/document/download/592008f6-3456-4afb-a13a-733a87da1b00_en?filename=mdr_proposal_extension-q-n-a_1.pdf.

European Medical Agency, Guidance on classification of medical devices; n.d. https://health.ec.europa.eu/system/files/2021-10/mdcg_2021-24_en_0.pdf.

European Medical Agency, High-risk medical devices: consultation procedures and advice; n.d. https://www.ema.europa.eu/en/high-risk-medical-devices-consultation-procedures-advice.

European Medical Agency, Market surveillance and vigilance; n.d. https://health.ec.europa.eu/medical-devices-sector/directives/market-surveillance-and-vigilance_e.

European Medical Agency, MDR – Article 61 – clinical evaluation; n.d. https://www.medical-device-regulation.eu/2019/07/11/mdr-article-61-clinical-evaluation/.
European Medical Agency, MDR Article 52, conformity assessment procedures; n.d. https://www.medical-device-regulation.eu/2019/07/11/mdr-article-52-conformity-assessment-procedures/.
European Medical Agency, Medical device expert panels; n.d. https://www.ema.europa.eu/en/human-regulatory-overview/medical-devices/medical-device-expert-panels.
European Medical Agency, Medical devices; n.d. https://www.ema.europa.eu/en/human-regulatory-overview/medical-devices.
European Medical Agency, Notified bodies; n.d. https://single-market-economy.ec.europa.eu/single-market/goods/building-blocks/notified-bodies_en.
European Medical Agency, Questions & Answers for applicants, marketing authorisation holders of medicinal products and notified bodies with respect to the implementation of the Regulations on medical devices and in vitro diagnostic medical devices (Regulations (EU) 2017/745 and (EU) 2017/746). https://www.ema.europa.eu/en/documents/regulatory-procedural-guideline/questions-answers-implementation-medical-devices-vitro-diagnostic-medical-devices-regulations-eu-2017-745-eu-2017-746_en.pdf.
European Medical Agency, Regulation (EU) 2017/745: clinical evidence needed for medical devices previously CE marked under Directives 93/42/EEC or 90/385/EEC; n.d. https://ec.europa.eu/docsroom/documents/40904.
FDA, Classify your medical device; n.d. https://www.fda.gov/medical-devices/overview-device-regulation/classify-your-medical-device.
FDA, Computer assisted surgical systems; n.d. https://www.fda.gov/medical-devices/surgery-devices/computer-assisted-surgical-systems.
FDA, De Novo classification process (evaluation of automatic class III designation); n.d. https://www.fda.gov/media/72674/download?attachment%20%5D%20%5B.
FDA, De Novo classification request; n.d. https://www.fda.gov/medical-devices/premarket-submissions-selecting-and-preparing-correct-submission/de-novo-classification-request#What_is_a_De_Novo_Classification_Request_.
FDA, Device classification under section 513(f)(2)(De Novo); n.d. https://www.accessdata.fda.gov/scripts/cdrh/cfdocs/cfpmn/denovo.cfm.
FDA. Establishment registration & device listing. Available from https://www.accessdata.fda.gov/scripts/cdrh/cfdocs/cfRL/rl.cfm.
FDA, How to determine if your product is a medical device; n.d. https://www.fda.gov/medical-devices/classify-your-medical-device/how-determine-if-your-product-medical-device.
FDA, Humanitarian device exemption; n.d. https://www.fda.gov/medical-devices/premarket-submissions-selecting-and-preparing-correct-submission/humanitarian-device-exemption.
FDA, Marketing Submission recommendations for a predetermined change control plan for artificial intelligence-enabled device software functions; n.d. https://www.fda.gov/media/166704/download.
FDA, PMA application methods; n.d. https://www.fda.gov/medical-devices/premarket-approval-pma/pma-application-methods.
FDA, Predetermined change control plans for machine learning-enabled medical devices: guiding principles; n.d. https://www.fda.gov/medical-devices/software-medical-device-samd/predetermined-change-control-plans-machine-learning-enabled-medical-devices-guiding-principles.
FDA, Premarket approval application and humanitarian device exemption modular review; n.d. https://www.fda.gov/media/73513/download.
FDA, Premarket Approval (PMA); n.d. https://www.fda.gov/medical-devices/premarket-submissions-selecting-and-preparing-correct-submission/premarket-approval-pma.
FDA, Premarket Approval (PMA) database; n.d. https://www.accessdata.fda.gov/scripts/cdrh/cfdocs/cfPMA/pma.cfm.
FDA, Requests for Feedback and meetings for medical device submissions: the Q-submission program; n.d. https://www.fda.gov/media/114034/download.
FDA, The 510(k) program: evaluating substantial equivalence in premarket notifications [510(k)]; n.d. https://www.fda.gov/media/82395/download?attachment%5D.

George EI, Brand TC, LaPorta A, Marescaux J, Satava RM. Origins of robotic surgery: from skepticism to standard of care. JSLS: Journal of the Society of Laparoendoscopic Surgeons 2018;22(4). Available from https://doi.org/10.4293/JSLS.2018.00039.

Guidance on Class I medical devices; n.d. https://www.gov.uk/government/collections/guidance-on-class-1-medical-devices.

Guidance on Classification rules for in vitro diagnostic medical devices under regulation (EU) 2017/746; 2025. https://health.ec.europa.eu/document/download/12f9756a-1e0d-4aed-9783-d948553f1705_en?filename=md_mdcg_2020_guidance_classification_ivd-md_en.pdf.

Health Canada, Guidance document - guidance on the risk-based classification system for non-in vitro diagnostic devices (non-IVDDs); n.d. https://www.canada.ca/en/health-canada/services/drugs-health-products/medical-devices/application-information/guidance-documents/guidance-document-guidance-risk-based-classification-system-non-vitro-diagnostic.html.

Health Canada, Guidance document: the management of drug submissions and applications; n.d. https://www.canada.ca/en/health-canada/services/drugs-health-products/drug-products/applications-submissions/guidance-documents/management-drug-submissions/industry/document.html.

Health Canada, Guidance on clinical evidence requirements for medical devices: additional obligations during post-market phase; n.d. https://www.canada.ca/en/health-canada/services/drugs-health-products/medical-devices/application-information/guidance-documents/clinical-evidence-requirements-medical-devices/additional-obligations-post-market-phase.html.

Health Canada, Incident reporting for medical devices: guidance document; n.d. https://www.canada.ca/en/health-canada/services/drugs-health-products/reports-publications/medeffect-canada/incident-reporting-medical-devices-guidance-2021.html.

Health Canada, Medical Device Single Audit Program (MDSAP); n.d. https://www.canada.ca/en/health-canada/services/drugs-health-products/medical-devices/activities/international/transition-medical-device-single-audit-program.html.

Health Canada, Medical devices; n.d. https://www.canada.ca/en/health-canada/services/drugs-health-products/medical-devices.html.

Health Canada, Medical Devices Active Licence Listing (MDALL); n.d. https://www.canada.ca/en/health-canada/services/drugs-health-products/medical-devices/licences/medical-devices-active-licence-listing.html.

Health Canada, Medical Devices Establishment Licence (MDEL) Listing; n.d. https://health-products.canada.ca/mdel-leim/.

Health Canada, Medical Devices Regulations (SOR/98-282); n.d. https://laws-lois.justice.gc.ca/eng/regulations/sor-98-282/.

Health Canada, Predetermined change control plans for machine learning-enabled medical devices: Guiding principles; n.d. https://www.canada.ca/en/health-canada/services/drugs-health-products/medical-devices/good-machine-learning-practice-medical-device-development/predetermined-change-control-plans-machine-learning-enabled-medical-devices.html.

Health Canada, Pre-market guidance for machine learning-enabled medical devices; n.d. https://www.canada.ca/en/health-canada/services/drugs-health-products/medical-devices/application-information/guidance-documents/pre-market-guidance-machine-learning-enabled-medical-devices.html.

Health Canada, Questions and answers regarding the implementation of a risk-prioritized periodic safety update report regulatory review pilot (PSUR-RRP) at Health Canada; n.d. https://www.canada.ca/en/health-canada/services/drugs-health-products/drug-products/applications-submissions/guidance-documents/product-vigilance/questions-answers-implementation-risk-prioritized-periodic-safety-update-report-regulatory-review-pilot-psur-drug-products.html.

Iftikhar M, Saqib M, Zareen M, Mumtaz H. Artificial intelligence: revolutionizing robotic surgery: review. Annals of Medicine & Surgery 2024;86(9):5401–9. Available from https://doi.org/10.1097/ms9.0000000000002426.

Johner Institut, Expert report-comparison of the regulatory systems in the USA and EU with respect to medical device safety; 2023 https://www.swiss-medtech.ch/sites/default/files/2024-01/230822_Gutachten_Johner_Systemvergleich_US-EU_FINAL%20EN.pdf.

International Medical Device Regulators Forum (IMDRF), Regulatory Update from PMDA – Japan; n.d. https://www.imdrf.org/sites/default/files/2024-03/Japan.pdf.

Japan PMDA, Consultations; n.d. https://www.pmda.go.jp/english/review-services/consultations/0002.html.

Japan PMDA, Post -marketing Safety Measures of medical devices; n.d. https://www.pmda.go.jp/files/000266885.pdf.

Japan PMDA, Regulations and approval/certification of medical devices; n.d. https://www.pmda.go.jp/english/review-services/reviews/0004.html.

Japan PMDA, Report on AI-based Software as a Medical Device (SaMD); n.d. https://www.pmda.go.jp/files/000266100.pdf.

Japan PMDA, Revision of Japanese Medical Device QMS requirements; n.d. https://www.pmda.go.jp/english/review-services/regulatory-info/0004.html.

Lee A, Baker TS, Bederson JB, Rapoport BI. Levels of autonomy in FDA-cleared surgical robots: a systematic review. Digital Medicine 2024;7(1). Available from https://doi.org/10.1038/s41746-024-01102-y, https://www.nature.com/npjdigitalmed/.

Medical Device Coordination Group (MDCG) Working Groups; n.d. https://health.ec.europa.eu/medical-devices-dialogue-between-interested-parties/medical-device-coordination-group-working-groups_en#one.

Medical devices: conformity assessment and the UKCA mark; n.d. https://www.gov.uk/guidance/medical-devices-conformity-assessment-and-the-ukca-mark.

Mennella C, Maniscalco U, De Pietro G, Esposito M. Ethical and regulatory challenges of AI technologies in healthcare: a narrative review. Heliyon 2024;10(4):e26297. Available from https://doi.org/10.1016/j.heliyon.2024.e26297.

National Health Service (England); n.d. https://gettingitrightfirsttime.co.uk/wp-content/uploads/2025/05/FINAL_NHS-England-and-GIRFT-implementation-of-robotic-assisted-surgery-in-England_06-19-2025-1.pdf.

National Health Service (England), Implementation of robotic assisted surgery (RAS) in England; n.d. https://gettingitrightfirsttime.co.uk/wp-content/uploads/2025/05/FINAL_NHS-England-and-GIRFT-implementation-of-robotic-assisted-surgery-in-England_06-19-2025-1.pdf.

Nüssler A. The new European medical device regulation: friend or foe for hospitals and patients? Injury 2023;54:110907. Available from https://doi.org/10.1016/j.injury.2023.110907.

RAPS, De Novo classifications: FDA drafts guidance on acceptance reviews, finalizes guidance from; 2014 https://www.raps.org/News-and-Articles/News-Articles/2017/10/De-Novo-Classifications-FDA-Drafts-Guidance-on-Ac.

RAPS, The evolution of Canada's medical device regulatory framework; n.d. https://www.raps.org/News-and-Articles/News-Articles/2023/6/The-evolution-of-Canada-s-medical-device-regulator.

Regulating medical devices in the UK, n.d. https://www.gov.uk/guidance/regulating-medical-devices-in-the-uk.

Sharma Y, Singh D, Singh A. An introduction to regulatory aspects related to medical devices in Japan and Canada. Current Indian Science 2023;01. Available from https://doi.org/10.2174/2210299x01666230720103101.

Study supporting the monitoring of the availability of medical devices on the EU market; n.d. https://health.ec.europa.eu/document/download/59b9d90e-be42-4895-9f6f-bec35138bb0a_en.

The Medical Devices (Post-market Surveillance Requirements) (Amendment) (Great Britain) regulations 2024: guidance on implementation; 2024 https://www.gov.uk/government/publications/medical-devices-post-market-surveillance-requirements/the-medical-devices-post-market-surveillance-requirements-amendment-great-britain-regulations-2024-guidance-on-implementation.

The Medical Devices Regulations; 2002 https://www.legislation.gov.uk/uksi/2002/618/contents/made.

UK MHRA, Chapter 2: Classification; n.d. https://www.gov.uk/government/consultations/consultation-on-the-future-regulation-of-medical-devices-in-the-united-kingdom/chapter-2-classification.

UK MHRA, Software and AI as a Medical Device Change Programme roadmap; n.d. https://www.gov.uk/government/publications/software-and-ai-as-a-medical-device-change-programme/software-and-ai-as-a-medical-device-change-programme-roadmap.

UK MHRA, UK conformity assessment; n.d. https://www.gov.uk/guidance/uk-conformity-assessment#combining-eu-and-uk-certification,%20Chapter%202:%20Classification%5D%20%5Bhttps://www.gov.uk/guidance/medical-devices-conformity-assessment-and-the-ukca-mark,%20Conformity%20Assessment.

Van Norman GA. Drugs, devices, and the FDA: part 2: an overview of approval processes: FDA approval of medical devices. JACC: Basic to Translational Science 2016;1(4):277–87. Available from https://doi.org/10.1016/j.jacbts.2016.03.009, http://www.journals.elsevier.com/jacc-basic-to-translational-science.

Van Norman GA. Expanded patient access to investigational new devices: review of emergency and nonemergency expanded use, custom, and 3D-printed devices. JACC: Basic to Translational Science 2018;3(4):533–44. Available from https://doi.org/10.1016/j.jacbts.2018.06.006, http://www.journals.elsevier.com/jacc-basic-to-translational-science.

Yang GZ, Cambias J, Cleary K, Daimler E, Drake J, Dupont PE, et al. Medical robotics-regulatory, ethical, and legal considerations for increasing levels of autonomy. Science Robotics 2017;2(4). Available from https://doi.org/10.1126/scirobotics.aam8638, http://robotics.sciencemag.org/content/robotics/2/4/eaam8638.full.pdf.

CHAPTER 4

Training healthcare professionals for artificial intelligence-driven medical robotics

Neetha S.S.[1], **Swati Sah**[2], **and Deepa S.**[1]
[1]Department of Computer Science, Christ University, Bangalore, Karnataka, India,
[2]School of Computing Science and Engineering, Sharda University, Greater Noida, Uttar Pradesh, India

4.1 Introduction

The convergence of artificial intelligence (AI) and medical robotics is a critical innovation in practice today. AI-assisted medical robotics is successfully transforming medicine to improve diagnostic accuracy, enhance procedural accuracy, and improve operational skills in medicine as a whole. Medical robotic systems with AI are now unavoidable aspects of surgery, rehabilitation, diagnostic procedures, and patient care with exceptional skills that exceed the boundaries of traditional medical practice (Faiyazuddin et al., 2025). Artificially intelligent robotic systems utilize advanced machine learning algorithms and complex computational processes, vision practices, and real-time data analysis in their efforts to assist medical professionals in the provision of quality patient-centered healthcare services. A good example is robotic surgery systems are able to process preoperative images, suggest optimal surgical paths, and provide intraoperative support, thereby decreasing complications and enhancing clinical results (Shen and Gu, 2024). Diagnostic robotic AI is able to detect subtle imaging or pathology slide anomalies with accuracy that is equally good or even superior to human experts, thus enabling early detection of life-threatening diseases (Ness et al., 2024). Even though the high-tech aids in the clinical environment do not necessarily ensure their efficient utilization. Efficient utilization of AI-based medical robots requires a healthcare professional's ability to comprehend, make use of, and interact with sophisticated technology systems. Conventional medical education is mainly anatomical, physiological, and pathological, and understanding often ignores the need for the interdisciplinary framework required to meet modern challenges and hazards triggered by robots and AI (Francis et al., 2023).

This highlights the need to re-engineer existing educational systems to include computational thinking abilities, ethics, control of robotic systems, and data analysis. Increasing use of AI within health settings has important implications for medical decision support systems. In principle, AI algorithms have the ability to merge enormous datasets and select the most effective clinical treatments, but they are limited. Algorithmic bias, data privacy issues, and system compatibility require an urgent need for serious consideration. Healthcare

AI-Powered Developments in Medical Robotics
DOI: https://doi.org/10.1016/B978-0-443-27783-2.00001-6

professionals must be trained not only to use such technologies but also to critically assess their results and assume ultimate accountability in clinical decision-making (Vaidya et al., 2024). The global health environment also emphasizes profound inequalities in access to AI-enabled robotic technology. Whereas the world's best healthcare facilities in the developed world are at the forefront of AI adoption, many healthcare environments, especially in low- and middle-income countries, are lacking in the facilities and training infrastructure to fully utilize these technologies at scale. This disparity exacerbates existing healthcare disparities and underscores the necessity for scalable and standardized training systems that can be deployed across different healthcare environments (Nimkar et al., 2024). In this regard, training healthcare professionals in the skills critical to AI-enabled robotic medicine is not a secondary concern; it is an issue of extreme priority. Such training programs need to be adaptive, multidisciplinary, and responsive to technological changes. They need to encompass conceptual learning and hands-on training with actual or simulated robotic systems, as well as extensive ethical training. Only through the creation of solid, forward-looking learning systems is the full potential of AI-enabled medical robotics achievable in clinical medicine.

This chapter critiques the prevailing paradigm of AI-enhanced medical robotics, specifies the technical and nontechnical competencies of healthcare professionals, and evaluates evidence-based pedagogies. It also considers issues of successful implementation of training systems and makes strategic recommendations for curricula development, policy support, and worldwide standardization. By providing required knowledge and skills to healthcare professionals to effectively communicate with AI, the medical practice can ensure that technological advances are translated into improved care and clinical outcomes (Joshi et al., 2025).

4.2 Fundamentals of artificial intelligence-driven medical robotics

The AI and robotics have created a new generation of intelligent machines that stand to revolutionize the very fabric of healthcare delivery. Medical robotics with AI is a prevalent phrase used to describe robotic systems augmented by cognitive computing capabilities that include, among others, machine learning, computer vision, and natural language processing (NLP). The advanced machines can handle complicated data, decide based on knowledge, and complete clinical responsibilities with some level of speed and precision never imagined by human professionals confined to operating in silos (Faiyazuddin et al., 2025).

4.2.1 Overview of artificial intelligence in healthcare robotics

At the heart of AI robotics in medicine lies the capacity of machines to learn from information and improve their functional effectiveness independently. In contrast to conventional rule-based systems, AI robotics technologies can learn how to transform themselves to fit new data, identify underlying patterns, and provide real-time assistance to clinical decision-making. These capacities make them extremely useful in scenarios where instant data-driven action is needed, such as operating rooms, intensive care units, and diagnostic laboratories (Shen and Gu, 2024). The application of AI is prevalent across many robotic systems, such as (but not limited to) surgical automation, diagnostic imaging, rehabilitation, pharmaceutical supply chains, and telehealth services. The algorithms based on AI, integrated into these robotic systems, process multimodal datasets consisting of medical imaging, sensor data, and electronic health records (EHRs), thereby enabling faster and more accurate decision-making processes. For example, sophisticated robotic assistants can assist surgeons by determining the most optimal incisions or highlighting areas of potential risks based on historical case data and predictive analytics (Ness et al., 2024).

4.2.2 Types of artificial intelligence-driven medical robots

Robotics enabled by AI can be generally categorized into different functional domains, each serving particular purposes in the field of patient care:

- Surgical robots: These advanced systems facilitate minimally invasive surgery using high-definition visualization, enhanced precision control, and less patient trauma, as indicated in Fig. 4.1 (Introducing NVIDIA Isaac for Healthcare, an AI-Powered Medical Robotics Development Platform, 2025). The da Vinci Surgical System, for instance, employs AI to convert a surgeon's hand movements into micro-movements of the instruments, thus increasing dexterity and eliminating tremors (Francis et al., 2023).
- Diagnostic robots: Diagnostic intent devices driven by AI can analyze radiological images, samples of pathology, or biometric information to identify diseases like

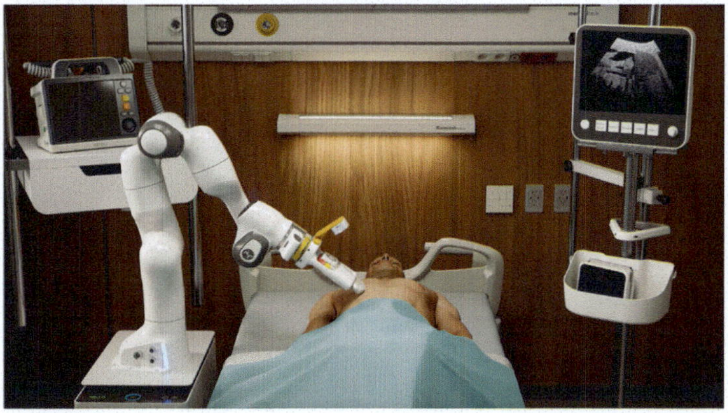

Figure 4.1 Surgical robot.

tumors, tuberculosis, or cardiovascular defects with high accuracy. Robotic systems work in collaboration with medical practitioners by indicating abnormalities for further investigation (Vaidya et al., 2024).

- Rehabilitation robots: Used in postsurgical rehabilitation and the treatment of chronic disease, these robots are designed with the capability of modifying treatment protocols in real-time, using AI to enhance speed, range of motion, and motor learning methods (Nimkar et al., 2024).
- Assistive robots: These are robots programmed for helping the disabled and aging population, which can be employed to assist in daily activities, track health indicators, and participate in basic NLP conversations, thereby ensuring independence and safety (Joshi et al., 2025).
- Pharmacy and logistics robots: Robotic systems driven by AI deployed in hospital pharmacies allow for the automation of dispensing drugs, perform strict testing of prescriptions, and track stock in an efficient manner. The systems minimize medication mistakes and enhance the supply chain efficiency (Yisa, 2024).

All these robotic system classes are based on the combination of AI subsystems actuation, reasoning, and perception to function well in actual healthcare settings. Fig. 4.2 shows these robotic classes.

4.2.3 Role of machine learning and computer vision

Machine learning, specifically deep learning, is the cognitive foundation of the majority of the robots falling under AI. The algorithms are given big sets of data and trained to pick up patterns, make predictions, and refine their actions in an iterative manner. Machine learning allows a variety of functions in the healthcare robots, including disease

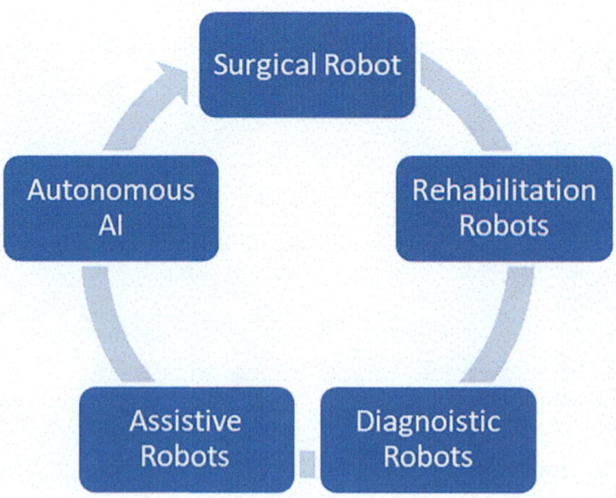

Figure 4.2 Types of robots.

diagnosis and predicting surgical motion (Barua, 2024). Computer vision, as a branch of AI that deals with the study of visual information, is particularly vital in areas of robotic surgery and diagnosis. In robotic laparoscopic surgery, for instance, AI software processes real-time video streams to outline tissue borders, identify abnormalities, and offer visual image enhancements that supplement the surgeon's field of view (Cruz et al., 2024). In diagnostic use, computer vision systems are able to automatically assess medical images for disease presence, thereby improving diagnostic accuracy and lessening observer variation.

In addition, coupling robotics with computer vision increases the possibility of autonomous decision-making, for example, the control of grip pressure in surgery or the evaluation of joint mobility in a patient during rehabilitation. AI allows the systems to learn over time, making the robots able to increase their accuracy and responsiveness as they are exposed to new information and provided with feedback (Integrated Surgical Robotics Training Program Overview [White paper]. Department of Surgery, 2024).

4.2.4 Natural language processing and data integration

NLP is a key and critical functionality to enable human–robotic system interaction and effective data interoperability. NLP technologies enable the robotic agents to comprehend and interpret unstructured clinical text, EHR notes, and voice commands. This ability is extremely useful in the use of documentation, voice-guided surgical procedures, and clinical decision support (AI-enhanced Diagnostic Robotics Training Initiative: Annual Report 2024, 2024). Coupled with NLP, machine learning enables AI robots to offer context-aware assistance, including translating speech between health workers and patients to identify symptoms or condensing lengthy patient histories into to-do lists. Not only do the capabilities increase the efficiency of processes, but they also minimize the cognitive burden on health workers, thereby opening up a larger window of interaction with patients.

4.2.5 Ethical and legal considerations

The development of medical robots that combine AI is also evoking fundamental ethical and legal issues simultaneously. It is critical to scrutinize extremely carefully issues like patient consent, accuracy of data, and risk of algorithmic bias. For example, machine learning algorithms that are being trained on biased data can reproduce health inequities by providing unequal outcomes to particular groups (Comprehensive AI and Robotics Integration in Multi-specialty Care, 2024).

Furthermore, liability is an issue. When an artificially motivated robot causes wrongdoing, the most important questions are: Who is to be held responsible: the software engineer, the hospital, or the attending physician? Legal systems will need to adapt to clearly outline the various degrees of responsibility, particularly as robotic systems become more autonomous (Global Guidelines on Artificial Intelligence Training for Healthcare Professionals, 2025).

In addition, adherence to data protection laws such as the Health Insurance Portability and Accountability Act (HIPAA) and the General Data Protection Regulation (GDPR) is required to protect sensitive patient data processed by AI robots. This responsibility justifies the inclusion of cybersecurity training in the education of healthcare personnel managing AI robotic systems (Health Insurance Portability and Accountability Act of 1996, 1936).

4.3 Skill requirements for healthcare professionals

Effective implementation of AI-based medical robotics in clinical processes requires a ground-level shift in the competencies required of healthcare professionals. Conventional knowledge of human anatomy, diagnostic assessment, and procedural skill is increasingly being found wanting when taken in isolation. With AI and robotics being increasingly embedded in healthcare settings, it is important that medical professionals acquire a blend of technical proficiency, clinical integration skills, and interpersonal soft skills to operate effectively in these sophisticated, technology-driven systems (Faiyazuddin et al., 2025).

4.3.1 Technical competencies

Technical proficiency is the key to AI–robotic readiness. Clinicians need to gain a thorough grasp of fundamental AI concepts, such as machine learning, neural networks, and data-driven algorithms. Knowledge of such areas allows clinicians to critically analyze the rationale behind AI-derived outcomes and understand when such outcomes need to be trusted or need to be intervened (Shen and Gu, 2024). This includes robotic system knowledge. This includes robotic operating system knowledge, motion control interfaces, haptic feedback systems, and sensor data processing. For a surgeon, this involves gaining hands-on robotic console knowledge and robotic instrument manipulation nuances, as well as experience with advanced 3D visualization technology (Ness et al., 2024). Computer vision is an essential area of expertise. Experts should be familiar with the mechanisms by which AI equipment processes visual inputs from areas such as radiology, endoscopy, or live surgery video. This allows experts to critically evaluate the output produced by AI, identify false positives or negatives, and work with the system in the decision-making process (Francis et al., 2023). Just as important is the idea of data literacy, or understanding how to interpret the outputs generated by AI systems, to read performance metrics, and to interact with predictive analytics interfaces. This skill will allow healthcare providers to tailor treatment protocols, streamline processes for operations, and detect system holes through the use of real-time feedback or historical trend analysis (Vaidya et al., 2024).

Briefly speaking, since these systems are likely to rely on cloud connectivity and digital integration, it is essential that they be secure as far as their cybersecurity awareness is concerned. Healthcare professionals need to be informed of common vulnerabilities, be confidential with information, and adhere to regulations such as HIPAA and GDPR. Cybersecurity awareness not only ensures confidentiality of patients but also generates trust among AI systems (Nimkar et al., 2024).

4.3.2 Clinical integration competencies

Apart from merely possessing technical expertise, healthcare professionals need to acquire the skills required to successfully implement AI-based robotic technologies in healthcare infrastructures. This must include an in-depth knowledge of interfacing robotic systems with EHRs, decision-support systems, and telemedicine systems. Healthcare professionals need to use judgment in the right moments to apply AI-based suggestions for diagnosis or treatment protocols, as contrasted with the use of human judgment, particularly in cases of high risk or uncertainty (Joshi et al., 2025). One of the key skills in this field is workflow optimization. Clinical workflows are capable of being revolutionized by AI robots, from steps that vary from preoperative preparation to discharge processes. Healthcare professionals need to acquire the required knowledge in order to redesign workflows in a manner that incorporates AI tools but does not compromise on safety, patient satisfaction, and teamwork (Yisa, 2024). A second key competency is the identification and correction of errors. AI systems, though precise, can be subject to technical malfunctions or skewed outcomes. Medical professionals need proper training so that they can identify when an abnormality occurs within a robotic system, comprehend the probable causes, and implement fail-safes or emergency overrides to protect patients (Barua, 2024).

Additionally, healthcare professionals need to possess the ability to engage actively in the evaluation and ongoing refinement of AI systems. This entails activity with performance metrics, detection of anomalies, and engagement in feedback loops that lead to software or algorithmic changes. Such an iterative process ensures that pragmatic issues of clinical practice are adequately represented in the development of AI technologies (Cruz et al., 2024).

4.3.3 Interpersonal and ethical competencies (soft skills)

As AI algorithms assume more intermediary functions in the clinician–patient relationship, the function of communication and empathy remains largely paramount. Clinicians ought to communicate AI-generated decisions and robotic interventions to patients in clear and human-oriented language. Clinicians need to alleviate patients' anxieties regarding automation while protecting patients' autonomy in health decisions (Integrated Surgical Robotics Training Program Overview [White paper]. Department of Surgery, 2024). Interdisciplinary groups are very significant in such a situation. The application of AI robotics requires coordination between data scientists, engineers, IT experts, and ethics advisers. Health professionals need to have the capacity to handle more than one discipline, converting clinical demands into technical demands and conveying technical limitations in clinical environments (AI-enhanced Diagnostic Robotics Training Initiative: Annual Report 2024, 2024). The importance of ethical thinking is that it can be applied to guide responsible use of AI. Practitioners must deal with intricate issues, including data bias issues, inequities in access to AI tools, and the possibility of dehumanizing medical care. They must be educated to ready them to see these issues, conduct ethical thinking, and support technology deployments that emphasize patient-centered care (Comprehensive AI and Robotics Integration in Multi-specialty Care, 2024). Leadership as a soft skill in change management is essential. Clinicians, being the champions and drivers of innovation and technological advancements, are required to lead their teams through times of change. This involves breaking down resistance, driving consensus, and establishing a culture of ongoing learning (Global Guidelines on Artificial Intelligence Training for Healthcare Professionals, 2025).

4.4 Training strategies and educational frameworks

To effectively prepare healthcare professionals for the complex issues of AI-enabled medical robotics, institutions of education have a mandate to develop pedagogic models that combine requisite theoretical training with experiential, hands-on learning modalities. Lecture-based teaching methodologies, when isolated and viewed in isolation, fall short of addressing the evolving demands of this new field. Consequently, contemporary training programs are compelled to embrace an inclusive and flexible model encompassing simulation, practical application, certification, and continuing education in varied healthcare settings (Faiyazuddin et al., 2025).

4.4.1 Curriculum design for artificial intelligence and robotics in medical education

The integration of AI and robotics into the curriculum of medical and nursing education is no longer a choice; it is a necessity. The courses must be significantly overhauled to incorporate interdisciplinary material that integrates the principles of AI, robotic technologies, algorithmic ethical principles, clinical uses, and data protection (Shen and Gu, 2024).

Fundamental components of a rewritten curriculum are

- A foundational course focused on AI and machine learning, tailored specifically for use in clinical settings.
- NLP and computer vision solutions in diagnosis and EHR integration.
- Special robotic control systems and kinematics for surgery and rehabilitation settings.
- Human factors engineering and patient safety considerations.
- Ethical and legal implications of AI-led healthcare interventions.

In addition, interdisciplinary modules for facilitating the collaboration between medical, engineering, and computer science students can enhance understanding and facilitate interworking, that is, mirror the innovative ecosystems present in real-world healthcare settings (Ness et al., 2024).

4.4.2 Simulation-based training and virtual reality

Simulation has also been embraced as a primary component in AI and robotics education. Simulation with high-fidelity simulators precisely replicates actual robotic surgical settings, enabling experts to practice procedures without compromising patient safety. The simulations often include AI-based feedback systems, where the system analyzes performance, identifies errors, and provides machine learning-based suggestions based on past user performance (Francis et al., 2023). Virtual reality (VR) technology offers interactive learning that enables the user to conduct activities like laparoscopic navigation or diagnostic image interpretation in an entirely virtual and controlled environment. The technique enables learning of cognitive aptitudes as well as psychomotor skills and offers repetitive drills without the related cost factors or safety issues (Vaidya et al., 2024). Augmented reality (AR) enhances physical training immensely in the context wherein virtual signals, body models, or system alerts are projected on top of real-world environments. As an example, an AR head-mounted display may assist a trainee in calibrating robotic tools by providing live feedback and fault correction functionalities (Nimkar et al., 2024). Simulation-based training also allows for a tiered architecture in learning, with beginners starting with system orientation and then progressing toward advanced, scenario-based problem-solving. The scalable architecture supports a wide range of learners, from medical students to fully qualified experts learning to work with AI-aided technologies.

4.4.3 Hands-on training with real and simulated robots

Though simulations may offer a basic framework for exposure, practice with actual AI-integrated robotic equipment is essential for the growth of clinical confidence and functional skills in actual conditions. Schools must invest in initiating robotic labs or collaborating with vendors to expose students to advanced surgical, diagnostic, and rehabilitation robots (Joshi et al., 2025).

Effective practical training schemes have a transparent and sequential approach:

- Stage 1: Getting accustomed to system interfaces and following safety procedures
- Stage 2: Supported participation in nonclinical activities (e.g., suturing and navigation)
- Stage 3: Observation and support in real-life clinical practice
- Stage 4: Autonomous practice under the guidance of an experienced mentorship

Case-based learning greatly enhances this learning approach by incorporating the application of robotics in real clinical scenarios, thereby allowing active learners to identify the respective contexts and approaches to the effective deployment of robotic technologies.

For example, a clinical decision would be whether to use classical or AI-assisted diagnostic tools, based on the case complexity of the patient and the trust in the system output (Yisa, 2024).

The training sessions must be complemented by objective measuring instruments such as performance checklists, real-time AI feedback, and video analysis. These must be put in place to guarantee accountability and consistency in the skill-acquisition process.

4.4.4 Certification programs and continuing medical education

Formal certification and credentialing are also necessary to guarantee that healthcare professionals who work with AI-based robots are kept at the same levels of competence. These certifications, ideally harmonized with global regulatory bodies such as the WHO, FDA, or EMA, must examine technical skills as well as clinical judgment skills in AI-facilitated settings (Barua, 2024).

Certification courses can include:

1. Internet-delivered theoretical modules with associated monitored tests.
2. Simulation-based assessments
3. Real-time clinical assessments
4. Ethics and patient safety modules
5. AI system troubleshooting workshops

As technology is also increasing rapidly, training has to be ongoing. Continuing medical education (CME) has to include yearly updates on:

1. System and interface changes
2. Updated clinical guidelines for the use of AI in robotics
3. New emerging ethical and legal frameworks
4. Performance indicators based on real-world usage

This educational content can be presented in the form of microlearning modules (e.g., 10-minute bundles of AI updates) or modular CME courses (e.g., "AI in Radiology 2025").

In addition, the creation of fellowships with specializations in AI-integrated diagnostics or surgery will assist highly motivated clinicians in assuming leadership positions within this field. The fellowships will enable comprehensive education, clinical innovation, and the production of future academic faculty (Cruz et al., 2024).

4.5 Challenges in training healthcare professionals

In spite of the gigantic change-making potential that resides with AI-based medical robots, the path to ubiquity is fraught with monumental challenges. These challenges traverse the technological, institutional, and financial as well as the cultural fronts. These need to be addressed to ensure healthcare professionals are adequately equipped to integrate these technologies into clinical settings safely and effectively (Faiyazuddin et al., 2025).

4.5.1 Resistance to artificial intelligence adoption

One of the biggest impediments to the application of AI technologies within the healthcare sector is the resistance that has been voiced by medical professionals, on the grounds of practical concerns and psychological apprehensions. The professionals fear as to whether the widespread application of AI systems would jeopardize their employment prospects or diminish their professional autonomy. Surgeons and radiologists are particularly apprehensive about how much AI would supplant human expertise rather than augment it (Shen and Gu, 2024). Besides, the acquisition of skills in relation to learning robotic systems may be highly demanding, particularly for less experienced individuals when dealing with digital instruments. Such fear is even more plausible with older professionals who may be less familiar with the use of advanced technological systems (Ness et al., 2024).

One of the major enablers of skepticism is the lack of trust in AI algorithms. Insufficient transparency, reliability, and algorithmic data skewness problems fuel additional skepticism regarding AI-generated recommendations' performance and safety. Unless these issues are successfully addressed by targeted educational initiatives and experiential training, they can deter people from taking part in training and restrict technology adoption (Francis et al., 2023).

4.5.2 Lack of standardized training modules

Inconsistency of training models between institutions results in inconsistency in skill development and system implementation. Most healthcare organizations establish internal programs of their own, and there are significant variations in the breadth of curriculum, difficulty of tests, and amount of exposure to the clinical environment. This inconsistency adversely impacts the quality of care provided as well as the portability of credentials across institutions (Vaidya et al., 2024). The issue is further aggravated by the fast-paced development of AI and robotics technologies. Curricula soon become outdated, and schools grapple with the dilemma of having to revise their curricula to conform to newly developing systems, software updates, and regulatory requirements (Nimkar et al., 2024).

Further, the diversity of robotic systems offered by various firms necessitates specialized education, so it becomes difficult to obtain universal certification. A surgeon who is experienced with one robotic system might not be adequately qualified to use a different system without another training course, resulting in duplicated effort and wastage of resources (Joshi et al., 2025).

4.5.3 Cost and resource constraints

Creation of training facilities for robotics and AI entails a significant financial outlay. Technical assistance, license fees, robotic surgery systems, and high-fidelity simulators account for astronomical fees, which are likely to be out of reach for poorly funded or small institutions (Yisa, 2024). Apart from infrastructure requirements, human resource challenges remain. The instructors need to prove competence not just in clinical procedures but also in AI technology, which is unusual and costly to acquire. The continuous upkeep of training facilities and equipment contributes to the continuous cost burden, further making sustainability a point of concern (Barua, 2024). Healthcare professionals also bear opportunity costs. Training time means time spent away from clinical service, creating vacancies and reduced productivity. Organizations have to plan training time without interrupting service delivery, especially where healthcare systems are already stressed (Cruz et al., 2024).

To solve these problems, it is important to look at possible solutions such as regional simulation centers, vendor partnerships, and government agency funding. These could reduce costs by allowing the sharing of resources and facilitating equitable access to more sophisticated training (Integrated Surgical Robotics Training Program Overview [White Paper]. Department of Surgery, 2024).

4.5.4 Bridging the gap between medicine and artificial intelligence/robotics

The other major challenge is the interdisciplinary gap between clinicians and AI engineers. Clinicians, such as physicians and nurses, typically lack sufficient knowledge of computational methods, whereas engineers typically lack sufficient knowledge of clinical procedures and ethical issues. This kind of gap greatly impedes the design and deployment of robot systems that are both clinically meaningful and user-oriented (AI-enhanced Diagnostic Robotics Training Initiative: Annual Report 2024, 2024). Educational courses tend not to create interdisciplinary discussion, thus keeping isolated cognitive systems. Medical professionals may be alienated due to the over-specialization of subject matter, while developers find it hard to understand intricate clinical needs (Comprehensive AI and Robotics Integration in Multi-specialty Care, 2024).

To remedy this gap, it is critical that training highlights the significance of AI literacy within the healthcare field while, at the same time, developing experiential clinical skills for engineers. Combined workshops, hackathons, and jointly created simulation platforms can create a common ground and improve system design, implementation, and trust (Global Guidelines on Artificial Intelligence Training for Healthcare Professionals, 2025).

4.6 Case studies and best practices

Research on successful applications of AI-based medical robotics training produces hard data on successful models, reproducible approaches, and actual results in real applications. Leading health innovation hubs have formalized programs that merge technical expertise and clinical use, thus providing successful blueprints for wider application. The following case studies illustrate how evidence-based training models, multidisciplinary partnerships, and institutional engagement can transcend typical challenges and augment the preparedness of health practitioners (Faiyazuddin et al., 2025).

4.6.1 Johns Hopkins Hospital—integrated surgical robotics training program

Johns Hopkins Hospital has developed an extensive program in surgical robotics training, which combines theoretical foundations with the acquisition of practical skills. What is distinctive about this program is its new competency-based system, with foundation modules in robotics and AI that encompass machine learning, haptic feedback systems, and system use safety procedures (Shen and Gu, 2024). Participants advance through high-fidelity simulation training, which involves VR-based modules and simulated surgery on robotic systems like the da Vinci System. These simulations yield AI-based performance analytics, which encompass accuracy-related metrics, decision-making latency, and error rate metrics. Most significantly, the course is integrated with clinical mentorship via exposure between junior trainees and advanced robotic surgeons. This model of delivery not only enhances technical competence but also enhances decision-making in the operating room in AI-assisted settings. The last step involves supervised practice in real clinical settings and a comprehensive multidomain assessment that tests robotic competence, clinical decision-making skills, and ethics (Ness et al., 2024). Outcome measurement has demonstrated dramatic declines in the incidence of surgical errors, reduced procedure time, and enhanced satisfaction scores among graduates of the program. The education model that has been developed at Johns Hopkins has become globally accepted as the gold standard for teaching robotic surgery (Francis et al., 2023).

4.6.2 Mayo Clinic—artificial intelligence-enhanced diagnostic robotics training initiative

Mayo Clinic has initiated a landmark program for radiologist and pathologist training in the context of AI-powered diagnostic robotics. The program underscores the collaboration between human professionals and AI in image examination, cancer detection, and diagnosis decision support. The course begins with training components devoted to algorithm literacy that will provide healthcare professionals with knowledge of the various procedures used in training, validating, and deploying deep learning models. Students learn to identify the strengths and limitations of algorithms and the possibility of embedded bias.

One of the most important elements of the program is case-based learning, where the trainees are working on real diagnostic cases both through conventional and AI-aided techniques. The AI systems give suggestions that are to be critically assessed by practitioners, hence a cautious response to decision-making. (Vaidya et al., 2024). It further includes visualization tools demystifying the logic of AI-generated outputs, which in itself enhances transparency and instills confidence. Quality assurance sessions have been instituted to train professionals on continuous monitoring of system performance, identifying model drift, and application of interventions as needed. Below is a representative diagram, as Fig. 4.3 (Lopez-Jimenez et al., 2020), showing how AI enables possible diagnosis and prognostic calculations for an individual patient.

More diagnostic accuracy and reliability have been established in research, especially in oncology and imaging of the musculoskeletal fields. Additionally, patient satisfaction has been reaffirmed in terms of increased confidence in clinical evaluation, where health workers explain the diagnostic role of AI (Nimkar et al., 2024).

4.6.3 Cleveland Clinic—multispecialty artificial intelligence robotics integration program

Cleveland Clinic has taken a broad and interdisciplinary strategy in the training of robots powered by AI, taking into account the expanding utilization of robotic machinery across various clinical settings ranging from surgical operations, and rehabilitation to the automation of drug processes. The educational model is both modular and designed to

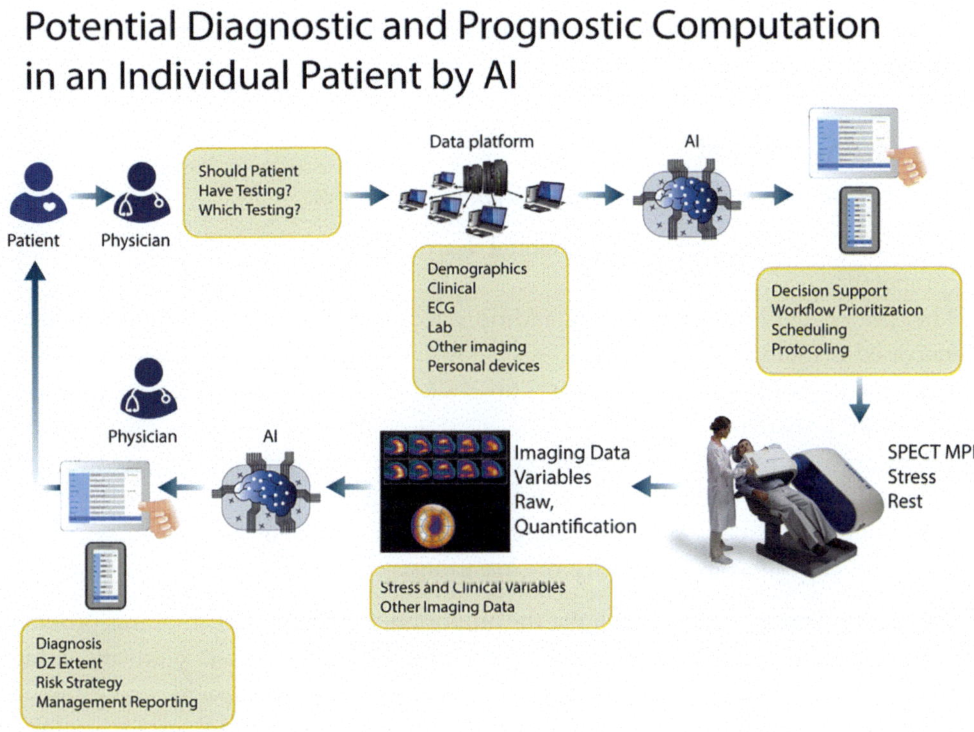

Figure 4.3 AI analysis of patient.

fulfill some specific requirements. All students start training with a basic curriculum that is common to all and addresses the ethical aspect of AI, cybersecurity concerns, and the overall principles of robotics. Later, specialized modules branch out into specialized streams, such as orthopedics, neurology, and pharmacy. One of the most outstanding aspects of this program is the focus on strong team training from different backgrounds. Professionals in healthcare, information technology, biomedical engineering, and data science meet and interactively participate in training exercises under simulated clinical environments. This leads to shared understanding, enhanced human operator–automated system interaction, and enabling of productive integration of robotic technology (Joshi et al., 2025).

The project also consists of live pilot projects, under which experienced professionals oversee the deployment of robotic systems within the concerned departments. All of the projects include a feedback loop of students, system vendors, and patients, aimed at refining operational workflows continuously and identifying areas requiring more training. Internal assessments indicate a notable increase in collaboration, reduced errors in implementation, and increased satisfaction levels among patients and staff. Cleveland

Clinic's project is an example of total institutional integration and integration of robotics into a cooperative system (Yisa, 2024).

4.6.4 Lessons learned

Across these case studies, several similar themes are present:

- Simulation and practice in experiential learning are at the core of safe clinical competency acquisition.
- The presence of mentorship in addition to cross-disciplinary training adds to increased readiness.
- Transparency of AI decision-making processes leads to trust between patients and healthcare providers.
- Continuous perforormance evaluation and periodic revisions are required in order to keep pace with the fast innovation in technology.
- Use of modular and adaptive designs supports customized adaptations in various specialties as well as in healthcare organizations.

The above best practices highlight the importance of extensive training programs beyond minimum technical training. Training programs must cultivate ethical reasoning, systems thinking, and cooperative problem-solving capacities to empower healthcare professionals to deal with the intricate realities of AI-enhanced healthcare provision (Barua, 2024).

4.7 Practical implications of artificial intelligence in healthcare

The introduction of AI-facilitated medical robots to clinical settings is more than a technology shift; it triggers radical transformations affecting all aspects of healthcare provision. These technologies have shown impressive advancements in diagnosis, rehabilitation, personalized therapy, and administrative processes, improving clinical accuracy, patient protection, operational effectiveness, and accessibility to healthcare globally. Grasping these implications is essential to formulate training programs aligned with reality-driven requirements and priorities of healthcare policy (Faiyazuddin et al., 2025).

4.7.1 Enhanced diagnostic accuracy and clinical decision-making

One of the major benefits of AI-based systems is improved diagnostic accuracy. By proper development with large-scale data, including radiographic images, pathology slides, and entire patient histories, machine learning algorithms can detect patterns that would be missed by clinicians, particularly in difficult or unclear cases (Shen and Gu, 2024).

AI-based diagnostic robots are used today to:

1. Identify early-stage neoplasms with greater sensitivity.
2. Employ imaging methods to identify diabetic retinopathy and cardiovascular malformations.
3. Cancer classification based on genomic and phenotypic characteristics.

These systems support rapid, evidence-based decision-making, thus optimizing diagnostic speed and improving outcomes. Significantly, AI does not replace human judgment but enhances it—serving as a clinical copilot that increases confidence and reduces variation in decision-making (Ness et al., 2024).

4.7.2 Personalized treatment and adaptive therapies

Robotics with AI supports highly individualized care paths. In rehabilitation, for instance, exoskeleton robots can change resistance, timing, and pattern of movement based on real-time feedback gathered with sensors installed on the patient. Surgical robots can also change trajectory or force application based on patients' unique anatomical characteristics (Francis et al., 2023). By integrating data collected using wearable technology, imaging diagnostic tests, and EHRs, AI systems prepare patient-specific recommendations based on a patient's comorbidities, treatment history, age, and lifestyle. This allows healthcare professionals to personalize interventions, avoid complications, and improve the accuracy of recovery predictions.

This degree of accuracy occurs in light of the shift toward value-based healthcare, whose metric is outcomes, not the number of services provided. As such, health professionals with AI analytics and robotic system calibration training are best suited to provide evidence-based, patient-focused care (Vaidya et al., 2024).

4.7.3 Operational efficiency and resource optimization

AI-based robotic systems significantly enhance the efficiency of operation in clinical and administrative sectors. In surgical scenarios, AI systems can optimize case scheduling based on the projection of procedure length, resource allocation optimization, and downtime between procedures minimization, respectively. Diagnostic robotics reduces imaging and pathology department delays by having the capability of processing massive amounts of data rapidly (Nimkar et al., 2024).

Robotic drug dispensing systems remove drug and inventory errors by computerizing drug compounding, dosage checking, and dispensing. AI telepresence robots are being used in the same way to enable remote consultations and patient monitoring, especially in areas with geographical isolation or in rural communities, and therefore reduce unnecessary hospitalization (Joshi et al., 2025).

These advances in technology create:

- Decreased costs per patient
- Reduced clinician workload and risk of burnout
- Increased access to specialist medical services in low-resource settings.

Use of AI systems minimizes cognitive and administrative workload on healthcare professionals, thus making room for direct patient interaction, one of the essentials for providing high-quality care (Yisa, 2024).

4.7.4 Impact on patient safety and quality standards

Patient safety is one of the top priorities in all healthcare settings. AI-driven robots are made to protect patients through continuous monitoring, error detection through computerized means, and advanced predictive analysis. As an example, robotic surgical systems can track vital signs, instrument location, and environmental conditions and alert when certain thresholds are achieved (Barua, 2024). Besides this, AI-driven algorithms also reflect the capability of detecting trends that indicate a worsening in the condition of patients at an earlier stage than traditional chart review processes. AI-supported surveillance systems in intensive care units and postoperative recovery units are increasingly being used in an effort to decrease the chances of sepsis, respiratory failure, or falls by anticipating high-risk events and alerting healthcare workers in advance (Cruz et al., 2024).

Through the implementation of standardized performance measurement and data collection, such systems take on a key position in enabling continuous improvement in quality. Hospitals utilizing AI-facilitated robotic technologies are expected to have greater clinical guideline compliance, improved documentation accuracy, and reduced negative events.

4.7.5 Global health and accessibility

One of the wide-ranging implications of robotics based on AI is that they can democratize access to healthcare services. Where there is a shortage of specialized medical attention, robotics based on AI can:

1. Improve teleoperation or telementoring remote surgery
2. Optimize diagnostic processes in mobile clinics using portable imaging and AI
3. Offer automated rehabilitation treatment or support to older adults in home settings.

Such technologies are of immense value in low- and middle-income nations, where there are chronic shortages of trained healthcare workers and requisite infrastructure. With proper education and policy settings, AI robotics may be essential to closing the global gap in the delivery of healthcare (Integrated Surgical Robotics Training Program Overview [White paper]. Department of Surgery, 2024).

4.7.6 Ethical and professional implications

Whereas AI systems provide special advantages, they also raise new professional and ethical obligations. It is essential that healthcare professionals recognize the limitations of AI-based suggestions, particularly in ethically challenging scenarios concerning end-of-life care, mental illness, or social determinants of health.

Additionally, professionals must tackle data ownership, informed consent, and fairness in algorithms. Patients must be informed when AI is used in their treatment and how decisions are made. Training sessions should, therefore, include modules on clear communication and shared decision-making in AI-assisted settings (AI-Enhanced Diagnostic Robotics Training Initiative: Annual Report 2024, 2024).

4.8 Future trends and recommendations

As medical robotics is being enhanced by AI, faster than at any time in the history of the world, the trajectory of healthcare education and service delivery will be influenced by a number of emerging trends. They signal a worldwide transition to intelligent systems that, in addition to enhancing clinical practice, reshape the education, accreditation, and support of professionals throughout their careers. To gain maximum rewards from these advances, institutions of learning need to embrace visionary strategies that emphasize innovation, equity, customization, and policy alignment (Faiyazuddin et al., 2025).

4.8.1 Integration of artificial intelligence-driven robotics in medical curriculum

Perhaps the most promising trend is the deliberate incorporation of AI and robotics into the core medical and nursing school curriculum. Future healthcare professionals need to be exposed to machine learning concepts, robotic system operation, and ethical thinking in their professional training—not when they enter practice.

Large academic centers are now incorporating:

- Undergraduate health sciences degrees with AI literacy modules
- Interdisciplinary education in medicine, computer science, and engineering
- Simulation-based robotic surgical training in anatomy and diagnostics
- Project-based learning models where students codesign AI tools with developers

These curricula train graduates to be competent clinicians with the technical skills necessary to comprehend and work with advanced medical systems (Shen and Gu, 2024).

Also, the trend is increasing toward adopting modular systems of micro-credentialing that enable learners to acquire specialist skills in areas such as AI-assisted radiology or robot-assisted surgical methods. These learning modules can be stacked to address elective credit, ongoing continuing professional development, or certification (Ness et al., 2024).

4.8.2 Role of augmented reality and artificial intelligence-powered tutors

The evolution of AR technologies can potentially have a significant impact on procedural training and patient education. Unlike VR, which replaces the environment, AR enriches the real world by adding digital information, thus providing real-time feedback during procedures or simulation (Francis et al., 2023).

AR-enhanced training platforms can:

- Display robot system statistics during simulation
- Mark anatomical landmarks or surgical risk sites
- Show in real-time AI-driven alerts or decision support cues
- Enable remote proctoring and performance evaluation by instructors

At the same time, AI-based personalized tutoring system development is underway. They learn from student information to:

- Trace individual learning and performance curves
- Provide adaptive content based on user strengths and gaps
- Recommend remediation strategies or upper-level modules
- Mimic “smart mentoring” while using robotic systems

This data-driven personalization has the potential to raise learning achievements considerably, particularly in advanced skill areas such as robotic surgery or image-guided diagnosis (Vaidya et al., 2024).

4.8.3 Policy recommendations

Policy making is a necessary basis for the successful, equitable, and secure utilization of AI and robotics within the healthcare industry. In relation to current trends and institutional practice, the following policy interventions are proposed:

- Develop international standards for AI and robotics literacy among the health workforce.
- Incorporate AI–robotics training into medical school accreditation standards and professional board regulations.
- Support research on AI effectiveness, bias mitigation, and job impact.

- Fund national and regional simulation centers to provide training equity.
- Foster cross-sector partnerships among academia, industry, and government institutions.
- Control certification of AI systems and postmarket surveillance to ensure safety and trust.
- Promote ethical AI deployment through the requirement of algorithm development and deployment transparency (Nimkar et al., 2024).

Governments, accrediting agencies, and ministries of health need to collaborate to ensure that policy is not lagging behind innovation.

4.8.4 Future research directions

As the field develops, several research priorities are emerging that need a priority focus:

- Explainable AI (XAI) model architecture enabling clinicians to comprehend and trust algorithmic outcomes.
- Studies assessing the effect of AI robotics on clinical outcomes, operational efficiency, and healthcare economics.
- Research examining the temporal adaptation of professionals to robotic systems through longitudinal studies.
- Human–robot interaction studies of building trust, emotional intelligence, and cooperation.
- Culturally and linguistically responsive AI tool design for worldwide application in global health.
- Research on simulation fidelity to identify the most beneficial features for robotics training platforms (Joshi et al., 2025).

Additionally, more interest is in bio-integrated AI systems, such as neural-linked prosthetics and closed-loop robotic rehabilitation systems that are real-time-adapted in accordance with neurological signals. Training of such systems will require a new system of education in clinical neuroengineering that incorporates medicine, bioelectronics, and AI (Yisa, 2024).

4.9 Conclusion

AI-powered healthcare robots are no longer fiction; they are a changing, dynamic force in the delivery of healthcare. From the operating theater to the diagnostic lab, from rehabilitation centers to telehealth facilities across the nation, smart robotic platforms are revolutionizing the delivery, tracking, and personalization of care. Such a radical shift holds much promise but demands an integrated analysis of the approaches used in the delivery of education, credentialing, and continuous support to healthcare professionals throughout their career's journey (Faiyazuddin et al., 2025).

This book has unmasked the sophisticated essence of the training dilemma: it is not merely a matter of machinery operation but also of interdisciplinary proficiency, ethical judgment, data awareness, and a cooperative spirit. It challenges clinicians to transition from being exclusive decision-makers to informed collaborators with technology, knowing when to trust AI-generated results, when to intervene, and how to justify AI-enhanced decisions to patients and colleagues in an open and compassionate manner (Shen and Gu, 2024). Broad-based training programs via simulation-based training, hands-on training, AR/VR platforms, and credentialing paths must be installed to prepare the next generation of workers. These programs should be dynamic, customized, and revised at regular intervals to keep pace with the fast-evolving advances in AI and robot technologies. As the nature of medical science is ever-evolving, so too must the processes by which we train workers in technological proficiency (Ness et al., 2024). Johns Hopkins, Mayo Clinic, and Cleveland Clinic case studies show that when training is deliberate, iterative, and institutionally backed, success is possible. Their models highlight the importance of simulation, mentorship, interdisciplinary collaboration, and continuous assessment factors that need to become the fabric of any successful training model. No less important is the development of policy and regulatory structures that enable safe, equitable, and responsible AI use in medicine. Domestic and international stakeholders must reach a consensus on training requirements, encourage open-access educational settings, and ensure that new systems are tested, understandable, and ethically sound. Without regulation, disparities in access and proficiency will increase exponentially, endangering quality along with equity in healthcare (Francis et al., 2023).

In the future, AI-powered education itself, such as personalized learning systems and intelligent tutoring, will become a routine part of training pipelines. Clinicians will not only apply AI in practice, but they will also learn with AI, opening up a new era of technology-enabled professional development. Finally, the success of AI-backed robots in the healthcare industry depends not just on development but also on people's readiness. Current and future clinicians' education and training are the most crucial elements; they are the deciding factor that will make this revolution realize its full potential to improve patient care, save lives, and improve global health equity (Vaidya et al., 2024). The path to a human–AI collaboration in medicine is not far away. With the right training, policy backing, and cultural transformation, healthcare systems will be able to position themselves to make the path to such collaboration improve, not supplant, the values of compassion, accuracy, and professionalism that make quality medicine.

A comparison of AI-driven methods has been given below as Table 4.1

Table 4.1 Comparison of artificial intelligence (AI)-driven robotics training methods.

Training method	Key features	Advantages	Limitations	Ideal use case
Simulation-based training	Uses high-fidelity VR/AR environments to replicate robotic procedures	– Safe and risk-free environment – Enables repeated practice – Offers AI-driven performance feedback	– Lacks tactile realism of real procedures – High cost of development for realistic simulations	Early skill acquisition, decision-making practice, preclinical exposure
Hands-on training with real robots	Direct interaction with live robotic systems under supervision	– Builds real-world operational confidence – Enables system troubleshooting skills – Reflects clinical complexity	– Requires expensive equipment and infrastructure – Limited access for large groups	Skill refinement in advanced learners, surgical residency, and fellowship training
Augmented reality (AR) assisted learning	Overlays digital data onto real environments during procedures or training	– Real-time guidance – Enhances situational awareness – Allows blended learning environments	– May cause visual distraction – Requires advanced AR hardware and custom software	Enhancing procedural precision and intraoperative teaching
Online AI and robotics courses	Structured theoretical modules, sometimes with simulations or interactive tasks	– Flexible and self-paced – Scalable for global learner – Cost-effective for institutions	– Limited practical exposure – Lower retention without hands-on practice	Introduction to AI fundamentals, CME for working professionals

(Continued)

Table 4.1 (Continued)

Training method	Key features	Advantages	Limitations	Ideal use case
Certification and fellowship programs	Formal credentialing with multimodal learning (theory, practical, assessment)	– Recognized qualification – Enhances career progression – Structured, tiered learning paths	– Time-intensive – Limited availability in some regions – May require prior clinical experience	Specialist upskilling, board certification, and career transition into robotics
Interdisciplinary team simulations	Collaborative scenarios involving clinicians, engineers, and AI developers	– Improves cross-functional communication – Mirrors real deployment environments	– Complex logistics – Requires participation from multiple disciplines	Deployment training, system testing, workflow integration
AI-powered personalized training platforms	Adaptive learning platforms using performance data to personalize content	– Individualized feedback – Tracks progress over time – Recommends targeted remediation	– Still emerging in medical education – Requires high data security compliance	Longitudinal learning, residency programs, and institutional LMS integration

AI disclosure

During the preparation of this work, the author(s) used SCI SPACE in order to review the literature. After using this tool/service, the author(s) reviewed and edited the content as needed and take(s) full responsibility for the content of the publication.

References

AI-enhanced diagnostic robotics training initiative: Annual report 2024; 2024.

Barua R. An In-Depth Exploration of AI and Humanoid Robotics' Role in Contemporary Healthcare. IGI Global; 2024. p. 42–61. Available from https://doi.org/10.4018/979-8-3693-2238-3.ch003.

Comprehensive AI and robotics integration in multi-specialty care; 2024.

Cruz EMGNV, Oliveira S, Correia A. Robotics application in the hospital domain: operator–patient–robot interface. Preprints 2024:202406. Available from https://doi.org/10.20944/preprints202406.1161.v1.

Faiyazuddin M, Rahman SJQ, Anand G, Siddiqui RK, Mehta R, Khatib MN, et al. The impact of artificial intelligence on healthcare: a comprehensive review of advancements in diagnostics, treatment, and operational efficiency. Health Science Reports 2025;8(1) Available from https://doi.org/10.1002/hsr2.70312, onlinelibrary.wiley.com/journal/23988835.

Francis J, Varghese JV, Thomas A. Impact of artificial intelligence on healthcare. International Journal of Advances in Medicine 2023;10(10):737–43. Available from https://doi.org/10.18203/2349-3933.ijam20232839.

Global guidelines on artificial intelligence training for healthcare professionals. WHO Press; 2025.

Health Insurance Portability and Accountability Act of 1996. Pub. L. No. 104; 1936.

Integrated surgical robotics training program overview [White paper]. Department of Surgery. Johns Hopkins Medicine; 2024.

Introducing NVIDIA Isaac for Healthcare, an AI-Powered Medical Robotics Development Platform; 2025.

Joshi R, Pandey K, Kumari S. Healthcare robots enabled with IOT and artificial intelligence in healthcare applications. Artificial Intelligence Technologies for Engineering Applications. India: CRC Press; 2025. p. 104–29. Available from https://doi.org/10.1201/9781003565529-8, https://www.taylorfrancis.com/books/edit/10.1201/9781003565529.

Lopez-Jimenez F, Attia Z, Arruda-Olson AM, Carter R, Chareonthaitawee P, Jouni H, et al. Artificial intelligence in cardiology: present and future. Mayo Clinic Proceedings. Mayo Clinic 2020;95(5):1015–39. Available from https://doi.org/10.1016/j.mayocp.2020.01.038.

Ness S, Xuan TR, Oguntibeju OO. Influence of AI: robotics in healthcare. Asian Journal of Research in Computer Science 2024;17(5):222–37. Available from https://doi.org/10.9734/ajrcos/2024/v17i5451.

Nimkar P, Kanyal D, Sabale SR. Increasing trends of artificial intelligence with robotic process automation in health care: a narrative review. Cureus 2024;16(9). Available from https://doi.org/10.7759/cureus.69680.

Shen R, Gu J. The transformative impact of artificial intelligence in healthcare: innovations, applications, and future prospects. SHS Web Conference 2024;207:03018. Available from https://doi.org/10.1051/shsconf/202420703018.

Vaidya R, Kulkarni S, Gaikwad T, Jadhav S. Advancing healthcare through artificial intelligence: opportunities, challenges and future directions. International Research Journal of Advanced Engineering and Management (IRJAEM) 2024;2(05):1302–8. Available from https://doi.org/10.47392/irjaem.2024.0177.

Yisa MI. Leveraging Artificial Intelligence in healthcare to optimize patient outcomes, with specialized staff training programs. World Journal of Advanced Research and Reviews 2024;24(3):3078–93. Available from https://doi.org/10.30574/wjarr.2024.24.3.4041.

CHAPTER 5

Strategies for clinical adoption of artificial intelligence-powered systems

Anil Pise[1,2,3] and Sowmya Bachigari[4]
[1]Sr. Data Scientist, X-idian, Johannesburg, Gauteng, South Africa
[2]Technical Partner, Cumulus Solutions, Johannesburg, South Africa
[3]Senior Consultant, Deloitte SA, Johannesburg, South Africa
[4]Woxsen University, Hyderabad, Telangana, India

5.1 Introduction

Artificial intelligence (AI) swiftly reshapes the healthcare landscape, delivering transformative benefits nationwide across patient management, quick diagnosis, and many novel therapeutic interventions. AI-enhanced robotic surgery redefines precision standards utterly and boosts clinical efficiency remarkably with utmost consistency in every operation. Sophisticated algorithms remarkably facilitate precision in intricate surgical procedures, often eclipsing traditional methods with swift patient recuperation afterward. Widespread integration of AI into clinical settings encounters substantial obstacles, such as bureaucratic red tape and statutory ambiguities virtually everywhere nowadays.

Barriers pose significant challenges to both adoption and scalability despite the clear advantages AI brings, rather swiftly within the surgical domain. Robotic-assisted surgery has been further empowered by AI integration, representing a major breakthrough in minimally invasive surgical techniques nowadays. Surgeons wield precise control over instruments and glean highly detailed 3D visuals with AI-assisted surgical systems, such as the da Vinci Surgical System. AI algorithms bolster surgical systems with rapid interpretation of data and automated support for decisions alongside predictive analytics based on past surgical records and info specific to patients. Such capabilities pretty effectively minimize human error and lead to markedly improved clinical outcomes with great optimism intraoperatively. AI-driven surgical systems morph quickly, analyzing metrics alongside patient responses after each op, enabling adaptive refits and steady accumulation of knowledge. Over time, this evolution lays a foundation for more personalized surgical care and significantly contributes to better results simultaneously (Fig. 5.1).

The journey from technical innovation toward clinical adoption proceeds tortuously and not straightforwardly under most circumstances anyway. Healthcare organizations

AI-Powered Developments in Medical Robotics
DOI: https://doi.org/10.1016/B978-0-443-27783-2.00005-3

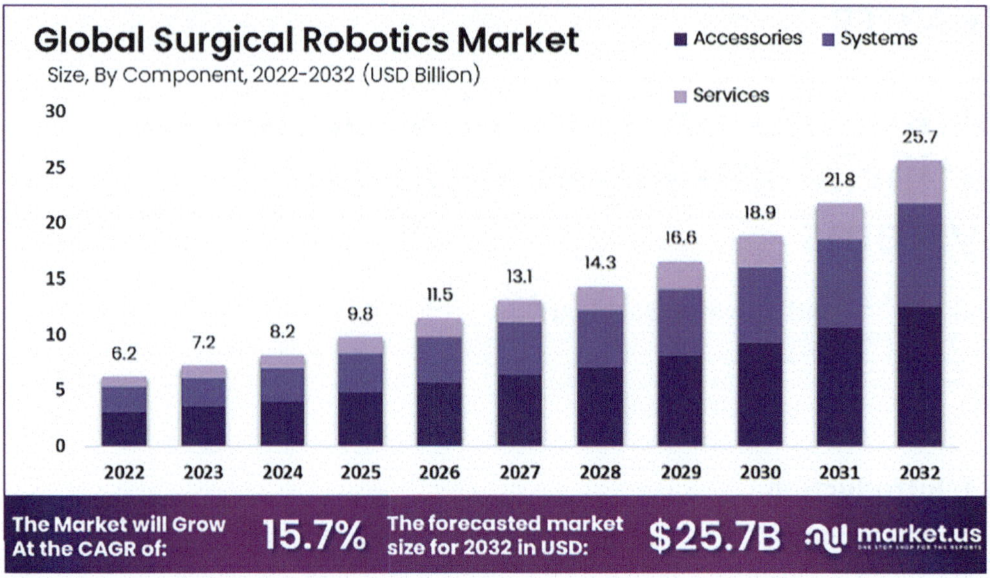

Figure 5.1 Projected growth of the surgical robotics market (2022–32), showcasing the increasing adoption of AI-driven robotic systems in medical procedures. *Based on Tajammul Pangarkar, 2025 market.us.*

must squarely confront multifaceted challenges, such as dodgy system interoperability and regulatory red tape, amidst wavering clinician confidence. Strategies for adopting AI-driven robotic systems in modern surgical practice are explored here, tackling key obstacles and offering very practical guidance.

5.2 The role of artificial intelligence in medical robotics

Robotic surgery has transformed surgical procedures significantly, offering advantages in precision and patient recovery with greatly enhanced control over operations. Advanced systems like the da Vinci Surgical System empower surgeons with enhanced dexterity, performing complex operations pretty effectively under high-definition 3D visualization. These features mitigate physical constraints inherent in traditional laparoscopic tools and significantly extend human hand capabilities, allowing for delicate maneuvers even deep within bodily regions (Chawda and Fatima, 2023). Robotic-assisted procedures frequently correlate with diminished blood loss and curtailed hospital stays alongside markedly lower infection risks and swifter recovery times postoperatively.

Robotic systems currently serve mostly as elaborate appendages of surgeons operating under stringent manual control with considerable precision nowadays. Mostly, they stay inert, lacking discretion or situational savvy under many circumstances, and offering

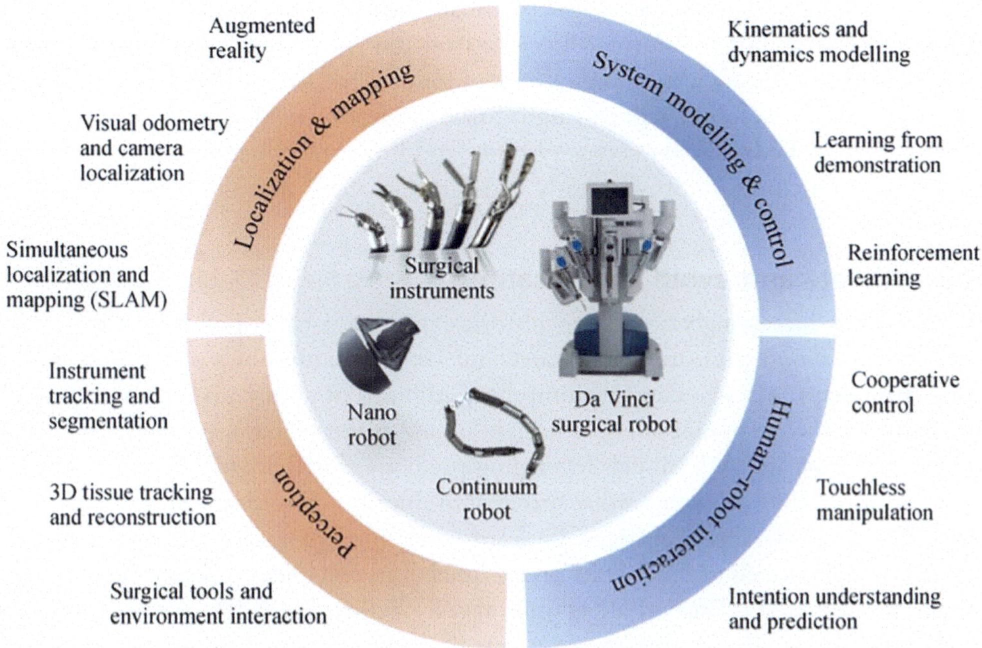

Figure 5.2 AI techniques for surgical robotics. *Based on Erh-Ya (Asa) Tsui, 2020.*

virtually no keen insight whatsoever. Surgeons' expertise and finesse play a crucial role in determining the outcome of procedures undertaken with utmost precision and care daily. Reliance on existing robotic surgery setups glaringly highlights a yawning deficit in adaptive acumen needed for informing high-pressure medical decisions. Robotic systems have markedly elevated surgical performance, but their full potential gets unlocked when somehow infused with AI-driven context-aware smarts (Fig. 5.2).

Robotic surgery evolves into a highly precise and minimally invasive approach, enabling surgeons to perform complex procedures with enhanced accuracy, flexibility, and improved patient outcomes.

5.2.1 Enhance intraoperative cognitive processes

Proprio Vision's AI-powered surgical navigation system remarkably exemplifies the transformative role of computer vision in surgery nowadays with great precision. Advanced AI algorithms can automatically recognize tumors and label critical anatomical structures like nerves and blood vessels in real-time endoscopic video feeds. These systems can flag anomalies like bleeding hazards or spots susceptible to inadvertent trauma, enabling surgeons to tweak their methods with increased finesse. Real-time guidance significantly enhances situational awareness during complex surgical procedures or those that are minimally invasive in nature. Surgical teams consequently make rather informed decisions on the spot, thereby minimizing nasty complications and greatly improving outcomes for patients. Studies published in Annals of Surgery 2023 indicate AI-driven surgical navigation can reduce tissue damage by as much as 30% markedly. Faster recovery times and lower healthcare costs are achieved while clinical outcomes improve greatly and surgical safety increases substantially.

5.2.2 Estimate and avoid complications

ML algorithms increasingly scrutinize historical patient data alongside real-time vital signs, such as blood pressure, heart rate, and oxygen saturation, in various clinical settings nowadays. Massive datasets comprising millions of anonymized patient records train algorithms to recognize subtle correlations and patterns that are usually missed by experienced clinicians. ML models leveraging predictive analytics can detect early signs of complications like sepsis or cardiac arrest with uncanny precision remarkably. These algorithms assess dynamic changes in a patient's physiological state, continuously adapting to evolving clinical contexts pretty quickly in real-time. The system generates bespoke risk scores and fires off alerts quite rapidly before overt deterioration becomes visibly apparent, enabling rather proactive medical intervention. Early warning systems can significantly improve patient safety and reduce admissions to the ICU, thereby enhancing hospital efficiency noticeably overall.

5.2.3 Make adaptive learning possible

AI-driven surgical systems evolve dynamically over time, with each new procedure performed very meticulously and accurately almost always. Refining algorithms with data from past surgeries makes each new operation an utterly transformative chance to boost system intelligence remarkably. Intuitive Surgical's SkillFire platform offers surgeons personalized feedback on performance largely based on data mined from numerous prior surgical procedures (Salunkhe et al., 2022). AI systems continually get better in real-time through this feedback loop, customizing suggestions and decision support for each unique surgery context swiftly. Continuous learning significantly enhances the accuracy of robotic-assisted surgeries and reduces the likelihood of human error substantially during complex operations. AI identifies anomalies and suggests corrections rapidly during procedures and adapts quickly to unexpected gnarly challenges with real-time data analysis. Algorithmic refinements informed by clinical outcomes eventually contribute significantly to much safer surgeries over a fairly long period. Surgical efficiency increases markedly, and patient outcomes improve substantially alongside shorter recovery periods and greater trust in AI-powered healthcare tech (Fig. 5.3).

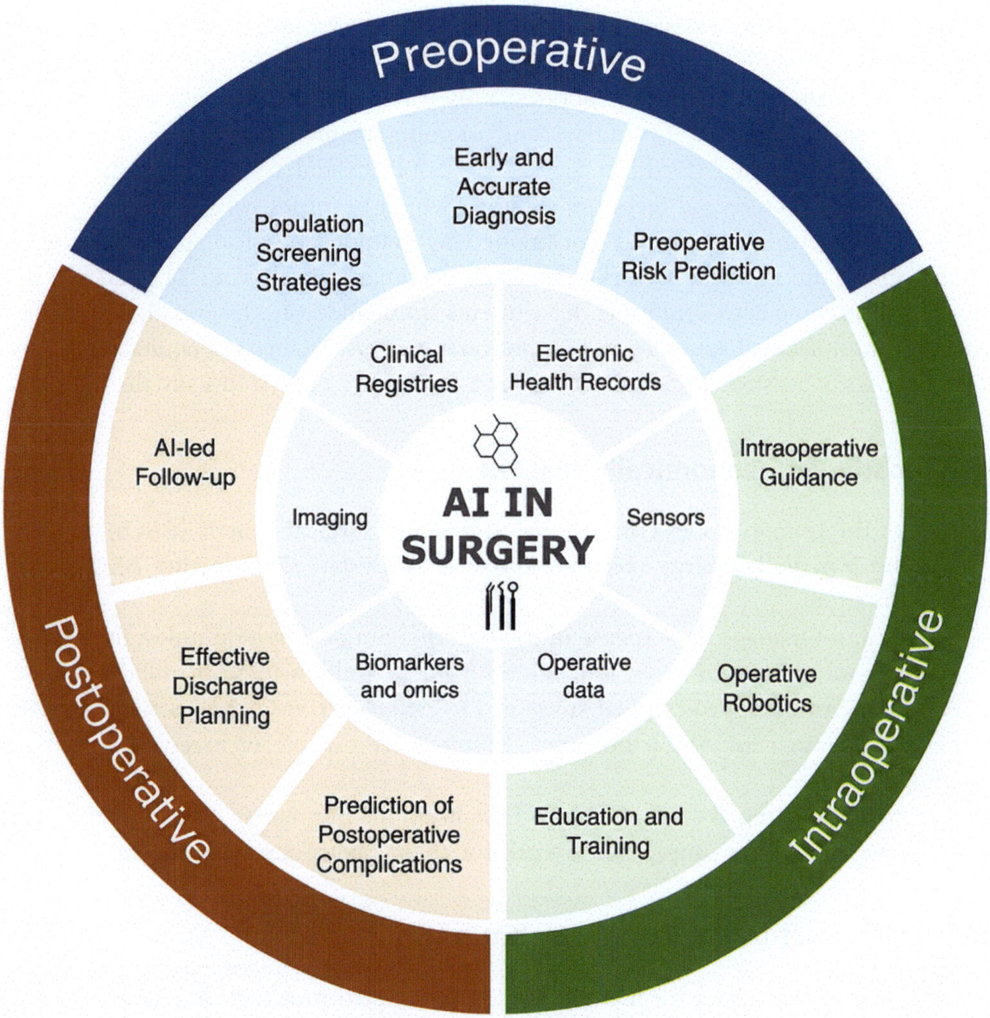

Figure 5.3 Chart depicting the applications of AI in surgery and the data sources. Applications are divided into preoperative, intraoperative, and postoperative stages. *Based on Guni, 2024.*

5.3 Challenges in clinical adoption

Robotic surgical systems inject sleek, transformative capabilities into modern clinical practice, offering unrivaled precision and greatly enhanced patient outcomes, suddenly. Such technologies facilitate procedures that are minimally invasive with significantly reduced blood loss and a pretty rapid patient recovery afterward. Several barriers continue to impede the widespread adoption of surgical robotics, especially in resource-sensitive environments such as ambulatory surgery centers and smaller community hospitals, nevertheless. Advanced robotics, poised to redefine surgical capabilities, are often stymied by hefty capital outlays and ongoing maintenance costs, necessitating highly trained personnel (Orthi et al., 2022). Sophisticated robotic systems demand substantial infrastructure and technical support, which becomes woefully inadequate in frantic clinical settings mercilessly constrained by cramped physical space. Integrating robotics into existing surgical workflows often demands radical protocol overhauls and intense staff training across multiple departments simultaneously. Essentially, globally, it is tackling multifaceted systemic and operational barriers, facilitating equitable deployment of robotic-assisted surgery across diverse healthcare ecosystems sustainably.

5.3.1 Costly yet economically feasible

- One of the primary issues confronting ambulatory surgical centers (ASCs), as stated by John Van Valkenburg, executive director of Upstate Orthopedics ASC, is slim profit margins.
- ASCs cannot manage the expense of the da Vinci Surgical System unless they handle a significant volume of cases, as it ranges from $1 million to $2.5 million in cost.
- According to orthopedic spinal surgeon Dr. Michael Burdi, the cost-benefit ratio poses a dilemma for ASCs; however, hospitals are capable of investing in robotic equipment.
- According to Greg Roche, CEO of Distalmotion, numerous ASC operators find that "the economics simply don't work for them" with current solutions.

5.3.2 Infrastructure and space limitations

- Finding dedicated space for many existing robotic platforms is challenging since they are oversized for ASC operating rooms.
- "The size of these robots has traditionally been a significant barrier," states Greg Roche, because ASCs cannot afford to dedicate an entire room for a robotic system.

5.3.3 Surgical adaptation and workflow integration

- "The surgeon must adapt to the workflow of the robot," as Greg Roche points out, in numerous robotic platforms, which conflicts with the more streamlined team setups present in ASCs.
- In contrast to hospitals that have larger surgical teams, ASCs need robots that conform to the surgeon instead of the other way around.

5.3.4 Limited market uncertainty and adoption

- In outpatient environments, the incorporation of robotic systems is still in its early stages. According to Greg Roche, this transformation is "inevitable" but will require time.
- Despite the sector still facing financial and technological challenges, Tenet Healthcare CEO Saum Sutaria identifies opportunities for the growth of robotics in ASC services.

5.3.5 Training and readiness gap for surgeons

- Although many ASCs do not have the necessary equipment to support this skill, younger surgeons obtain most of their training on robotic systems.
- According to Greg Roche, new surgeons possess "limited laparoscopic experience" and are seeking opportunities to continue using robotic platforms.

5.3.6 System restrictions versus patient demand

- In outpatient environments where recovery durations are shorter, patients prefer minimally invasive robotic surgery.
- However, ASCs struggle to meet this growing demand because of logistical, financial, and regulatory obstacles (Fig. 5.4).

5.4 Strategies for clinical adoption of artificial intelligence-powered systems

AI- and ML-infused robotic surgery revolutionize healthcare pretty rapidly by elevating surgical precision and facilitating real-time decision-making under pressure.

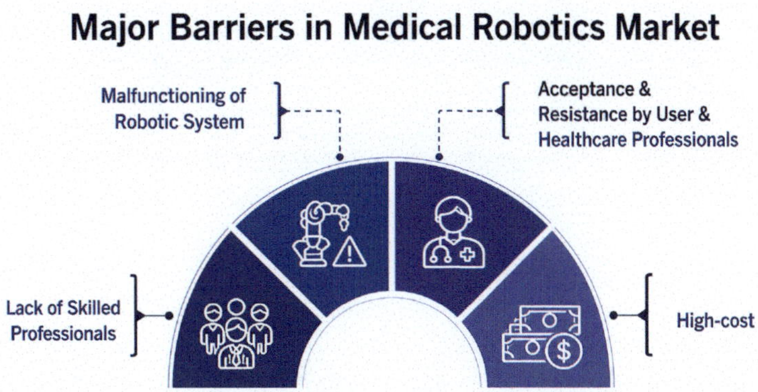

Figure 5.4 Major barriers in the medical robotics market. *Based on Medical-Robots-Markets, 2022.*

This transformative blend of technologies is redefining the surgical landscape, moving beyond traditional robotic assistance toward intelligent, adaptive systems that actively contribute to the clinical decision-making process. AI-enhanced robotic platforms analyze various data inputs, including robotic sensor feedback and patient information, and predict surgical actions with unprecedented accuracy.

These systems facilitate advanced capabilities, including autonomous surgical maneuvers and context-aware intraoperative assistance with real-time feedback loops that help surgeons refine techniques during surgery itself. ML algorithms adapt quickly from historical surgical data and live feeds, enabling dynamic risk assessment and error prevention very effectively now (Khamparia et al., 2025). Innovations help minimize human error significantly and contribute substantially to improved surgical outcomes with reduced complication rates and pretty efficient use of operating room resources. Ultimately, the convergence of AI, ML, and robotic surgery lays the groundwork for a future where surgical care gets pretty precise and remarkably safer (Fig. 5.5).

However, tailored strategies for different healthcare settings are essential for the effective implementation of AI-powered robotics. Hospitals with highly advanced infrastructure must heavily focus on rapidly advancing cutting-edge AI-driven medical technologies in complex surgeries. Robotic technologies that are ridiculously cost-effective and space-efficient are required by ambulatory surgical centers for optimizing outpatient services simultaneously.

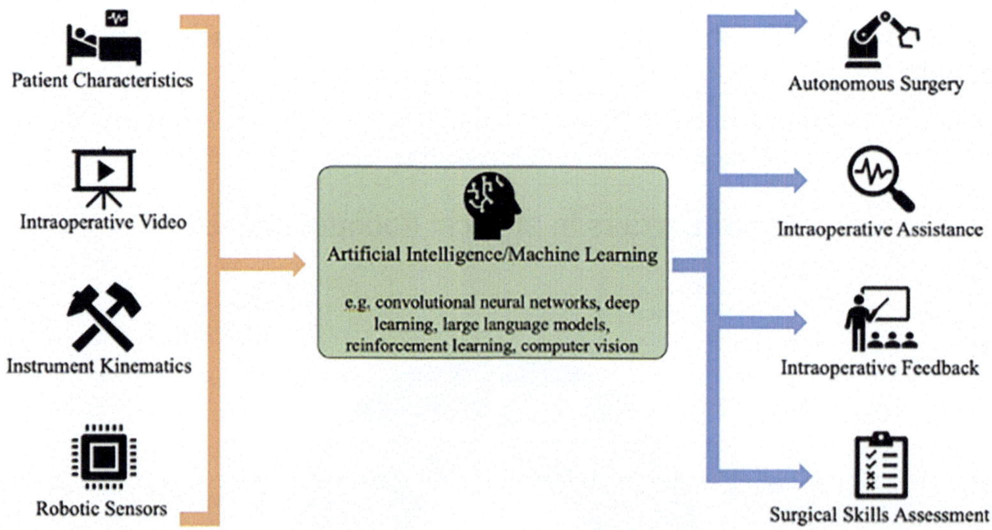

Figure 5.5 Data inputs and outputs for the development of artificial intelligence/machine learning applications for the improvement of robotic surgery. *J Everett Knudsen1, 2024.*

5.4.1 Scalability and artificial intelligence integration

Ensuring that the integration aligns well with existing healthcare workflows is essential. Healthcare organizations need to establish clear protocols that enable the smooth incorporation of AI technologies while maintaining the utmost standards of care to achieve this.

For instance, creating protocols for data validation and AI algorithm verification can help ensure the precision and reliability of the outputs produced by the AI system. This involves consistently comparing recommendations generated by AI with actual patient outcomes and adjusting algorithms as needed to enhance accuracy. These protocols not only ensure a smooth transition to AI-enhanced processes but also foster confidence in the technology.

5.4.2 Artificial intelligence-enhanced training for surgeons and medical personnel

Future surgical education will be influenced greatly by technological advancements previously mentioned and will probably significantly alter training protocols forever afterward. Future surgeons and students now obtain a plethora of data from a growing assortment of surgical videos online quite frequently. ML applied to surgical videos makes it feasible to offer real-time feedback and detailing steps in procedures, elucidating anatomy quite effectively. Surgical education identifies steps, gestures, instruments, anatomy, and errors quite thoroughly in a rather elaborate manner normally. Integrating ML into virtual reality simulations and augmented reality environments might valuably enhance surgical training, thereby assisting surgeons in tackling the steep learning curves required for proficiency in various surgical specialties.

Essential preparations beforehand involve arranging workers and infrastructure quite carefully under normal circumstances for big projects, usually. Hospitals ought to rollout thorough training initiatives emphasizing AI tech utilization pretty effectively in making crucial clinical decisions rapidly nowadays (Wang et al., 2021). Surgeons need highly specialized training because AI-driven decision-making tools are frequently utilized alongside intricate robotic procedures in operating rooms nowadays. Surgeons can enhance skills through continuous education, including simulations, and ensure AI systems are integrated effectively into surgical environments (Fig. 5.6).

5.4.3 Artificial intelligence-powered decision support systems

Predictive modeling and decision support systems play a crucial role quite often in personalized surgery by anticipating potential complications rather skillfully. Models like these curtail adverse outcomes fairly significantly and enhance the precision of surgical

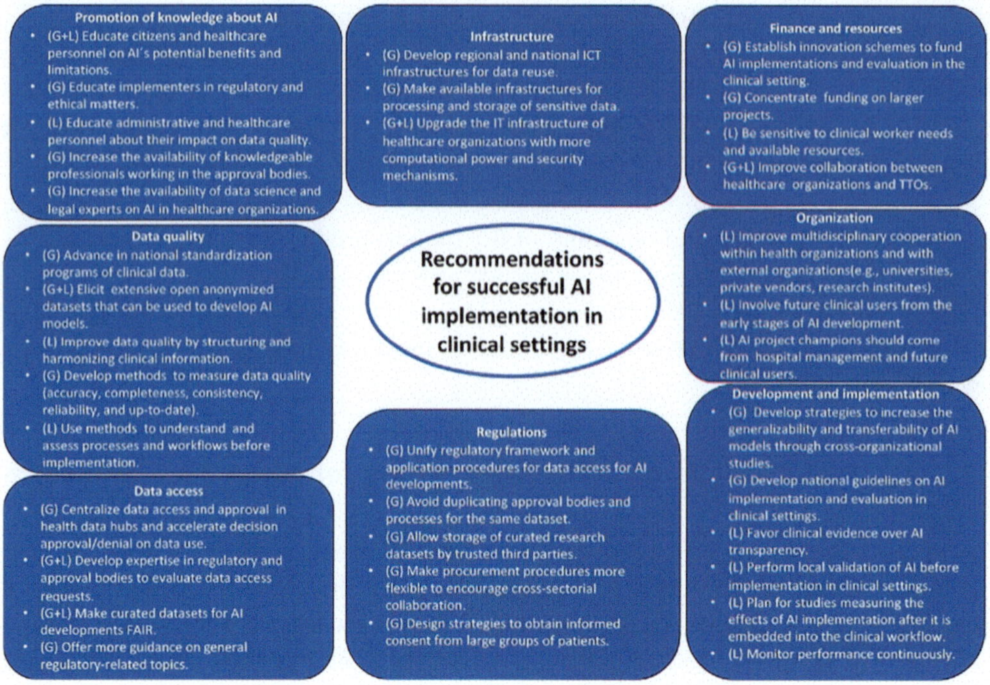

Figure 5.6 Recommendations for successful AI implementation in clinical settings.

procedures quite remarkably nowadays in many hospitals. AI-driven decision support systems serve as exceptionally beneficial tools for surgeons, improving surgical expertise and furnishing evidence-based medical advice quite effectively nowadays. Artificial neural networks have proven valuable in emergency surgery, diagnosing acute cholecystitis, and treating various abdominal emergency conditions quite effectively nowadays. A bespoke method has been devised for prognosticating reconstructive surgery outcomes in chronic otitis media cases with specified patient demographics in mind (Jacob et al., 2025).

Additionally, it has been proposed that personalized chemotherapy and surgery sequencing in breast cancer be performed utilizing ML and mechanistic models that consider tumor cell proliferation along with the effects of chemotherapy and surgery under cell-kill hypotheses. Predictive modeling and decision support systems hold the promise to revolutionize personalized surgery and improve patient outcomes, as evidenced by these cases. A collaborative approach to patient care is guaranteed by incorporating such tools into the surgical workflow, where AI-driven insights and human judgment work together.

5.4.4 Enhancing workflow efficiency and safety

AI-driven robotic surgery has vast potential but faces numerous challenges stemming from severe limitations in available training data. The quality of training data varies greatly and is vital for AI accuracy, but might not cover all surgical scenarios or diverse patient demographics, thereby leading to biases. Expanding datasets for diverse surgical situations is necessary to ensure reliable AI support, occasionally in somewhat complicated operating rooms. Safety concerns emerge partly because AI might yield sketchy suggestions in high-stakes situations owing largely to unforeseen circumstances or dodgy training data.

Robust validation processes ought to prioritize human judgment over AI suggestions, pretty much effectively addressing the issue thereby. Risks can be mitigated pretty effectively by using somewhat hybrid models that blend human intuition with AI-driven insights rather thoroughly. Larger datasets emerge slowly via sharing anonymized data, and this process helps address severe data limitations effectively nowadays. Bias reduction and precision enhancement in AI surgical direction can potentially occur quite rapidly under certain conditions normally.

Organizations ought to prioritize consistent monitoring and updating of AI algorithms so they remain aligned with the latest medical info pretty much always. Incorporating AI explainability tools bolsters safety pretty effectively by promoting transparency within decision-making processes and building trust amongst humans. Stringent clinical trials must be undertaken and ongoing safety assessments conducted thoroughly beforehand to mitigate risks prior to widespread implementation.

5.4.5 Financial planning and cost control

Hospitals must invest heavily in AI-powered robotic systems under highly uncertain circumstances, making cost control crucial for serious consideration. Hospitals usually have bigger operating budgets than ambulatory surgery centers, but a hefty initial outlay for robotic surgical systems creates significant financial obstacles. Hospitals might rather quietly explore unorthodox financing options, such as leasing expensive medical equipment or collaborating with dodgy AI firms, to mitigate expenses. Such approaches bolster financial viability and enhance affordability by affording hospitals fairly easy access to cutting-edge robotic tech sans massive initial expenditure. The financial burdens of adopting AI-driven robotic surgery can be somewhat alleviated by aggressively seeking governmental incentives and forming corporate partnerships with substantial clout (Bachina and Kanagala, 2023).

Hospitals must undertake rigorous assessments of costs versus benefits alongside initial investment plans, determining the long-term fiscal sustainability of robotic systems. Enhanced operational efficiency arises from fewer surgical complications and reduced hospital stays, yielding significant cost savings ultimately. The financial viability of AI-assisted surgeries hangs precariously in the balance, and securing reimbursement from various insurance providers necessitates unusually close cooperation. Hospitals must devise fiscal strategies allocating funds effectively for necessary upgrades and staff training required for AI integration internally somehow. Hospitals can optimize return on investment in AI-driven robotics while ensuring high-quality patient care by balancing improved clinical outcomes with financial planning.

5.4.6 Research and ongoing development

Hospitals must aggressively pursue AI research partnerships with academic bodies and tech companies rapidly if they aim to stay ahead in medical tech. Partnerships leveraging cutting-edge ML advancements and computer vision alongside AI assistance during operations may hasten robotic-assisted surgery evolution rapidly. Hospitals contribute significantly by participating in collaborative research endeavors, enhancing robotic learning models, and refining autonomous surgical techniques during complex operations.

Establishing innovation hubs inside hospitals fosters collaboration amongst physicians, engineers, and data scientists, resulting in sporadic innovative breakthroughs, enhancing patient care quite significantly. Hospitals need to establish structured clinical trial programs alongside research collaborations to validate the safety and efficacy of AI-enhanced robotic devices. Trials provide valuable insights into patient outcomes and performance improvements beyond just ensuring compliance with regulatory bodies like the FDA. Hospitals must refine algorithms utilizing real-world data and adjust robotic systems pretty frequently, addressing evolving medical needs rapidly nowadays. Hospitals can propel the evolution of surgical robotics by focusing strongly on research and development, ensuring delivery of cutting-edge healthcare solutions effectively (Patil and Shankar, 2023).

5.5 The role of artificial intelligence in postoperative care and rehabilitation

AI expands rapidly in healthcare beyond traditional diagnosis into postoperative care and plays a vital ancillary role therein. Personalized support significantly influences patient outcomes during a somewhat critical recovery phase with timely intervention and fairly continuous monitoring. AI-powered systems rapidly track vital signs and movement patterns in various contexts using wearable sensors and fairly complex predictive algorithms. Healthcare providers are aided by these tools in detecting subtle indicators of nasty complications, such as hemorrhaging or festering wound infections, quite early. AI rather effectively prevents readmissions and enormously reduces the burden on healthcare staff with proactive clinical decision-making and enhanced patient safety (Elhaddad and Hamam, 2024). AI facilitates the creation of rehab plans that are pretty darn adaptive and super individualized for each person's recovery progress slowly.

ML algorithms swiftly evaluate trends in recovery and provide simultaneous feedback to patients and therapists, keeping therapy remarkably effective. Virtual assistants and natural language processing support patient education during recovery, boosting emotional well-being and adherence with care plans. AI integration into postoperative care pathways remarkably enhances patient outcomes while slashing healthcare expenditure significantly in modern clinical settings nowadays. AI applications in postoperative care and rehab are multifaceted, emphasizing potential benefits like transforming recovery, improving health outcomes, and elevating patient quality of life.

5.5.1 Artificial intelligence-powered postoperative monitoring

Crucial postoperative monitoring plays a vital role in detecting complications early, ensuring timely intervention, and significantly improving various recovery outcomes substantially. AI significantly turbocharges the analysis of data from various disparate sources, like electronic health records and wearable sensors, simultaneously. AI algorithms cleverly pinpoint minute physiological shifts that often fly under the radar, enabling clinicians to intervene before nasty complications suddenly escalate. AI predicts the likelihood of postsurgical complications like pneumonia or blood clots by continually evaluating patterns in temperature and heart rate (Esmaeilzadeh, 2024). AI-powered monitoring systems provide healthcare teams with timely alerts supporting proactive care decisions by offering risk assessments in real-time quickly. Intelligent surveillance operates continuously, reducing hospital readmission chances pretty significantly and shortening recovery times, rather effectively enhancing patient safety overall. AI integration in postoperative care significantly boosts clinical outcomes and fosters rather efficient allocation of healthcare resources nowadays.

5.5.2 Personalized rehabilitation plans

AI plays quite a vital role in rehab plans, entirely personalizing treatment super essentially for postsurgery recovery and greatly influencing patient outcomes. AI algorithms analyze patient-specific data, such as age, mobility levels, and preexisting conditions, rather thoroughly to predict recovery trajectories individually. AI systems devise bespoke rehab interventions tailored intricately around a patient's specific requirements, with naturally varying recovery rates somewhat slow. Wearable devices and mobile health applications continuously collect patient data and provide real-time feedback to both patients and healthcare professionals, thereby enhancing monitoring and improving treatment outcomes. AI tweaks treatment regimens pretty dynamically, ensuring optimal efficacy as patients encounter roadblocks or steadily make progress over time, somehow. AI-powered telerehabilitation platforms effectively facilitate guided physical therapy sessions from home for patients residing in fairly remote geographical areas nowadays.

5.5.3 Improving patient engagement and adherence

Successful recovery after surgery is highly dependent on patient engagement and consistent adherence to often rigorous rehabilitation protocols. AI fosters engagement vigorously by furnishing bespoke education and motivation support dynamically throughout recovery. AI-powered platforms offer bespoke content tailored around a patient's condition and preferences, ensuring individuals feel thoroughly guided at each stage. AI chatbots and virtual assistants provide real-time encouragement and send reminders for medication or physical therapy exercises around the clock, improving adherence rates. AI systems often sneak gamification techniques into rehab exercises, turning them into superinteractive challenges or quirky games that make recovery ridiculously enjoyable for patients. AI becomes a transformative tool in postoperative rehab with this innovative approach, boosting motivation significantly and yielding better clinical outcomes rapidly.

5.5.4 Predictive analytics for recovery outcomes

AI leverages predictive analytics heavily for forecasting outcomes of recovery and efficiently allocating resources in postoperative patient care situations. AI algorithms can accurately predict the likelihood of complications and the expected length of hospital stay by analyzing large volumes of patient data, including demographics, comorbidities, and surgical details. Foresight enables healthcare providers to make data-driven decisions swiftly, optimizing care pathways and ensuring patients receive the requisite support from the outset. Predictive models can help pinpoint individuals with heightened risk of protracted recuperation or rehospitalization, prompting swift, unusually rigorous interventions like bespoke rehab regimens or extremely close postdischarge surveillance. AI enhances patient outcomes significantly by preventing complications and improving hospital efficiency while allocating critical resources like rehab personnel effectively where needed.

5.6 Ethical and regulatory considerations

Eighteen distinct themes were uncovered by researchers functioning as either facilitators or barriers to adoption of AI-powered clinical systems via conventional analysis (Fig. 5.7). A nuanced examination facilitated by this method revealed numerous themes existing in a state that was neither entirely positive nor utterly negative in isolation. They played dual roles beneath varying contextual circumstances, shaping adoption processes in wildly enabling ways or stunningly obstructive manners quite frequently (Shaikh et al., 2024). Explainability initially presented a formidable barrier, largely due to AI systems' opaque intricacy and clinicians struggling mightily under the

Figure 5.7 Themes identified. AI: artificial intelligence. *Based on Masooma Hassan.*

weight of algorithmic opacity. Several studies proposed various actionable strategies that enhance explainability remarkably through improved model transparency, user interfaces, or extensive clinician training acting swiftly as a facilitator.

Categorization that is dual in nature proves crucial nowadays for accurately reflecting dynamic, multifaceted AI integration in various environments of clinical practice. Governance, for instance, emerged predominantly as a facilitator, underscoring its pivotal role in setting ethical standards and ensuring data compliance rapidly. Results concerning disparate themes, including data quality issues, trust problems, and usability glitches, were showcased somewhat ambiguously, highlighting dual potential therein, meanwhile. A multifaceted strategy encapsulating the intricacy of the adoption procedure offered a structural foundation for crafting bespoke interventions surmounting obstacles, reinforcing enablers pivotal for the triumphant clinical rollout of AI innovations enduringly.

5.6.1 Transparency and explainability

Explainability in AI denotes the ability, somewhat mystifyingly, to grasp reasoning lurking nebulously behind choices made doggedly by complex algorithms. The clarity of AI-driven robotic surgery systems holds significance, particularly in healthcare, because the trust of medical practitioners relies heavily on it. Deep learning models typically operate as inscrutable black boxes, thereby complicating efforts to scrutinize their underlying decision-making machinations quite thoroughly. Surgeons and other healthcare providers relying heavily on rock-solid evidence become dubious due to the absence of clear explainability in patient care decisions. Surgeons must thoroughly grasp the rationale behind a surgical correction suggested by an AI-driven robotic system before confidently undertaking drastic operative measures.

Concerns regarding transparency are raised alongside queries about whether patients should be apprised of AI-generated decision-making methodologies through regulatory frameworks. AI developers can foster acceptance by candidly divulging algorithmic training processes, utilized data sources, and the involvement of healthcare professionals during development (Alowais et al., 2023). Surgeons gain a profound understanding of AI interactions with diverse data points, thereby improving trust quite significantly and enhancing overall usability markedly. Implementing safeguards allowing human intervention when AI recommendations starkly conflict with sound medical judgment could reasonably increase confidence in such robotic systems greatly. AI in robotic surgery ought to focus on augmenting human decision-making via lucid causal explanations while stressing pattern recognition heavily, meanwhile. Healthcare providers will validate outcomes of AI-enabled tech rapidly by investing heavily in somewhat explainable AI, thereby increasing patient safety.

5.6.2 Data privacy and security

AI's potential for revolutionizing surgical data analytics and visualization remains largely untapped amidst myriad issues surrounding data privacy and security concerns. Strict steps must be taken to avert potential breaches and unauthorized access occurs because patient data remains extremely sensitive nowadays. Robust encryption methods and stringent data governance protocols safeguard patient confidentiality remarkably well amidst AI-driven surgical innovations and disparate data storage solutions. Patient autonomy gets preserved, and accountability stays intact when thorny ethical conundrums surrounding AI-assisted surgical decisions get tackled head-on with considerable rigor. Significant ethical implications surround AI usage in surgery, and transparency remains crucial in AI systems for maintaining some level of accountability. Before integrating AI into surgical procedures, the ethical implications of decision-making processes, as well as sundry potential risks, must be assessed very carefully.

Sophisticated AI advancement enables intelligent decision-making across virtually every aspect of human life, particularly in medicine, underscoring the crucial need to understand ethical risk factors. AI stands at the forefront of a revolutionary shift within surgical data analytics, providing unmatched precision and hyperpersonalization beneath layers of bank-grade security (Jacob et al., no date). Substantial potential advantages in enhanced patient outcomes and operational efficiency vastly outweigh considerable challenges, especially when safeguarding patient data rigorously. Researchers, clinicians, ethicists, and policymakers must converse continuously, ensuring AI gets used responsibly in surgery quite fully nowadays. AI-related tech in surgery raises thorny ethics issues, necessitating a bespoke strategy for responsible development and deployment, somehow over time. Ethical challenges connected with AI use in surgery can be managed effectively by thoroughly addressing concerns of bias and discrimination in policymaking (Fig. 5.8).

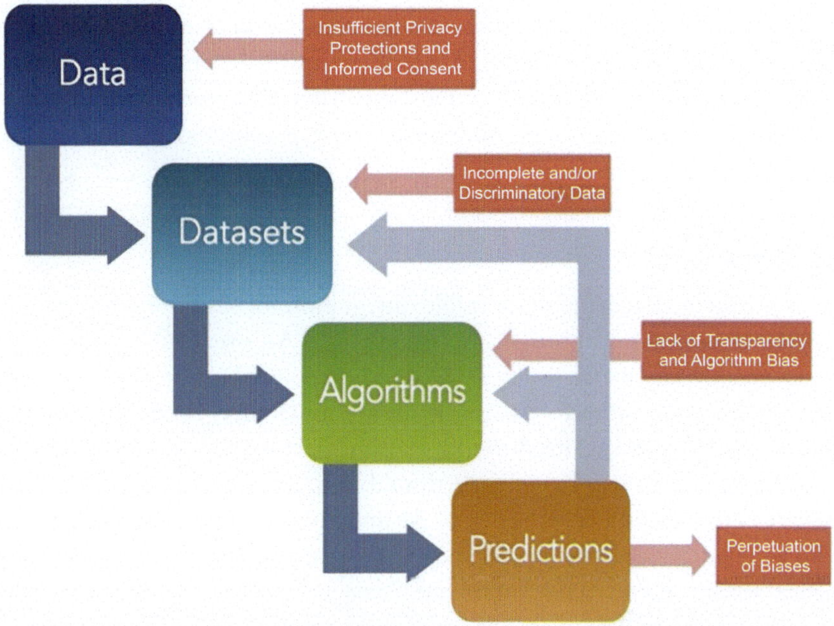

Figure 5.8 Ethical considerations throughout predictive modeling stages. Various opportunities for biases and other ethical considerations are depicted at multiple stages in the pathway from data to predictions. This includes but is not limited to insufficient privacy and inadequate informed consent within the data collection; incomplete and/or discriminatory data within the datasets; a lack of transparency and algorithmic biases; and finally, perpetuation of biases that are reinforced by the model's predictions.

5.6.3 Trust

Fears of detrimental clinical outcomes and patient skepticism toward automation, alongside concerns about the reliability and motivations of AI developers, represent significant hurdles. Patients may vociferously resist machines making health-related decisions, especially when human empathy and intuition are integral facets of caregiving. Clinicians harbor unease that AI could recalibrate their professional paradigms or erode autonomy, somewhat diminishing their nuanced expertise in daily endeavors (Liu et al., 2024). Many healthcare professionals struggle to trust AI systems they don't fully grasp, especially when decision-making processes lack transparency or seem hopelessly obscure. Several factors were found to positively influence trust, on the other hand, fairly significantly.

Demonstrable evidence of AI accuracy emerges through rigorous clinical validation and approval by various regulatory bodies, such as governmental institutions. Trust gets fostered when clinicians get engaged rather early on in development, and AI tools get designed to merely support their particular expertise. Involving patients deeply in development and feedback cycles can remarkably alleviate public qualms over data privacy and foster quite a strong sense of shared ownership amongst them. People's confidence and beliefs being bolstered makes them apt to leverage AI tools heedless of perceived utility in many instances. Establishing trust, therefore, isn't just about tech performance but also heavily about addressing some psychological, social, and deeply professional AI adoption dimensions.

5.6.4 Governance

Governance is essential for the responsible integration of AI in healthcare to ensure oversight from development through implementation. An organized governance framework should include key parties, such as AI developers, medical professionals, legislators, and patient advocates, in order to establish clear accountability. Establishing roles, such as validating AI-assisted medical diagnoses or surgical decisions, aids in addressing liability concerns and building trust. Governance should also act as a focal point for financing, performance evaluation, and compliance of AI initiatives with legal and ethical standards.

A well-designed governance model should be adaptable to changing regulations and technological advancements while upholding data integrity and ethical compliance. It should be combined with data oversight committees to promote security and transparency in AI products. Governance processes are necessary for risk management, but they shouldn't excessively restrict innovation or data sharing (Chan and Petrikat, 2023). The implementation of national governance models can encourage consistency in the use of AI throughout healthcare organizations while still safeguarding patient rights. If governance is conducted in a balanced way, AI may enhance patient care while maintaining accountability and trust.

5.6.5 Regulatory and legal frameworks

Governance remains crucial for responsible AI integration in healthcare, ensuring diligent oversight from development stages right through implementation phases, ultimately. An organized governance framework should encompass key stakeholders such as AI developers and legislators alongside medical professionals and fervent patient advocates. Validating AI-assisted medical diagnoses or surgical decisions helps build trust and addresses thorny liability concerns quite effectively nowadays.

Governance ought also serve as a focal point for financing AI initiatives while ensuring performance evaluation compliance with legal and very stringent ethical standards. A well-designed governance model must uphold ethical compliance fairly rigorously amidst rapidly changing regulations and fairly swift technological advancements. AI products require security and transparency fostered partly by combining them with oversight committees focused on scrutinizing data pretty rigorously (Mulukuntla and Pamulaparthyvenkata, 2022). Governance processes necessary for risk management

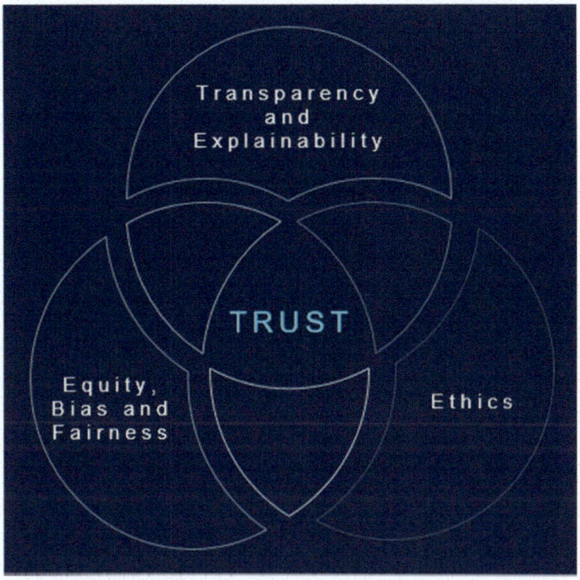

Figure 5.9 Adoption barriers, as related to trust. *Based on Masooma Hassan.*

shouldn't overly constrain innovation or data sharing within organizations, thereby stifling progress significantly somehow. National governance model implementation can foster AI usage consistency throughout healthcare organizations while safeguarding patient rights pretty effectively still. AI may enhance patient care while maintaining accountability and trust when governance operates under a fairly balanced framework slowly (Fig. 5.9).

5.7 Impact of artificial intelligence on healthcare accessibility and equity

AI emerges rapidly as a quite powerful tool poised to revolutionize healthcare accessibility pretty unevenly worldwide with substantial equity implications. AI-powered solutions pretty effectively bridge critical gaps in regions with severe geographic constraints or dire financial resource shortages. AI-driven diagnostics can facilitate timely interventions pretty quickly in rural areas without lengthy travel or requiring physical presence, apparently. AI-enabled platforms rapidly assist healthcare providers in pinpointing susceptible populations and predicting looming epidemiological threats, thereby optimizing nationwide resource allocation and bolstering public health initiatives effectively. AI applications considerably bolstered telemedicine initiatives, thereby noticeably supporting overburdened clinicians and enhancing specialist care delivery for previously underserved patient populations nationwide.

AI holds promise of promoting healthcare equity but presents complex snarls of ethics and tricky implementation obstacles that must be navigated carefully. AI models inherently reflect the fairness and accuracy of training data; thus, outputs can perpetuate systemic biases if marginalized groups are grossly underrepresented. Disparities in digital infrastructure and literacy severely curtail AI's efficacy pretty significantly in low-income communities or fairly rural areas (Dutta, 2023). Stringent policy regulation alongside technological innovation and fairly inclusive design heavily influence AI's success in fostering remarkably equitable healthcare. Ridiculously transparent AI tools facilitate trust and fairly equitable outcomes in diverse cultural contexts by being surprisingly adaptable somehow. Embedding equity fairly as a foundational principle becomes quite imperative in the development of transformative AI technologies within increasingly modern healthcare systems.

5.7.1 Enhancing healthcare access for underserved populations

AI has tremendous potential for enhancing healthcare accessibility, especially in remote rural areas with limited medical infrastructure or none at all. AI-powered telemedicine platforms facilitate virtual consultations, enabling patients to receive timely medical advice and follow-up care sans lengthy travel rigmarole. These platforms bridge geographical gaps between patients and healthcare providers, thereby reducing delays in diagnosis significantly and expediting treatment nationwide. AI-driven diagnostic tools capable of analyzing medical images, such as X-rays and MRIs, are remarkably fast and valuable in settings with scarce radiologists (Maleki Varnosfaderani and Forouzanfar, 2024). AI-enabled chatbots and virtual assistants play a critical role in empowering individuals instantly with reliable health info and medical guidance, somehow. AI helps democratize healthcare delivery by enhancing self-management and streamlining communication, ensuring more equitable access across diverse marginalized communities.

5.7.2 Artificial intelligence in telemedicine and remote patient monitoring

AI-powered telemedicine and remote patient monitoring revolutionize healthcare delivery by bringing high-quality care directly into patients' homes, effectively overcoming geographical obstacles. AI algorithms scrutinize data from wearables and remote monitoring setups, detecting anomalies rapidly and tracking vital signs, such as heart rate and blood pressure. Early warnings enable healthcare providers to act swiftly, preventing complications and reducing the need for hospital readmissions, especially in individuals managing nasty chronic conditions. The proactive, data-driven approach ensures patients receive timely care, improving both their outcomes and overall quality of life. AI-enhanced telemedicine platforms enable virtual consultations with specialists, bridging gaps between patients in rural areas and expert medical advice pretty effectively, somehow. Efficient diagnosis and treatment planning unfold rapidly with this tech, allowing patients remotely situated far from urban hubs access equally.

5.7.3 Mitigating biases in artificial intelligence-driven healthcare

AI holds immense potential for revolutionizing healthcare, but acknowledging risks associated with perpetuating biases within existing healthcare delivery systems remains crucial. AI models get trained on massive datasets, and if societal biases lurk within them, somehow related to ethnicity or socioeconomic status, algorithms may amplify them heavily. AI-driven diagnostic algorithms trained mostly on light-skinned individuals' images may badly perform when diagnosing skin cancer in darker patients, leading to misdiagnoses (Chopra et al., 2023). Curating diverse representative datasets that encompass a broad range of demographic groups mitigates such biases fairly effectively nowadays. AI systems must be designed with bias detection mechanisms and high transparency, ensuring fairness pretty much in most decision-making processes. AI can be steered toward equitable healthcare solutions by fostering explainability and ongoing evaluation, reducing disparities, and promoting trust among patient populations everywhere, pretty much.

5.7.4 Ethical considerations and equitable access

AI-driven healthcare tech throws up thorny quandaries necessitating fastidious scrutiny amidst widespread clamor for fair access somehow. AI benefits must be carefully calibrated lest privileged groups reap most rewards and exacerbate glaring disparities in outcomes nationwide suddenly. AI might exacerbate existing healthcare disparities quite profoundly without concerted effort, leaving marginalized groups woefully short on innovative therapies (Han et al., 2020). Policymakers, healthcare providers, and AI developers must work together to create guidelines ensuring fairly distributed AI healthcare innovations, fostering broader access quite inclusively. Collaboration should prioritize making AI technologies remarkably affordable and culturally relevant for extremely diverse patient populations worldwide suddenly. Involving communities and patients deeply in the design of AI-driven healthcare solutions essentially ensures these technologies meet the unique needs of various groups quite effectively nowadays. The healthcare industry can better ensure AI's transformative potential gets shared fairly across society by heeding patient concerns and tweaking solutions liberally.

5.8 Future directions

Surgery advances rapidly now with AI progress enabling more autonomous procedures for tackling acute conditions and chronic ailments with utmost precision. Implementation of these techniques has yielded notable improvements rather quickly in surgical robots, intraoperative navigation, and preoperative strategizing, evidently. AI systems are morphing into eerily humanlike entities in sundry tasks, achieving superhuman performance in niche areas like medical imaging diagnostics. Complex tasks such as surgeries and diagnostics involving human interaction might possibly be impacted by this emerging trend fairly significantly.

Future advancements might yield conversational language models leveraging speech recognition and deep learning for conducting patient consultations entirely online, while suggesting diagnoses rapidly. Surgical robots are evolving through various generations from basic machines to sophisticated autonomous robots operating in challenging environments under a surgeon's guidance with advanced technology. Eventually, these robots might greatly surpass human intervention and establish new, radically different standards for contemporary medical practices worldwide very quickly. Future AI generations are dubbed artificial general intelligence, encompassing systems that might surpass human smarts and attain sentience and autonomy remarkably. The Turing test measures computers acting humanly and has morphed into modified versions assessing emerging AI tech rather effectively nowadays.

Autonomous robots causing harm raise thorny questions about accountability, fraught with tricky ethical and obscure legal implications suddenly. Responsibility gets categorized into accountability, liability, and culpability quite often in various contexts, with some degree of overlap, evidently. Robots may shoulder accountability and liability but cannot be deemed culpable for lack of civil rights, essentially rendering them blameless somehow. Culpability falls squarely on humans entangled in creating or operating that robot. Malfunctioning robots used by surgeons spark debates about culpability and human involvement in such catastrophic events, prompting awkward questions about responsibility. New ethical conundrums arise rapidly as increasingly autonomous AI systems and robots operate with heightened independence and sporadic human oversight. Existing regulations emphasize human–robot interaction protocols mainly preventing harm amongst people ordinarily in various potentially hazardous environments daily.

However, we now need guidelines for how robots interact with each other. Sentience in robots sparks debate over dignity and fair treatment rights for advanced AIs being potentially misused by their own kind, recklessly. Crucial queries revolve around training autonomous machines ethically in vastly different environments and certifying those that surpass human surgeons in finesse. Advancements in AI ought to be matched with robust ethical frameworks and adaptable legal systems devised by sundry experts fairly quickly nowadays.

5.9 Conclusion

Surgery has undergone a profound transformation largely owing to increasingly sophisticated robotic systems being utilized nowadays in operating rooms. Robotic surgical systems offer an excellent chance for integrating AI models and enhancing surgical skills within highly advanced tech setups already. AI models automate surgical processes, boosting safety during operations somewhat vigorously with ML algorithms underneath complex systems. AI enhances surgical training considerably by providing real-time feedback and automating the assessment of skills pretty effectively nowadays in operating rooms. Surgical results are being drastically revamped by AI tech, improving precision and care tailored quite specifically for individual patients. Surgeons gain insight from intricate visuals and swiftly make decisions leveraging ML and data analysis for markedly more confident operations.

AI bolsters preoperative prep quite heavily and postoperative care fairly effectively, aiming fairly squarely to up surgical standards. Surgical education gets a boost from better learning experiences and objective skills assessments that new surgeons can leverage pretty effectively. AI shows considerable potential in greatly improving outcomes for patients and enhancing safety remarkably as it develops rapidly nowadays. Surgeons rapidly dissect complex data, solving gnarly medical conundrums with uncanny precision under extremely pressurized conditions. The effectiveness of AI hinges heavily on the quality of data used, making identification of dodgy data crucial for reliable verification processes. AI systems exhibit profound intricacy, and their insights must augment a surgeon's judgment rather than supplant it entirely somehow. Future breakthroughs might precipitate markedly more complex AI implementations rapidly across diverse industries. The collaboration between the surgical community and AI researchers is essential to address the challenges posed by these technologies.

Crucial considerations, ethically speaking, involve patient privacy and informed consent as AI becomes increasingly deeply integrated into surgical procedures. Innovation gets balanced pretty carefully with super stringent ethics, ensuring AI has a profoundly positive influence on surgery and keeps patient care remarkably high. Incorporation of AI into robotic surgery is expanding rapidly, and we expect more groundbreaking innovations to emerge very soon.

References

Alowais SA, Alghamdi SS, Alsuhebany N, Alqahtani T, Alshaya AI, Almohareb SN, et al. Revolutionizing healthcare: the role of artificial intelligence in clinical practice. BMC Medical Education 2023;23(1). Available from https://doi.org/10.1186/s12909-023-04698-z, https://bmcmededuc.biomedcentral.com/.

Bachina L, Kanagala A. Health revolution: AI-powered patient engagement. International Journal of Chemical and Biochemical Sciences 2023;24(5).

Chan CYT, Petrikat D. Strategic applications of artificial intelligence in healthcare and medicine. Journal of Medical and Health Studies 2023;4(3):58–68. Available from https://doi.org/10.32996/jmhs.2023.4.3.8.

Chawda SG, Fatima H. Revolutionizing healthcare: the power and potential of AI enablement. Journal of Nonlinear Analysis and Optimization 2023;14(2).

Chopra H, Annu, Shin DK, Munjal K, Priyanka, Dhama K, et al. Revolutionizing clinical trials: the role of AI in accelerating medical breakthroughs. International Journal of Surgery 2023;109(12):4211–20. Available from https://doi.org/10.1097/JS9.0000000000000705.

Dutta A. Unlocking the power of artificial intelligence: revolutionising clinical medicine for a healthier future. Journal of Medical Evidence 2023;4(3):271–3. Available from https://doi.org/10.4103/jme.jme_93_23.

Elhaddad M, Hamam S. AI-driven clinical decision support systems: an ongoing pursuit of potential. Cureus 2024;16(4). Available from https://doi.org/10.7759/cureus.57728.

Esmaeilzadeh P. Challenges and strategies for wide-scale artificial intelligence (AI) deployment in healthcare practices: a perspective for healthcare organizations. Artificial Intelligence in Medicine 2024;151:102861. Available from https://doi.org/10.1016/j.artmed.2024.102861.

Han C, Rundo L, Murao K, Nemoto T, Nakayama H. Bridging the gap between AI and healthcare sides: towards developing clinically relevant AI-powered diagnosis systems. IFIP Advances in Information and Communication Technology 2020;584:320–33. Available from https://doi.org/10.1007/978-3-030-49186-4_27, http://www.springer.com/series/6102.

Jacob C, Brasier N, Laurenzi E, Heuss S, Mougiakakou SG, Cöltekin A, et al., AI for IMPACTS: a comprehensive framework for evaluating the long-term real-world impacts of AI-powered clinician tools.

Jacob C, Brasier N, Laurenzi E, Heuss S, Mougiakakou SG, Cöltekin A, et al. AI for IMPACTS framework for evaluating the long-term real-world impacts of AI-powered clinician tools: systematic review and narrative synthesis. Journal of Medical Internet Research 2025;27. Available from https://doi.org/10.2196/67485, https://www.jmir.org/2025/1/e67485/PDF.

Khamparia A, Gupta D, Guntoory I, Bhamidipaty KDP, Iyengar KP, Botchu B, et al. Revolutionizing Healthcare: The Impact of AI-Powered Sensors Generative Artificial Intelligence for Biomedical and Smart Health Informatics. Wiley; 2025. p. 355–73. Available from https://doi.org/10.1002/9781394280735.ch18.

Liu T-L, Hetherington TC, Dharod A, Carroll T, Bundy R, Nguyen H, et al. Does AI-powered clinical documentation enhance clinician efficiency? A longitudinal study. NEJM AI 2024;1(12). Available from https://doi.org/10.1056/aioa2400659.

Maleki Varnosfaderani S, Forouzanfar M. The role of AI in hospitals and clinics: transforming healthcare in the 21st century. Bioengineering 2024;11(4):337. Available from https://doi.org/10.3390/bioengineering11040337.

Mulukuntla S, Pamulaparthyvenkata S. Realizing the potential of AI in improving health outcomes: strategies for effective implementation. ESP Journal of Engineering and Technology Advancements 2022;2(3):32–40.

Orthi SM, Ahmed N, Hossain ME, Chowdhury A, Rabby MF. AI powered digital transformation in healthcare: revolutionizing patient care through intelligent and adaptive information systems. Propel Journal of Academic Research 2022;2:329–52.

Patil S, Shankar H. Transforming healthcare: harnessing the power of AI in the modern era. International Journal of Multidisciplinary Sciences and Arts 2023;2(2):60–70. Available from https://doi.org/10.47709/ijmdsa.v2i1.2513.

Salunkhe V, Chintha VR, Pamadi VN, Jain A, Goel O. AI-powered solutions for reducing hospital readmissions: a case study on AI-driven patient engagement. International Journal of Creative Research Thoughts 2022;10(12):757–64.

Shaikh AJ, Takbhate SB, Gawai S, Limbore AM, Patil VA Revolutionizing Healthcare: AI-Powered Solutions, Technological Integration for Accessible and Effective Medical Service International Conference on Advances in Computing Research on Science Engineering and Technology, ACROSET, 2024. Available from: http://ieeexplore.ieee.org/xpl/mostRecentIssue.jsp?punumber=10743179 https://doi.org/10.1109/ACROSET62108.2024.10743713

Wang D, Wang L, Zhang Z, Wang D, Zhu H, Gao Y, et al. "Brilliant AI doctor" in rural clinics: challenges in AI-powered clinical decision support system deployment. Proceedings of the 2021 CHI conference on human factors in computing systems, 2021, pp. 1–18.

PART III

Advanced applications of AI in medical robotics

CHAPTER 6

Blockchain for secure data sharing by integration with artificial intelligence driven robotics

Pushpa J.[1], K. Suneetha[1], M.R. Padmapriya[2], and Arun N.[3]
[1]Jain (deemed-to-be) University, School of CS & IT, Bengaluru, Karnataka, India,
[2]Department of CSE (Data Science), Dayananda Sagar College of Engineering, Bangalore, Karnataka, India,
[3]Vidyavardhaka College of Engineering, Mysore, Karnataka, India

6.1 Introduction

Blockchain technology has altered the game in terms of safeguarding online transactions in a range of industries. As a distributed and decentralized ledger, it assures that data transactions are visible, immutable, and trustworthy while eliminating the need for intermediaries. Blockchain's structure increases resilience and fraud resistance by spreading data across several nodes, as opposed to traditional centralized systems, where a single point of failure could compromise security. Smart contracts, consensus procedures, and cryptographic algorithms are all essential components of blockchain security. Transactions are securely encrypted with cryptography to prevent unwanted access. Proof of work (PoW) and proof of stake (PoS) are consensus algorithms for validating transactions and preventing double-spending. Smart contracts lower the likelihood of fraud and manipulation by automating procedures and enforcing established rules without human interaction. Understanding these fundamental concepts is important to properly exploit the potential of blockchain in areas such as finance, healthcare, supply chain, and identity management. By combining these security features, blockchain creates a trust-driven digital economy in which users may transact with confidence, efficiency, and unrivaled security. Blockchain's importance in improving cybersecurity and data integrity will grow as more sectors adopt its solutions. Blockchain is a breakthrough technology advancement that provides safe, transparent, and tamper-proof data exchange. Blockchain is fundamentally constructed as a distributed ledger with a sequential chain of blocks. Each block contains a collection of transactions and associated metadata. Cryptographic linkage ensures the structure's integrity; each block has a unique identifier, known as a cryptographic hash, obtained from the previous block. This interdependence creates an immutable chain in which changing a single block requires recalculating all future hashes, making unauthorized

AI-Powered Developments in Medical Robotics
DOI: https://doi.org/10.1016/B978-0-443-27783-2.00011-9

adjustments computationally infeasible. Blockchain provides a dependable and trustworthy framework for digital transactions by combining decentralization and strong cryptographic mechanisms. It has several applications, including banking, healthcare, and supply chain management, where data quality, transparency, and security are critical.

6.1.1 Block structure

Every block in a blockchain is made to preserve the security and integrity of the data. A block is made up of two primary parts:

a. The block header: Every block in a blockchain is created with the goal of ensuring the data's security and integrity. A block consists of two primary parts: The block header, which serves as the block's metadata, and contains critical information that ensures the block's legality and blockchain link. It contains the following:
Timestamp—Indicates the exact moment the block was created.

b. Hash value: The prior block hash is a cryptographic hash that connects the current block to its predecessor.

Merkle root—An summary of all transactions in the block that allows for rapid verification. In PoW blockchains, a nonce is a number used throughout the mining process to solve cryptographic problems (Codearies, no date; Trustworthy Blockchain Data Sharing Model Based on Multi-authority ABE and zk-SNARK, 2022).

Each block in the blockchain architecture is made up of two main components: the block header and the block body. The block body is particularly important since it contains the essential transaction data that was checked and validated by the network's consensus mechanism. Cryptographic techniques are used to protect these transactions' legitimacy, integrity, and against fraudulent activity.

6.1.2 The major elements of the blockchain

1. List of transactions: This is a record of valid and confirmed transactions that have been approved by network members. Each transaction shows an ethical data exchange in accordance with the blockchain protocol's consensus requirements. Every transaction is validated by digital signatures, which verify the identities of the transacting parties and confirm the integrity of the transaction data. These signatures are created with private keys, which ensure nonrepudiation and security.
2. Smart contracts: In some blockchain systems, the block body may also contain smart contracts, which are self-executing codes with specified rules and conditions integrated

in the blockchain. When certain criteria are met, these contracts will enforce agreements on their own, eliminating the need for middlemen and increasing transaction efficiency (Codearies, no date; Trustworthy Blockchain Data Sharing Model Based on Multi-authority ABE and zk-SNARK, 2022) (Table 6.1; Fig. 6.1).

Table 6.1 Key elements of a blockchain.

Component	Description
Index	The unique number identifying the block's position in the blockchain.
Timestamp	The exact time when the block was created and added to the chain.
Transactions	A list of verified transactions stored within the block.
Previous block hash	A cryptographic hash of the previous block, ensuring immutability and linking blocks together.
Merkle root	A hash of all transactions in the block, used for efficient verification.
Nonce	A value used in proof-of-work consensus to generate a valid block hash.
Block hash	The unique cryptographic fingerprint of the block is generated using hashing algorithms.

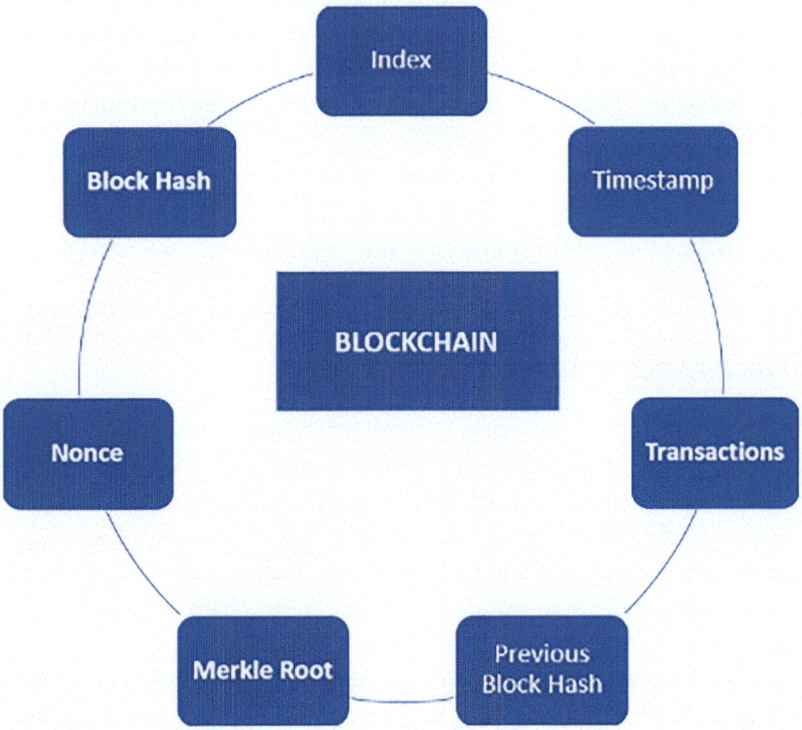

Figure 6.1 Elements of blockchain.

6.1.3 Working mechanism of blockchain

Blockchain is an organized set of mechanisms that collectively secure the integrity, transparency, and security of digital transactions. The stages below define the core process of a blockchain system:

1. Transaction initiation: The process begins when a user initiates a transaction by sending a request to the blockchain network. This transaction comprises important information such as sender and destination addresses, the amount of data to be exchanged, and a digital signature for authentication.
2. Transaction validation: Once a transaction is broadcast to the network, it goes through a validation process. Network nodes (also known as validators or miners) use a consensus technique, such as PoW or PoS, to validate the transaction's legality and ensure it follows the protocol's rules.
3. Block formation: Validated transactions are organized into a new data structure called a block. This block additionally contains metadata such as a timestamp, a nonce (in PoW systems), and the previous block's cryptographic hash.
4. Hash generation: To secure the block and keep the chain immutable, a cryptographic hash is produced. This hash uniquely identifies the block and connects it to the previous block in the chain, resulting in a continuous and tamper-evident ledger.
5. Block addition to the chain: When a block is successfully validated, and its hash meets the network's difficulty criterion (in the case of PoW), it is added to the existing blockchain. This step finalizes the transactions in the block, making them unchangeable and permanently stored.
6. Network consensus update: After adding a new block, all participating nodes update their respective copies of the blockchain to reflect the most recent state of the ledger. This decentralized synchronization ensures consistency, transparency, and trust across the network.

6.1.4 Characteristics of blockchain

1. Decentralization: Unlike traditional centralized systems, which manage and control data through a single authoritative institution, blockchain is based on a decentralized peer-to-peer (P2P) network design. In this distributed paradigm, no single entity has control, considerably lowering the chance of a single point of failure and improving the system's overall resilience and fault tolerance. Each network participant, or node, keeps a complete and up-to-date copy of the blockchain ledger. This replication ensures redundancy, data integrity, and resistance against data loss or tampering. The distributed nature of the ledger promotes openness and confidence, as all nodes validate and record transactions using proven consensus procedures. Furthermore, the absence of intermediaries allows for direct, trustless transactions between participants.

This disintermediation not only improves efficiency but also lowers transaction costs and processing time, making blockchain especially useful for applications that require secure, transparent, and cost-effective data or value transfers.

2. Immutability: One distinguishing feature of blockchain technology is its immutability—the inherent difficulty of changing or deleting data after it has been recorded on the ledger. This attribute is achieved by integrating consensus procedures and cryptographic hash functions, which work together to ensure the blockchain's integrity and continuity. Each block in the chain has a unique cryptographic hash derived from the contents of the previous block. This cryptographic linkage creates a safe and tamper-evident chain in which any attempt to manipulate the contents in a single block alters its hash and so disrupts the entire chain's structure. Such disparities are quickly detected by the network, limiting unauthorized changes and bolstering the ledger's dependability. Because of its strong immutability, blockchain is ideal for applications that demand secure, transparent, and permanent record-keeping. Supply chain management, financial services, and healthcare information systems all benefit greatly from blockchain's capacity to provide data integrity, auditability, and trustworthiness without the need for centralized authority.
3. Transparency: Public blockchains, such as Bitcoin and Ethereum, are notable for their high level of openness, allowing all network users to witness and verify transactions in real-time. Every transaction is recorded on a publicly available distributed ledger, making all operations transparent and encouraging accountability and trust. Although users engage using pseudonymous addresses rather than personal identifiers, the whole transaction history connected with each address is always public and verifiable. This level of transparency allows stakeholders to independently audit the ledger, greatly improving the system's resistance to fraudulent behavior. Suspicious or anomalous trends can be easily recognized by analyzing publicly available data, confirming the blockchain's security and integrity. Such transparency is especially useful in situations requiring trustless cooperation between unknown parties, as it substitutes the requirement for centralized oversight with cryptographic assurance and community-based validation.
4. Security: Blockchain technology uses a variety of complex cryptographic and architectural measures to assure transaction security, integrity, and authenticity, successfully preventing unauthorized access and harmful behavior. The basic security features that support blockchain systems include the following:

- Cryptographic hashing: Each block in the blockchain has a unique cryptographic hash, which is often created using powerful algorithms such as SHA-256. This hash acts as a cryptographic fingerprint for the block's content. Because cryptographic hashes are sensitive to input changes, even tiny modifications in the block data produce a completely different hash, making any attempt at tampering easily apparent.

- Consensus mechanisms: Blockchain networks use consensus algorithms such as PoW and PoS to validate transactions and establish agreements across distributed nodes. These procedures are critical to ensuring the ledger's integrity and consistency, avoiding double-spending and unauthorized changes.
- Decentralization: One of the most significant security advantages of blockchain is its decentralized architecture. Without a central point of control, the network is intrinsically more resilient. To successfully manipulate the blockchain, an attacker would need to obtain control of more than 50% of the total network power, which is referred to as a 51% attack. In big, well-established blockchain networks, such an endeavor is both computationally and economically impractical.
- Encryption: Blockchain data can be securely protected using public- and private-key cryptography. This method allows individuals to verify their identities and authorize transactions with confidence, ensuring privacy and trust across the network.

These features make blockchain highly resistant to hacking, fraud, and data manipulation, making it suitable for applications like digital identity management and financial transactions (Naidu and Mishra, 2018).

6.1.5 Cryptographic techniques in blockchain transactions

Blockchain transactions are largely secure thanks to modern cryptographic methods, including hashing, digital signatures, and encryption.

These strategies are critical for safeguarding data on the blockchain. They help to ensure that information is correct (integrity), private (confidentiality), and trustworthy (authentication). The following is an overview of the primary cryptographic methods used in blockchain systems, backed by relevant research and studies:

1. Hashing: Blockchain uses encryption algorithms to protect sensitive information by transforming it into unreadable code, which can only be decoded by those with the correct decryption key. This ensures that data remains confidential and accessible only to authorized users. Hashing algorithms play a key role in maintaining data integrity. They take input data and generate a unique, fixed-length string of characters known as a hash. Even the slightest change in the original data will produce a completely different hash, making it easy to detect tampering. A commonly used algorithm for this purpose is SHA-256, which helps ensure that the information stored on the blockchain remains accurate and unaltered (Ahmad et al., 2023).
2. Digital signatures: Digital signatures are required to validate users' identities and ensure the reliability of blockchain transactions. They have two important security features: authentication (confirming that the sender is who they claim to be) and nonrepudiation. Typically, the sender signs the transaction with their private key, and the recipient (or anyone else on the network) uses the sender's public key to

verify the signature. This technology, based on public-key cryptography, promotes confidence in a decentralized system by assuring that only authorized users can initiate valid transactions.

6.1.6 Popular cryptographic algorithms

- ECDSA (Elliptic Curve Digital Signature Algorithm) is widely used in blockchain networks due to its high security and efficiency, particularly for systems with limited computational capacity.
- RSA (Rivest–Shamir–Adleman) is a foundational method in public-key cryptography noted for its robust security, albeit it often requires bigger key sizes than ECDSA to achieve similar strength.

These algorithms are the foundation of safe and trustworthy blockchain systems, making it extremely impossible to falsify transactions or impersonate other users.

6.1.7 Encryption

Encryption is essential for maintaining privacy in blockchain systems since it ensures that only authorized individuals have access to transaction data. It functions as a digital lock, protecting critical information from unauthorized access.

Blockchain technology typically employs two primary types of encryption:

1. Symmetric encryption

This approach uses a single secret key to encrypt and decrypt data. It is fast and efficient, making it ideal for processing massive amounts of data.

Common examples include AES (Advanced Encryption Standard) and DES (Data Encryption Standard).

Symmetric encryption employs the same key for both encryption and decryption, which makes it efficient but necessitates secure key sharing. The most common symmetric encryption techniques are:

- AES is commonly used in blockchain for safeguarding transaction data due to its robust encryption and efficiency.
- DES: Although currently regarded as less secure, DES was previously used for encryption but has been mostly supplanted by AES (Singh et al., 2022).

2. Asymmetric encryption

This technology, also known as public-key encryption, encrypts data with a public key and decrypts it with a private key. This method is appropriate for secure communication between parties that have not shared a common key in advance.

Common examples include RSA and ECC (Elliptic Curve Cryptography).

By combining these encryption methods, blockchain systems ensure that sensitive information is kept private, secure, and only accessible to authorized parties.

Asymmetric encryption, also known as public-key cryptography, uses two keys: a public key for encryption and a private key for decryption. This method is widely used in blockchain for secure key exchanges and digital signatures. Common asymmetric encryption algorithms include:

- RSA: RSA is used for securing blockchain transactions by enabling encrypted communication between users.
- ECC: ECC is favored over RSA in blockchain applications due to its ability to provide strong security with smaller key sizes, making it more efficient (Patel et al., 2024).

6.1.8 Consensus mechanisms for secure transactions

Blockchain networks rely on consensus mechanisms to validate transactions and maintain the integrity of the ledger—without the need for a central authority. These mechanisms ensure that all participants in the network agree on the state of the blockchain, making unauthorized alterations virtually impossible. Below are some of the most commonly used consensus methods:

1. PoW: One of the earliest and most widely known consensus algorithms. It requires network participants, known as miners, to solve complex mathematical problems in order to validate transactions and add new blocks to the chain. While PoW provides a high level of security, it is also computationally intensive and energy-consuming, which has led to concerns about its environmental impact.
2. PoS: It offers a more energy-efficient alternative to PoW. Instead of relying on computational power, PoS selects validators based on the amount of cryptocurrency they "stake" or lock up in the network. This model not only reduces energy usage but also improves security by making attacks costly—especially 51% attacks, where a single entity gains control of the majority of the network.

6.1.9 Smart contracts and secure data execution

Smart contracts are self-executing programs that run on blockchain platforms. They automatically enforce and execute predefined rules and conditions when certain criteria are met—removing the need for intermediaries and enhancing transaction security.

These digital contracts are central to many blockchain applications, especially in decentralized finance (DeFi), healthcare, and supply chain management.

1. Best practices for developing secure smart contracts: To prevent vulnerabilities and ensure the secure execution of smart contracts, the following best practices should be

adopted: Conduct thorough security audits to identify and fix bugs or loopholes. Use well-established libraries, such as those provided by OpenZeppelin, to implement secure and standardized contract components. Implement strict access control to restrict functions to authorized users and prevent misuse.

2. Applications of secure blockchain transactions: Secure blockchain transactions have widespread applications across multiple sectors, demonstrating the technology's versatility and impact:

- Financial services: Enables secure, transparent, and efficient cross-border payments. Cryptocurrencies like Bitcoin and Ethereum are prime examples of blockchain use in finance.
- Healthcare: Facilitates secure sharing and storage of sensitive patient data, ensuring data privacy and interoperability across institutions.
- Supply chain management: Enhances traceability and authenticity of goods by recording each transaction on an immutable ledger, helping reduce fraud and ensure product quality.

6.2 Data sharing model in blockchain

In the digital age, data sharing is pivotal for collaboration and innovation. Traditional data-sharing methods often rely on centralized systems, which pose challenges related to security, trust, and control. Blockchain technology offers a decentralized alternative, enabling secure and transparent data sharing without intermediaries (Singh and Yadav, 2021).

6.2.1 Blockchain-based data sharing architecture

A typical blockchain-based data sharing architecture comprises several key components (Paliwal and Saraswat, 2021):

Data providers: Entities that generate and share data on the blockchain network. Data consumers: Entities that access and utilize the shared data. Blockchain network: The decentralized ledger that records transactions and ensures data integrity. Smart contracts: Self-executing contracts with terms directly written into code, facilitating automated data access and permissions. Access control mechanisms: Protocols that ensure only authorized entities can access specific data. Encryption techniques: Methods to secure data, ensuring privacy and confidentiality during storage and transmission. The interaction among these components ensures a seamless and secure data-sharing process (Naidu and Mishra, 2018).

6.2.2 Layered data sharing architecture

To address the complexities of data sharing, a layered architecture is often employed (Ahmad et al., 2023):

1. Entity layer: Comprises all participants, including data providers and consumers.

2. Data repository layer: Handles data storage, often integrating decentralized storage solutions such as the interplanetary file system.
3. Service layer: Manages business logic, smart contracts, and access control policies.
4. Application layer: Provides user interfaces and APIs for interaction with the system. This modular approach enhances scalability, security, and manageability.

1. Secure data sharing mechanisms
   Ensuring security in data sharing involves several mechanisms: Attribute-based encryption: Allows data to be encrypted based on user attributes, enabling fine-grained access control (Singh et al., 2022). Zero-knowledge proofs: Enable verification of data possession without revealing the data itself, enhancing privacy. Smart contracts: Automate access control decisions and enforce data sharing agreements transparently (Patel et al., 2024). These mechanisms collectively ensure that data remains confidential, integral, and accessible only to authorized parties (Codearies, no date; Trustworthy Blockchain Data Sharing Model Based on Multi-authority ABE and zk-SNARK, 2022).
2. Applications and use cases
   Blockchain-based data sharing models have been applied across various domains such as
   a) Healthcare: Secure sharing of patient records among healthcare providers while maintaining patient privacy (Buterin, 2014).
   b) Supply chain management: Transparent tracking of goods and data across the supply chain, enhancing trust among stakeholders (Gupta and Sadoghi, 2019).
   c) Financial services: Facilitating secure and transparent transactions between financial institutions and clients (Tomorrow's Apps Will Come From Brilliant (And Risky) Bitcoin Code, 2014). These applications demonstrate the versatility and effectiveness of blockchain in data sharing (Song et al., 2025).

6.2.3 Challenges and considerations

Despite its advantages, blockchain-based data sharing faces several challenges:

1. Scalability: Handling large volumes of data and transactions can be resource-intensive (IBM).
2. Privacy: Ensuring data privacy while maintaining transparency requires sophisticated cryptographic techniques (Singh et al., 2022).
3. Regulatory compliance: Adhering to data protection regulations across different jurisdictions can be complex (Investopedia).

Addressing these challenges is crucial for the widespread adoption of blockchain-based data-sharing models.

Blockchain technology offers a robust framework for secure and transparent data sharing. By leveraging decentralized architectures, encryption techniques, and smart contracts, it addresses many limitations of traditional data-sharing methods. However, ongoing research and development are essential to overcome existing challenges and fully realize its potential (Singh and Yadav, 2021).

6.3 Data security model

Technology is blended with all fields to process, analyze, transform, and discover the new agents. AI is amalgamated with technologies to bring a drastic change toward bettering the prediction process, filtering, or strengthening the core models. At the same time, challenges span to fortify the data security, integrity, and privacy.

Let's strive through all the challenges and address technologies, perhaps in a centralized or decentralized manner.

6.3.1 What is data?

Data is the crucial parameter in the digital world, which has much sensitive and confidential information of any principal, such as applications, users, processes, or identity users. Data is a payload in any standard protocol, or data is the content/message in many communications. Managing/monitoring the data is a highly privileged strategy in all fields. DoD quotes in Singh and Yadav (2021) show that data is a strong weapon that should be treated and secured with a full security model.

Data in blockchain technology is an immutable ledger in a decentralized network, whereas data is a digital asset in cybersecurity, and data is a chain of information in the IT world. Data should be securely managed in all forms to avoid any kind of mismanagement or mishandling that may result in sabotage of data or exploitation of it. Before discussing the secure building blocks of data, let's understand how the data is categorized. Classification of data can change from different perspectives as per the requirement; in this case, we are considering it with respect to security aspects. Data is classified as Personal Identifiable Information (PII), confidential, and strictly confidential data. Based on the classification, the level of security and strategy to handle the data will also vary. Exploitation of data is one big challenge in IT technology, with the 51% attack, which takes ownership of data and enacts financial fraud. Adopting the advanced technology for managing the data is highly recommended, as in the past, medical data exploitation was one big seismic event that created a lot of threats and insecurity in IT. Blockchain is creating a new platform for managing medical data as immutable, decentralized, and more secure than the traditional approach (Fig. 6.2).

PII data: Data/Information used to identify the individual identity.

Confidential data: Data/Information is restricted to the team/project/application.

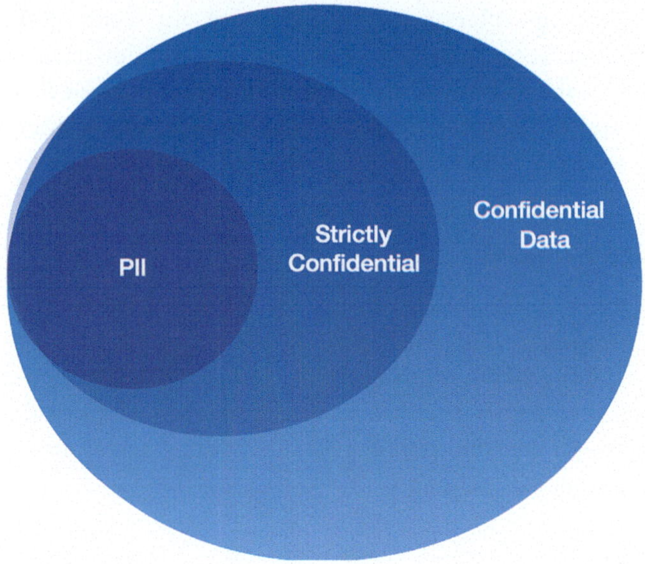

Figure 6.2 Classification of data.

Strictly confidential data: Highly restricted data of the organization with high-level security.

Sharing the data should also consider the nature of the data, as depicted in, according to the sensitivity, the governance rule will be applied. Data governance is the process of managing the availability, usability, integrity, and security of the data in enterprise systems, based on the internal data standards and policies that control data usage. Stated by the AWS Whitepaper (Paliwal and Saraswat, 2021). Data security models play a vital role in protecting the information while sharing the data. Many incidence (Naidu and Mishra, 2018) provide evidence that data is crucial and also sensitive blocks that are prime targets for attackers.

One standard protocol or rule cannot be applied to the data, as the data is accessible on the internet and stored in cloud or remote storage areas. The state of data is classified based on its form, such as data in rest, transit, and processing. Each state of difference should follow the different approaches/methodologies to secure the integrity of data. As per the HIPAA journal (Ahmad et al., 2023), the hackers are predominantly breaching the impermissible data of health information, which is a big threat to industries; on average, around 758,227 data were hacked/stolen per day. Paliwal and Saraswat (2021) show the gist of health data breached from 2009 to 2025, which are millions in number and create a big challenge in the security model (Fig. 6.3).

Cybersecurity is progressing toward incorporating fine-grained access control and strengthening the Identity and Access Management (IAM) process. Training the employees/business associates against attacks, building the security models, improving

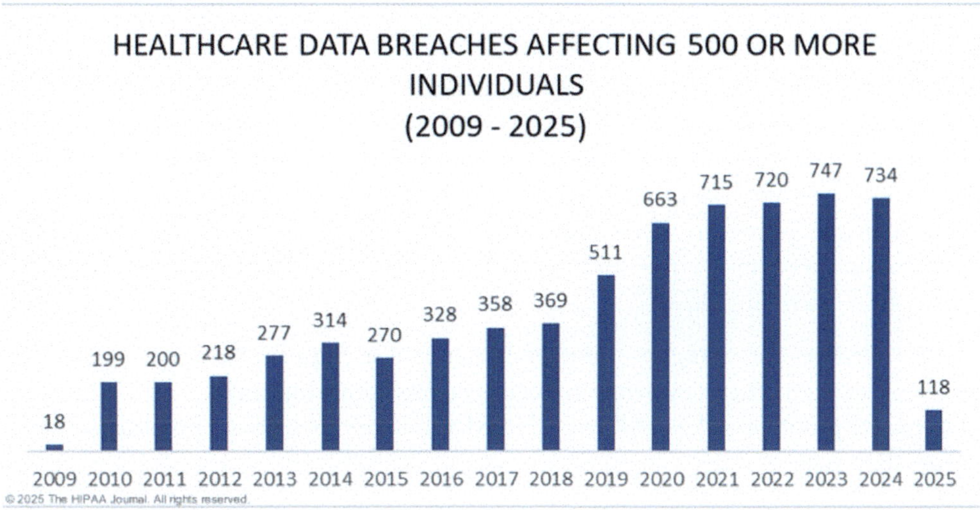

Figure 6.3 Data breach as per the HIPAA survey [From HIPAA Journal] (*HIPAA*, Health Insurance Portability and Accountability Act).

the technologies, and managing the isolation should be a major focus in an extended cybersecurity service.

6.4 Data governance model

Data governance in blockchain technology is about the practices, policies, and standards that manage the collection, storage, distribution, and management of data within decentralized networks. Blockchain technology represents different challenges and advantages compared to conventional data governance models, such as transparency, immutability, and decentralization. The governance model guarantees that data integrity, privacy, security, and compliance are ensured within blockchain environments. While blockchain-based applications extend their reach to financial, medical, and logistics sectors, there comes a compelling need for effective data governance, where regulatory compliance is guaranteed along with operational excellence.

6.4.1 Survey on data governance in blockchain

Blockchain data governance models have come a long way, with multiple frameworks formulated to govern data ownership, sharing, and compliance, as shown. The key aspects of

1. Decentralized identity management: Traditional systems focus on centralized authentication authorities, whereas blockchain facilitates self-sovereign identity

through decentralized identifiers and crypto keys, eliminating the risk of identity theft.

2. Permissioned versus permissionless governance: Permissioned blockchains (e.g., Hyperledger Fabric, Corda) permit targeted entities to oversee access control to ensure compliance with regulations, and permissionless blockchains (e.g., Bitcoin, Ethereum) permit open participation and therefore make governance more dependent on consensus mechanisms.
3. Consensus mechanisms for data integrity: Blockchain utilizes several consensus algorithms, including:

- PoW—Ensures data integrity using computational strength but is energy-expensive.
- PoS—Enhances efficiency while reducing energy consumption.
- Delegated proof-of-stake—Allows selected validators to manage governance decisions (Fig. 6.4).

1. Smart contracts to automate governance: Smart contracts enforce inevitable rules of governance, deriving a rich data permission chain on access to and compliance checks.
2. Regulatory and compliance frameworks: Various governments and international regulating agencies have developed frameworks to legalize blockchain-based data governance, including:

- GDPR (General Data Protection Regulation)—Data protection, privacy
- Health Insurance Portability and Accountability Act (HIPAA)—Controls how medical information is shared.
- Sarbanes-Oxley Act (SOX)—It promotes transparency and integrity in financial reporting.

6.4.2 Data governance model with technical implementation

The technical implementation of some critical aspects of a blockchain data governance model:

IAM: The type of IAM systems that constitute blockchain depends on public–private-key cryptography and ZKP to allow access without revealing sensitive information.

- Data integrity and transparency: Since a record is immutable on the blockchain, it can be audited, thus reducing fraud and ensuring compliance.

Actors verify transactions using on-chain governance models, creating a consensus-driven governance approach that doesn't rely on any single actor.

Audit trails and data provenance: Blockchain enables auditable logs that track alterations and ensure responsibility.

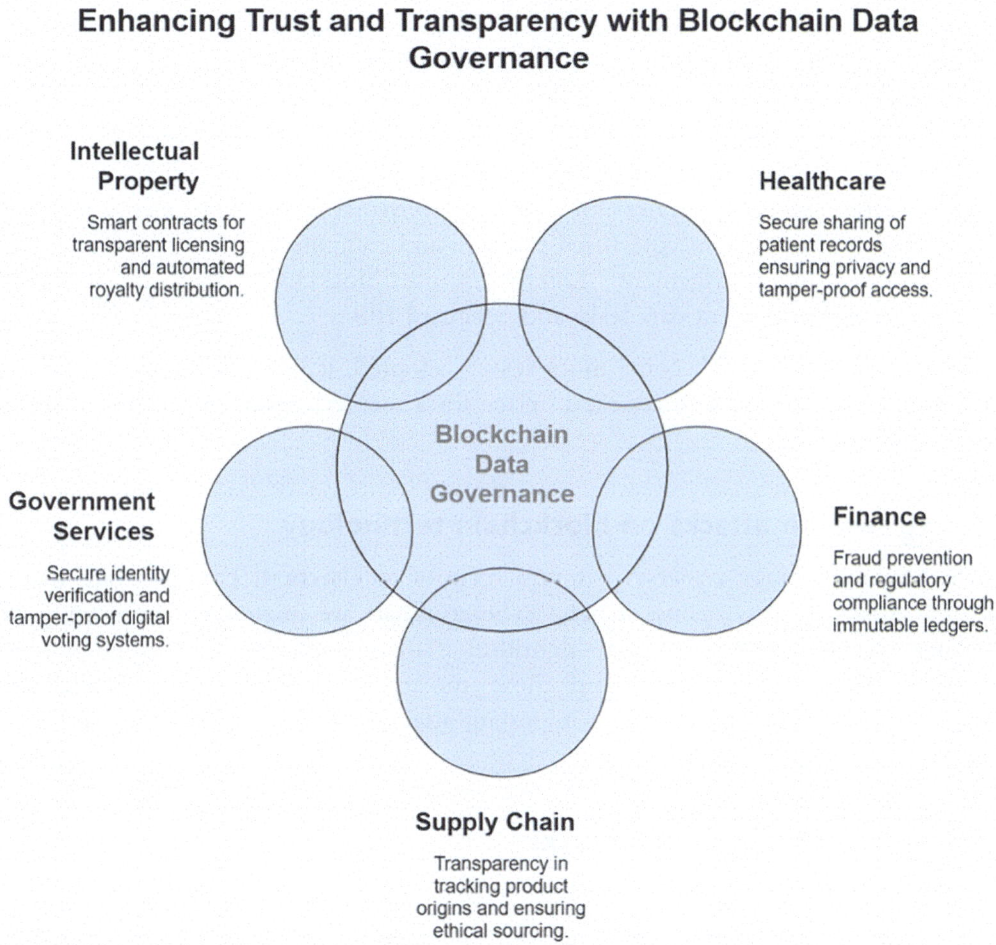

Figure 6.4 Blockchain data governance.

Interoperability standards: Cross-chain communication protocols such as Polkadot, Cosmos, and Chainlink allow for secure data transfer between such blockchain networks.

6.4.3 Applications in real-time

Good governance of blockchain data increases data security, privacy, and operational efficiency. Some real-world applications worth mentioning are:

1. Healthcare: Governance of blockchain data makes it secure, private, and efficient. Here are some of the most common real-world applications.

2. Financial sector: The banks and other financial institutions deploy blockchain governance frameworks to combat fraud and ensure they comply with antimoney laundering regulations.
3. Supply chain management: Blockchain provides traceability and proof of authenticity of goods, promoting ethical sourcing and reducing counterfeiting.
4. Governance and public services: Blockchain governance is used to fix clarity and ease the chance of intrigue in vote-based elections and computerized record frameworks.
5. Enterprise data management: Enterprises use blockchain-based digital rights management to prevent data leaks and safeguard IPR.

As these technologies become more widely adopted, it will still remain critical to have a solid data governance regime in place for a secure, transparent, and compliant digital environment.

6.5 Use case on attacks on blockchain technology

Blockchain technology possesses a sound security architecture, but there are many different attack vectors against it. The cybercriminals are unendingly innovating in utilizing weaknesses in consensus algorithms, smart contracts, and cryptographic protocols. We need to understand those threats to design resilient blockchain architectures. This section examines important attack scenarios, the resulting implications, and countermeasures.

6.5.1 Survey of blockchain attacks

The main attack vectors on blockchain security are:

1. 51% attack: This occurs when an entity controlling over half of a blockchain's processing power, known as a miner, assumes control, enabling them to alter transaction history, execute double-spending, and undermine the integrity of the network. In the past, PoW-based systems, such as Bitcoin and Ethereum Classic, have experienced 51% attacks. Use case: The Ethereum Classic 51% attack (2019): In January 2019, Ethereum Classic was compromised by a 51% attack, where attackers changed the structure of the blockchain and doubled-spent $1 million+ of ETC. PoW consensus was healed for this incident, but it did expose how weak this approach can be, where we have to rely on more secure measures like chain reorganization limits. Solution: Networks use hybrid consensus frameworks to deter 51% attacks like PoS or proof of authority. These protocols limit dependence on mining power, making it difficult for the attackers to obtain majority control. In addition, checkpointing mechanisms prevent deep chain reorganizations and thus provide security to the network.

2. Sybil attack: Malicious actors create lots of fake identities (nodes) to coerce network consensus. This attack is against the principle of decentralization that blockchain promotes, allowing malicious entities to take control over the processes involving decisions. Use case: Bitcoin network node Sybil attack: The researchers showed a way attackers could commandeer a large number of nodes in the Bitcoin network in order to isolate some users and make it difficult or impossible for them to get their transactions confirmed. Solution: Strong countermeasures are typically PoW/PoS in nature and are computationally or economically expensive, which makes a Sybil attack costly. In general, reputation-based systems and identity verification mechanisms enhance the security of a distributed database by allowing only valid nodes into the network.
3. Smart contract vulnerabilities: Smart contracts are vulnerable to attacks such as reentrancy, integer overflow, and denial-of-service if they are not coded securely. Use case: The DAO Hack (2016): The Decentralized Autonomous Organization (DAO) was a venture capital fund that lost over $50 million worth of Ethereum because of a reentrancy attack on the smart contract. Solution: Formal verification methods, automated security scans, and test environments such as MythX and OpenZeppelin to identify vulnerabilities predeployment. When employed, secure coding practices like reentrancy guards and access controls limited to restricted spaces further enhance contract security.
4. Phishing attacks: These are forms of inadvertently stealing private keys by using impersonated wallet interfaces and sending fake emails. Use case: MyEtherWallet DNS Spoofing attack, which manipulate the DNS resolver to divert the internet traffic dito the attacker nodes to steal the credentials (2018): MyEtherWallet suffered DNS hijacking, redirecting users to a phishing site where they unwittingly entered their private key. Solution: Phishing can be avoided by using hardware wallets, multifactor authentication (MFA), and domain name system security. It is important to teach users to avoid risks and how to spot phishing attempts.
5. Routing attacks: Blockchain relies on network infrastructure for communication (backbone of the network). This potentially leads to financial losses and disruptions of decentralized applications, while attackers can also intercept and slow down transactions. Use case: Border Gateway Protocol (BGP) Hijacking attack on Bitcoin network, 2018: Attackers exploited vulnerabilities in the BGP to intercept traffic from the Bitcoin network, effectively slowing down transactions and creating the potential for double-spending. Solution: Use encrypted channels of communication, redundant peer links, and traffic monitoring for outliers, all of which contribute to reducing routing attacks. Such features make a significant difference to security, as do advances in current blockchain-based DNS and decentralized communication protocols.
6. Eclipse attack: This attack targets individual nodes by bombarding them with malicious peers, such that they become isolated from the rest of the network, and

incorrect information is relayed to them. Use case: Bitcoin Eclipse attack experiment: Security researchers killed two birds with one stone, showing that an attacker could eclipse nodes from honest peers and feed them malicious transaction data. Solution: To mitigate Eclipse attacks, networks need to use random peer selection, implement per-IP connection limits, and offer continuous network monitoring. In addition, distributed node architectures offer redundancy, which poses a challenge for attackers to take down a given node.

7. Dusting attack: Attackers send tiny amounts of cryptocurrency (dust) to many addresses to analyze transactions and unmask users. Use case: Bitcoin Dusting attack (2019): Spammers simply had to send tiny amounts of BTC across hundreds of wallets in order to track and analyze their spending habits. Solution: Wallet providers must incorporate dust filtering features automatically in order to prevent users from being exposed to dust transactions. Privacy solutions such as CoinJoin and Wasabi Wallet also improve the anonymity of transactions, decreasing tracing dangers.
8. Front-running attack: In DeFi apps, hackers utilize the ordering of transactions by means of high gas prices to make profitable trades ahead of other users. Use case: Front-running in Uniswap decentralized exchange (DEX): Traders exploited large gas fees to front-run trade execution in the Uniswap DEX, exploiting regular users to their advantage. Solution: Secure mempools, batching transactions, and commit-reveal schemes avoid front-running by securing transaction privacy before execution. Also, layer-2 scaling solutions minimize congestion and render front-running less profitable.
9. Time-jacking attack: The attack manipulates the network time of a blockchain node to cause inconsistencies in block timestamps, causing possible double-spending and network instability. Use case: lightweight Bitcoin clients time manipulation: Attackers manipulated system time on lean Bitcoin clients to receive faulty timestamps so that they could take advantage of timing vulnerabilities in transaction authentication. Solution: Strict time synchronization policies, median time past facilities, and network time validation through multiple sources of information prevent time-jacking attacks. Validation rules based on timestamps must be enforced on blockchain networks as well.

6.5.2 Real-world applications

Security solutions for blockchain are essential to ensure confidence in distributed systems. Businesses that rely on secure blockchain solutions are:

- Cryptocurrency networks: Bitcoin, Ethereum, and DeFi platforms use sophisticated security features to resist attacks.
- Enterprise blockchain solutions: Firms utilize private blockchain networks with strong security architecture to safeguard business-sensitive information.
- Government and public sector: Secure blockchain-enabled voting systems and national identity verification schemes provide transparency and fraud prevention.

- Healthcare: Avoiding manipulation of data and unauthorized access to medical record storage systems.
- Supply chain and logistics: Reducing fraud risk through the guarantee of transaction authenticity for supply chain activities.

6.6 Secured architecture of blockchain

Data is a high-interest commodity in all the fields ranging from bank translation to the defense community; features to steward the data should be strengthened to make it tamper-proof.

Security models, as per the CIA, are based on three major triads, such as Confidentiality, Integrity and Availability. These CIA triads are guidelines for an information security model against many vulnerabilities to avoid attacks such as ransomware attacks. To protect data, many technologies are adopted, such as cybersecurity, information security, cloud security, and blockchain technology. Each of these technologies will provide features to secure the data either by encrypting the data or by using secure transactions.

A data security model is an outline of the business responsibilities involved in the ownership, management, and governance of an organization's data.

Features of these technologies should provide security and guard the complete life cycle of the data.

1. Encryption: Protecting the data by using the private key so the data looks like scrambled data, though accessed in an unreadable format.
2. MFA: Verifying the user identity should not be limited to user ID and password; adding a Hardware Security Model or pattern generation or multiverification enhances the security level.
3. Firewall: Protecting against data accessibility by a certain IP address/Protocol/Ports is a boundary wall to restrict the traffic toward the data storage.
4. Endpoint security: Accessing the data from the device or storage securely is one of the challenges, providing the firewall with respect to the network connectivity, API connectivity, and hardening the hardware devices.
5. Authentication: Providing the role-based access or attribute-based access control used to provision the permission for applications/services/users to access the data.
6. Data masking: Data masking is one of the encryption functionalities to mask sensitive data with different information.
7. Identity management: Policies or permissions on the set of resources are used to manage/monitor the service access.
8. Backup: Data may be lost or hacked, preventing the data, which can be achieved by replicating the data to enhance high availability, such as RAID. Multiregion availability or mirroring the data.

9. IDS/IPS: Many detection/prevention algorithms are used as early detector agents before intrusion services affect the resources, which should be enabled as a preventive measure, such as AWS Guard Duty, HIPS, NBA, and many more.

In Naidu and Mishra (2018), we try to synthesize the CIA recommended by the information security community and the functionality that should encompass the security model. A vibrant technology signifies the various approaches for securing the information.

Blockchain security model integrates ->traceability->accessibility->integrity and confidentiality (Fig. 6.5).

In this section, we understood the necessary features for securing the data; secure data sharing is the process of mechanisms, standards, protocols, framework, and technologies. Let's walk through the secure framework of blockchain and its benefits. Blockchain is an emerging technology for managing data securely to avoid fraudulent activities on any critical data, such as supply chain management, health records, or bank transactions. Blockchain technology builds security by creating a hash code from the previous node, which creates decentralized security system so that no centralized attack can be triggered.

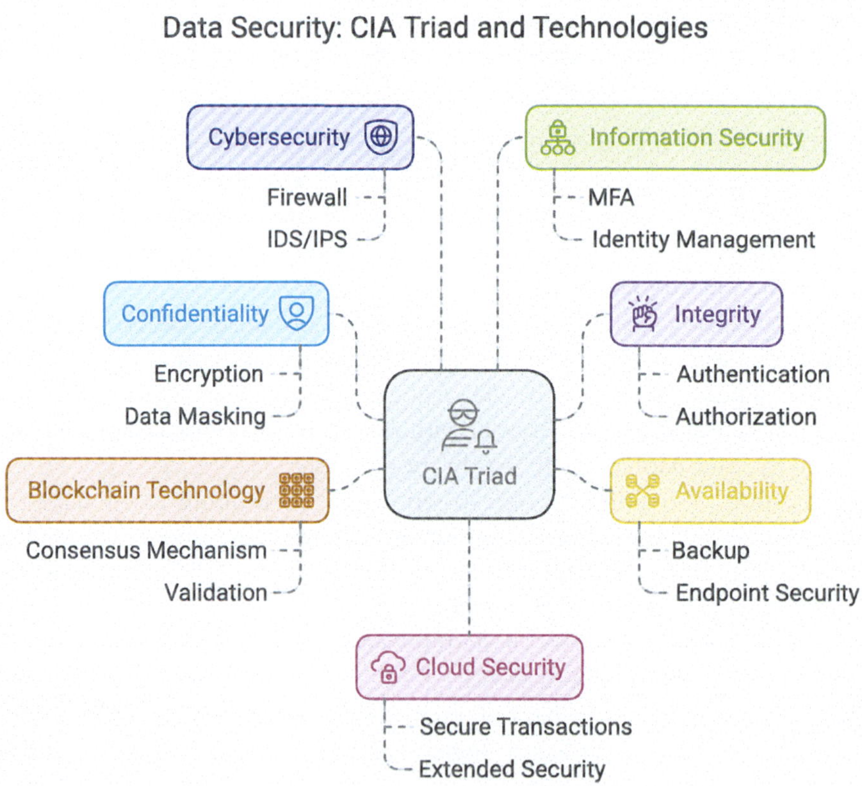

Figure 6.5 CIA traid in blockchain.

Trending blockchain technology is adopted by many organizations, such as healthcare, Bitcoin, supply chain, financial services, manufacturing, and fabric industries.

1. Why are industries attracted toward blockchain technology?
2. What are the secure services available in the blockchain?
3. Whether the architecture is constructed with more secure components in blockchain?

In this chapter, those questions will be addressed to understand the evolving blockchain in the field of IT to securely share data.

6.6.1 Why are industries attracted toward blockchain technology?

Data sharing is highly prioritized in any industrt that enforces the special stewardship of it along with privacy, where blockchain fits the requirement to bridge the gap between traditional and advanced technology. Blockchain is a cutting-edge technology that can be used for Benefits Data Trust, which is a nonprofit program to share information about nutritionist information, supplements, and vulnerable areas by the health organization. Security, integrity, scalability, and confidentiality are the key characteristics that attract toward blockchain.

IT industries also demand Privacy Enhancing Technology to secure data sharing.

Some of the features offered by blockchains are

1. Improved accuracy by eliminating human involvement in verification.
2. Cost savings from removing third-party verification
3. Decentralization makes it difficult to mess with.
4. Transactions are secure, confidential, and efficient.
5. Transparent technology.
6. Provides a financial alternative and a mechanism to secure personal information for inhabitants of nations with unstable or weak governments.

Due to these unique methodologies, industries adopt blockchain. Blockchain technology can be used by healthcare professionals to safely preserve patients' medical records. A medical record that has been created and signed can be entered onto the blockchain, giving patients assurance that the information is unchangeable. To ensure privacy, these private health records might be encrypted and kept on the blockchain using a private key, making them only accessible by designated users.

6.6.2 What are the secure services available in the blockchain?

Despite the fact that blockchains are cryptographically signed, blockchain security is required to ensure that only authorized users have access to nodes and the entire blockchain network. Blockchain networks utilize a variety of cybersecurity procedures and frameworks to keep critical user information private and hidden.

Blockchain technology encompasses new security mechanisms such as

1. Selective endorsement: Network connectivity is restricted based on the accessible permission as public and private by the process known as selective endorsement.
2. Byzantine fault tolerance (BFT): It is basically used against the large malicious nodes, and the replication strategy builds the fault tolerance node. The extended BFT, such as particle and federated BFT, provides fault tolerance in a large area network to support scalability.
3. Decentralization: Blockchain technology works against the single point of failure by following the decentralization architecture, which is highly fault-tolerant, distributing the control to multiple control access and fidelity of data.
4. Nonce: In cryptography, a nonce is a number that guards against replay attacks and protects confidential communications. Authentication techniques use nonces, which are random or pseudo-random numbers, to identify communications. In order to emphasize the transient nature of these interactions, these numbers occasionally carry a timestamp.
5. Block validation: PoW and PoS are two different mechanisms in consensus to validate the participants to prevent bad actors in the network.

Blockchain is incredibly hard to sabotage because of the anonymity and security features embedded within the technology.

6.6.3 Whether the architecture is constructed with more secure components in blockchain?

Yes, blockchain is embedded with a new process for securing the data sharing. Let's deep-dive into its platform. The following diagram is refered from (Ahmad et al., 2023) to understand more about the blockchain platform and its secure runtime. This diagram represents the usual node capabilities required to participate in a blockchain architecture on three networks: public, cloud, and enterprise. The capability locations are illustrative of industry best practice; however, any capability can be implemented in any network, depending on the solution needs. Let's understand the major secured features in the blockchain platform discussed below (Fig. 6.6).

6.6.4 Secured blockchain platform

1. Smart contract: A smart contract, also known as "chaincode" according to fabric, is a trustworthy distributed application that derives its security and trust from the peer consensus and the blockchain. It is a blockchain application's business logic.

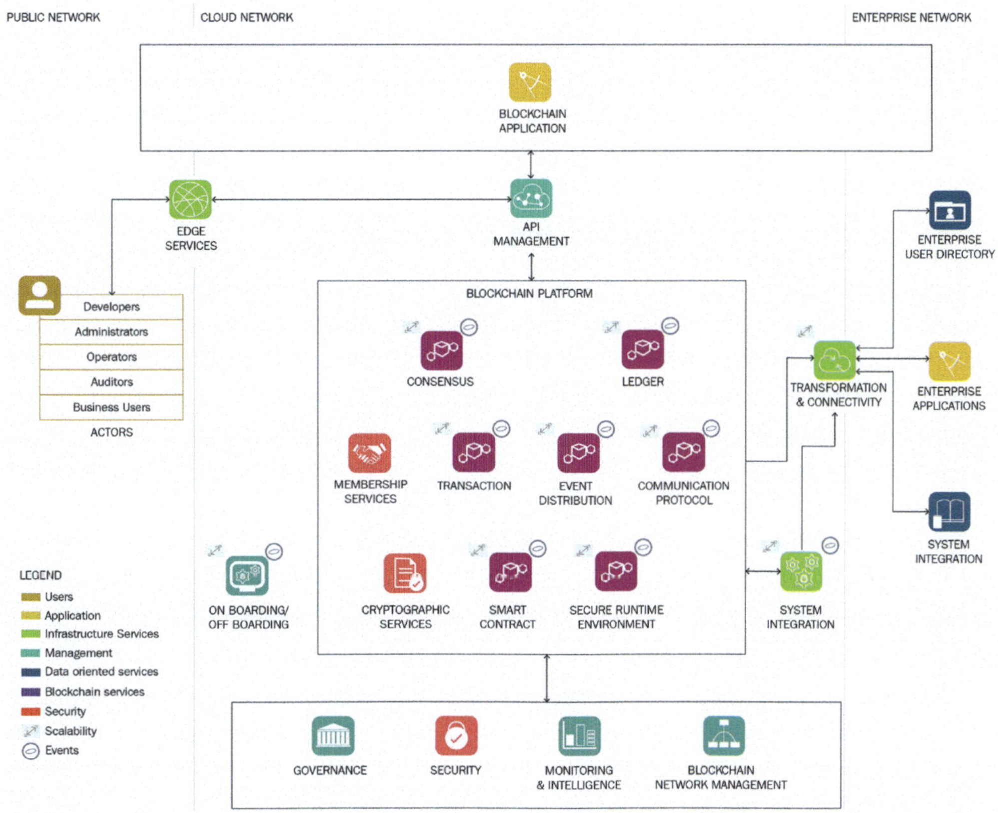

Figure 6.6 Secured blockchain platform.

2. Membership services: The task of assigning cryptographic identities to network entities falls to a pluggable membership service provider.
3. Consensus: By applying transactions that have been verified by a consensus procedure and arranged into blocks with a hash connecting each block to the one before it, nodes individually keep a copy of the ledger.
4. Secure runtime environment: A blockchain transaction may use smart contract features that need a secure environment during runtime. Server-side blockchain business logic is hosted in a secure runtime environment. Using a secure container with a collection of signed runtime components, including a secure operating system and libraries for programming languages supported by blockchain, and their corresponding runtimes (Patel et al., 2024).
5. Cryptographic services: The process of creating methods and procedures to stop a third party from accessing and learning about the information included in private communications throughout a communication process is known as cryptography.

Services included symmetric, asymmetric, and hashing to provide nonrepudiation, access control, and authentication.

Along with those features, Merkle trees and decentralized consensus work together to ensure the trustworthy transaction of data in a secure way.

6.7 Enhancing secure data sharing in blockchain

Blockchain technology has evolved a new approach in business transactions by creating a distributed ledger management system. Though blockchain provides secure transactions using the consensus mechanism, major attack such as the 51% attack, in which attackers control more than 50% of hashing power, phishing attacks, investment scam, and routing attacks.

Avoiding such incidents by amalgamation with confidential computing, quantum computing, microservices architecture, and machine learning.

Yes, integrating technologies or enhancing methodologies resolves many challenges, such as scaling, resilience, integrity, and security. Further, our discussion is on the fusion of other technologies and their proliferation to boost the capabilities of blockchain (Fig. 6.7).

1. Confidential computing: By guaranteeing that data is encrypted while it is in transit, at rest, and during active processing—a crucial component that is frequently disregarded in conventional security frameworks (MIT's Bitcoin-Inspired Enigma Lets Computers Mine Encrypted Data, 2015).

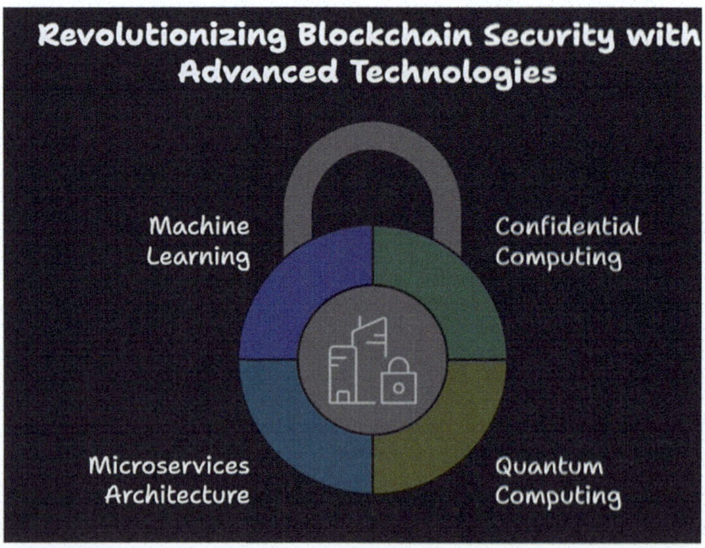

Figure 6.7 Enhancement of blockchain.

2. Tomorrow's Apps Will Come From Brilliant (And Risky) Bitcoin Code (2014)—Confidential computing tackles this problem. Confidential computing offers a safe environment where only authorized code can access sensitive data by enclosing it in a secure CPU is a micro processor enclave Trusted Exceution Environment(TEE). This guarantees that data is hidden from unauthorized users even during processing. Confidential computing and blockchain enable a new degree of security for transactions and data exchange. Because blockchain technology is immutable, every transaction is permanently documented, promoting transparency and trust. A confidence fabric, a trustworthy digital environment where sensitive data may be handled and verified safely without exposure, is created when blockchain technology and confidential computing are integrated.
3. Quantum computing: Quantum is a word emphasized in physics to state the small particles that form many forms of energy. Extending this concept, quantum computing has come up with the qubit, which is a unit of information that holds the superposition of all possible states. Quantum computing is applied to blockchain to anticipate the attack ("Quantum Secured Blockchain"). Improving the digital signature, hashing technique, and cryptographic methods in blockchain (Quantum Secured Blockchain, 2022).
4. Microservices architecture: Designing blockchain applications based on microservice architecture principles can also provide business advantages. Enhanced efficiency within the software system lowers infrastructure expenses and the likelihood of service outages related to capacity. These features hold specific importance for private blockchains, where cost efficiency and service reliability are essential demands for enterprises. Microservice architecture principles can facilitate the use of interchangeable components, thereby diminishing technical debt that may result in outdated, unpredictable environments. Solidity, the programming language utilized for smart contracts on Ethereum, includes a feature for defining the precise runtime version for each contract's execution. Over time, a smart contract's version number can serve to identify outdated blockchain-stored code that might be eligible for replacement. Just remember that in a blockchain, smart contracts that have already been executed (meaning, are included in a "mined" block) cannot be removed—a new version of a contract needs to be issued for subsequent transactions.
5. Machine learning: Machine learning concepts contain federated learning, data analytics mechanisms, and anomaly detection techniques to address the challenges of blockchain and its limitations. Blockchain-driven systems can attain improved security, efficiency, and customized user experiences by utilizing machine learning algorithms and methods. One of the key advantages of incorporating machine learning into blockchain lies in the area of security. Blockchain technology is renowned for its resilience owing to its decentralized and tamper-proof

characteristics. Nevertheless, the integration of machine learning's capacity to recognize patterns and irregularities can enhance security by detecting and addressing possible threats and attacks. In DeFi applications, machine learning plays a vital role in evaluating credit risk, determining interest rates, and facilitating automated market making.

6.8 Automation of artificial intelligence—blockchain

The use of technology and various techniques and methods indicates that Tech automation carries out the tasks automatically without requiring human intervention. The importance of tech automation is to automate the system with the help of IT processes like system monitoring, cloud management, and software testing using advanced tools and technologies.

Automation assists in bringing valuable insights into business processes that cover a wide range of multiple domains, removing various obstacles, adding speed, and increasing efficiency. Embedding the technology with another stream or domain plays a vital role.

Automation is employed to streamline and optimize various operations;, any other domain is utilized, such as computer systems, robotics, and artificial intelligence (AI). The necessity and demand for automation are increasing rapidly every year, in industries like manufacturing, logistics, healthcare, and customer service.

6.8.1 Benefits

By utilizing automation, we can enhance the efficiency of our resources and allow humans to focus on more creative tasks or those necessary for addressing edge cases and refining processes. Automation enables quicker handling of repetitive tasks and reduces turnaround durations. It also reduces the dependence on individuals being present to perform tasks or recalling to complete them. Automated processes are more reliable than manual ones, particularly when dealing with large amounts of data. Manual processing might also be more inconsistent, and some steps could be overlooked.

Automated tasks can keep operating even when team members change or are assigned different responsibilities.

6.8.2 Artificial intelligence

AI is a branch of computer science that allows machines to replicate human intelligence—such as learning, solving problems, and making decisions. AI encompasses areas like machine learning, deep learning, computer vision, and natural language processing. Intelligent automation (IA) encompasses the incorporation of AI technologies into automation processes, allowing systems to be self-learning, flexible, and able to make decisions rather than merely adhering to predefined rules. While conventional automation performs repetitive tasks precisely as instructed, intelligent automation can manage complexity, uncertainty, and change—similar to human capabilities.

6.8.3 Essential elements of artificial intelligence

1. Robotic process automation: Software bots that automate repetitive tasks based on specific rules (e.g., data input, form completion).
2. IA: Imparts cognitive functions—such as data learning, language comprehension, or decision-making.

6.8.4 Artificial intelligence with machine learning

1. Allows systems to enhance gradually without the need for explicit reprogramming.
2. Natural language processing: Enables machines to understand and react to human language (utilized in chatbots, virtual aides).
3. Computer vision: Enables systems to perceive and understand data, such as examining images or video recordings.

6.8.5 Artificial intelligence in blockchain

This pertains to employing blockchain technology to automate AI tasks, choices, or processes within a secure, decentralized, and tamper-resistant setting. It additionally involves incorporating AI models and decisions into blockchain smart contracts, enabling the automation, verification, and trust of AI-driven actions without any human involvement.

1. How these three technologies integrate together

Gains insights from data, forecasts outcomes, identifies trends, and streamlines smart choices. Please provide the text you would like me to paraphrase. Blockchain technology

A distributed digital registry that securely and transparently logs transactions.

Information on the blockchain is unchangeable and can be verified.Certainly! Please provide the text you'd like me to paraphrase. Automation (through Smart Contracts).Specified conditions are fulfilled Smart contracts are self-implementing codes kept on a blockchain that automatically execute actions whenBy automating AI decisions through smart contracts, you create a reliable, verifiable, and self-sufficient AI system.

6.8.6 Sample applications: automating artificial intelligence through blockchain

1 Supply chain automation: AI is more efficient, forecasting possible delays or risks based on real-time data, which includes weather forecasts and logistics data. The blockchain technology would provide visibility of all the deliveries, with the smart contracts automatically sending payments or messages according to AI predictions, such as in the case of a late delivery. Through this integration, mistakes are minimized, accountability is enhanced, and operations are simplified.
2 Robotized insurance requests: The AI will be capable of processing the data of insurance claims and determining their legitimacy in a short period of time and without errors. Smart contracts enable the authorization of payments in case of verified claims by implementing blockchain. All the steps of this process are documented, and this makes the whole system traceable, which minimizes fraud activities.
3 Sharing healthcare data: AI is used to analyze patient records and suggest possible diagnoses in order to assist healthcare providers. With blockchain, patient-sensitive data is disseminated safely and exclusively to the relevant people. Automation also ensures that no information is shared without authorization, and all the outcomes of the AI can be completely traced, holding everyone responsible.
4 Federated learning using blockchain: Federated learning enables the training of AI models on noncentralized devices, including smartphones, without uncovered data. Smart contracts on blockchain are used to automate the process of aggregating and validating model updates to create transparency. Such a strategy ensures the privacy of data, accountability, and equitable compensation to every stakeholder involved in the AI model development.

The technical functioning of automation

- Step 1: AI reaches a conclusion: A machine learning model identifies a transaction as suspicious.
- Step 2: Blockchain logs the outcome: The outcome is recorded on the blockchain, producing an unchangeable, verifiable record.
- Step 3: Smart contract initiates an action: Advantages of automating AI through blockchain.

6.8.7 Benefit overview

- Transparency blockchain offers an unalterable record of every AI decision made.
- Trust removes the "black box" aspect of AI; outcomes are documented and verifiable.

- Security blockchain guarantees data integrity and stops unauthorized modifications.
- Automated smart contracts minimize human involvement and accelerate workflows.
- Decentralization eliminates the necessity of a central authority to validate AI results.

6.9 Integration of artificial intelligence-driven robotics

The rise of AI is increasingly associated with software-driven robotic systems such as mobile robots, unmanned aerial vehicles, and, more recently, semiautonomous vehicles. However, a significant gap remains between the algorithmic (digital) world and the physical world. This disconnect presents a major challenge in achieving the goal of creating intelligent, intuitive, and user-friendly robots capable of effectively navigating and interacting with human-centric environments.

To address this, the emerging discipline of machine intelligence (MI)—an interdisciplinary fusion of robotics and AI—aims to develop robust, embodiment-aware systems. These systems are envisioned to possess situational awareness, self-understanding, and the ability to adapt to the dynamic bodies and environments in which they function.

The seamless integration of AI with robotics in areas such as control, perception, and machine learning is essential for the advancement of fully autonomous intelligent systems that can operate reliably in everyday settings.

This review explores the historical evolution of MI, tracing its conceptual roots back to the twelfth century. It further analyzes the current landscape of robotics and AI, highlighting key technologies, influential systems, and cutting-edge research trends. Finally, the article discusses ongoing challenges and reflects on the future possibilities of human–machine interaction, many of which are still on the horizon. The embedding of AI into robotics has attracted considerable attention in recent years. This evolving field of research focuses on enhancing robotic systems by leveraging AI to improve their intelligence, adaptability, and overall performance, as shown in Table 6.2

Table 6.2 Key components of a blockchain block.

Component	Description
Index	The unique number identifying the block's position in the blockchain.
Timestamp	The exact time when the block was created and added to the chain.
Transactions	A list of verified transactions stored within the block.
Previous block hash	A cryptographic hash of the previous block, ensuring immutability and linking blocks together.
Merkle root	A hash of all transactions in the block, used for efficient verification.
Nonce	A value used in proof-of-work consensus to generate a valid block hash.
Block hash	The unique cryptographic fingerprint of the block is generated using hashing algorithms.

Table 6.3 The evolution of robotics.

Year	Milestone	Description
2000	Programming	Introduction to basic robotic programming. (Placeholder text used in image)
2003	Automation	Development of automated robotic processes.
2007	Technology	Advancements in robotic technologies and integration.
2012	Electronic industry	Robots are being widely used in electronic manufacturing and automation.
2018	Robots	Emergence of more sophisticated and humanoid robots.
2022	Artificial intelligence	Integration of AI to make robots more autonomous and intelligent.

The evolution of robotics—timeline overview as shown in Table 6.3

The convergence of automation and AI is rapidly becoming a driving force behind the creation of new industries, breakthrough technologies, and increased productivity across existing sectors. As AI continues to emerge within the realm of robotics, its practical applications in real-world scenarios are gaining widespread recognition. From self-driving vehicles and customer service to healthcare and industrial automation, AI is playing a transformative role in enhancing various aspects of everyday life.

While concerns persist about the potential displacement of human labor due to automation, global forecasts suggest a net increase in employment opportunities as a result of this technological advancement. This growth presents a valuable opportunity for reskilling the workforce and investing in knowledge development aligned with current and future innovations.

The integration of AI and robotics holds immense potential for reshaping job roles across industries. It enables the automation of repetitive tasks in manufacturing while introducing adaptability and intelligence into routine operations. The diverse and expanding applications of AI in robotics make this an exciting and essential area of exploration, with vast opportunities for individuals to contribute to its future evolution and societal impact.

Robotics is a multidisciplinary field that involves the design, construction, operation, and application of robots. It merges principles from engineering and computer science to develop robotic systems capable of performing predefined tasks autonomously, reducing the need for direct human involvement. At its core, robotics leverages technological innovations to enhance the efficiency, safety, and automation of various operations.

Deployment to undertake tasks that are either physically demanding or hazardous for humans, such as lifting heavy equipment or working in dangerous environments. They have also been widely used for highly repetitive tasks, like automobile assembly, where precision and consistency are critical. By automating such roles, robotics not only improves productivity but also enhances workplace safety, allowing human workers to focus on more complex, creative, and value-added responsibilities.

An important distinction is that robots are not subject to the physical and mental limitations that affect humans. While repetitive tasks can lead to fatigue, boredom, or reduced concentration in humans, robots can perform these same activities with unwavering accuracy and efficiency. Robotics has already made a notable impact across various industries, from precision agriculture and automated delivery services to optimized manufacturing processes, demonstrating its vast potential to revolutionize the way we work and live.

6.9.1 Difference between artificial intelligence and robotics

Although AI and robotics are often used interchangeably, they are in fact distinct yet closely related disciplines. They have the power to significantly impact various industries and aspects of human life, but they aid different purposes and operate through different mechanisms.

At a conceptual level, AI can be compared to the "brain," involving neural network models that mimic the structure and function of human cognition. Robotics, on the other hand, is more akin to the "body," focusing on the physical structure and mechanics required for movement and interaction with the environment.

AI involves the progress of systems capable of performing tasks that traditionally require human intelligence, such as learning, reasoning, problem-solving, and decision-making. These systems are designed to operate autonomously, learning from data and adapting their behavior without the need for constant human input.

In contrast, robotics centers on the creation of machines that can achieve specific physical tasks. Robots can be programmed to carry out simple, repetitive actions, such as sorting objects or assembling miniature components. While the integration of AI into robotics can significantly enhance a robot's capabilities enabling more complex decision-making and adaptability—it is not always required. Some robotic systems are designed to function without any cognitive features, simply executing preprogrammed behaviors (Antonopoulos, 2017).

Ultimately, while AI and robotics are distinct fields, their integration creates powerful synergies. Together, they can drive innovation and bring about transformative benefits across a wide range of sectors, from manufacturing and healthcare to logistics and everyday consumer technology.

Artificial intelligence: A field of computer science focused on creating systems that can mimic human intelligence.

Key abilities:

1. Learning from data
2. Problem-solving

3. Decision-making
4. Adaptability

Analogy: Like the brain enables thinking and learning.

Examples: Chatbots, recommendation systems, AI-powered diagnostics, and self-learning algorithms.

Robotics: A branch of engineering and technology dedicated to designing and building physical machines (robots) that perform tasks.

Key features:

1. Physical structure and movement
2. Sensor-based interaction
3. Programmed behavior

Analogy: Like the body enables action and movement.

Examples: Industrial robots, delivery drones, robotic arms in factories, and Home cleaning robots.

AI + Robotics = Intelligent Robots

The synergy: When AI is integrated into robotics:

1. Robots can make smart decisions
2. Adapt to their environment
3. Perform more complex tasks

In summary, as stated in Table 6.4

Enhancing robotics through AI: applications, methods, and future potential.

6.9.2 Overview

Recent breakthroughs in AI have accelerated its integration into robotics, marking a natural evolution in both fields. While AI is not yet universally embedded in all robotic systems, its adoption is expanding rapidly due to increasing capabilities and the demand for smarter, more adaptive machines. The synthesis of AI and robotics is opening up

Table 6.4 Comparison of AI and robotics with their combined applications.

Aspect	AI	Robotics
Focus	Intelligence and learning	Physical task execution
Role	"Brain"	"Body"
Example	Virtual assistant	Robotic vacuum
Combined power	Self-driving cars, smart manufacturing, and autonomous drones	

(*AI*, artificial intelligence).

new possibilities across industries, improving productivity, safety, efficiency, and adaptability in a wide range of environments.

6.9.3 Key methods of artificial intelligence integration in robotics

1. Machine learning and behavior modeling: One of the most impactful methods of AI implementation in robotics is machine learning. Robots equipped with this capability can learn from observing human actions and replicate specific tasks, gradually improving their performance. This enables robots to move beyond repetitive, preprogrammed duties and become intelligent collaborators in dynamic workspaces.
2. Computer vision: AI enhances robotic perception by using computer vision; robots understand how to navigate in a complex environment and classify objects and understand the visual input to make context-sensitive decisions. This mental ability enables robots to be more flexible and interactive in the physical world, to become more useful in various uses.
3. Edge computing in real-time decision-making: In robotics, edge computing is important in real-time AI task processing. Robots with edge computing process sensor data onsite, rather than transmitting the data to remote cloud servers. This allows real-time decision-making, minimizes latency, and also improves situational awareness, which is necessary in time-sensitive tasks like autonomous driving and industrial automation.

6.9.4 Advanced sensory input for smarter robots

AI-driven robots leverage a variety of sensors to better perceive and interact with their environments, including:

1. Time-of-flight sensors: Measure the reflection time taken from surfaces to gauge distance.
2. Temperature and humidity sensors: Monitor environmental conditions in smart agriculture or climate control.
3. Ultrasonic sensors: To detect objects and measure proximity using sound waves.
4. Vibration sensors: Monitor movement or structural integrity in industrial equipment.
5. Millimeter-wave sensors: Enable high-resolution detection in automotive or security systems.

These sensory inputs, combined with AI, allow robots to continuously learn, adapt, and make more intelligent decisions in real-time.

6.9.5 Applications of artificial intelligence-driven robotics across industries

1. Customer service: AI-powered chatbots and service robots are revolutionizing customer support by handling routine queries, automating responses, and learning from interactions to provide more human-like assistance over time. These systems are increasingly being used across online platforms and physical retail environments.
2. Manufacturing and automation: In industrial settings, AI-enabled robots are streamlining production lines, handling quality control, and optimizing logistics. Their ability to learn and adapt ensures greater efficiency and precision, reducing human error and downtime.
3. Healthcare, agriculture, and beyond: From surgical assistance to precision crop harvesting, AI in robotics is making waves across sectors. Robots are becoming essential tools in areas where precision, repeatability, and safety are critical.

   AI in robotic assembly and visual systems: revolutionizing precision and efficiency.

6.10 Artificial intelligence-driven innovation in manufacturing and packaging

AI is proving to be a transformative force in robotic assembly, particularly within intricate manufacturing industries such as aerospace. The integration of AI with advanced visual systems enables real-time decision-making, allowing robots to perform intricate tasks with greater autonomy and precision.

By analyzing visual input and environmental feedback, AI allows robotic systems to self-correct and adapt during operations. This capability not only enhances performance but also supports robots in learning optimal pathways and task strategies through experience during real-time execution.

1. AI in the packaging industry: The packaging industry is one of the latest sectors to implement AI to improve efficiency in their operations, as well as their accuracy in handling and sorting and overall cost of operation. Moreover, AI will make it easier to install and move robotic units by training and storing the optimal motion paths. This allows quicker installation and deployment in cases of moving systems or reconfiguring a system so that operations can be more flexible and efficient.
2. Improved visual recognition and accuracy of imaging: Industries that require precision, for example, assembly lines and logistics, also need computer vision with AI. Vision systems that are enhanced with AI enable robots to recognize the slightest visual variations and defects, carry out very delicate operations such as aligning components or inspecting quality, and have better object recognition and spatial recognition. Therefore the visual acuity of the robotic systems improves and allows them to see and react to the slightest shifts in the workplace.

6.10.1 Core components of a robot

Robotic systems are composed of various interconnected hardware elements that enable motion, perception, energy supply, and control. The primary components include:

- Electric motors—Provide rotational force essential for motion and mechanical operations.
- Actuators—Convert electrical signals into mechanical movement, enabling interaction with the environment.
- Sensors—Gather data from —the surroundings, aiding perception and feedback mechanisms.
- Muscle wires—Shape-memory alloys that contract when electrically charged, simulating muscle movement.
- Power supply—Delivers the necessary energy to operate all robotic components.
- Pneumatic air muscles—Mimic human muscles using compressed air, ideal for soft robotics.
- Ultrasonic motors—Utilize ultrasonic vibrations for precise movement, especially in compact designs.
- Piezo motorsOffer fine control and minimal wear, commonly used in precision applications.

6.10.2 Toward artificial general intelligence

The future of robotics lies in the fusion of AI with physical systems, aiming toward artificial general intelligence (AGI) AI that possesses human-like comprehension. For this to materialize, robots must be equipped with mobility, sensory abilities (touch, vision, and hearing), and the capability to interact with their environment. These features create a dynamic feedback loop, allowing the system to learn from its actions and evolve through real-time interaction with the physical world.

Shift in AI focus in robotics: There is a growing shift in the robotics domain from asking "What can robots do for humans?" to "How can robots enhance AI cognition?" Allowing AI systems to physically interact with real-world objects fosters deeper learning, mimicking how human children learn through exploration. This embodied intelligence approach can lead to more intuitive and adaptive AI systems capable of real-world reasoning and decision-making.

Transformative potential across industries: The merging of AI with robotics has transformative implications for multiple sectors, including healthcare, manufacturing, defense, and space exploration. As robots gain more intelligent and responsive capabilities, they can take on increasingly complex roles. Ultimately, this convergence could accelerate innovation and bring us closer to realizing AGI, where machines not only assist but also learn, reason, and collaborate with humans at an advanced level.

6.10.3 Applicability of artificial intelligence in blockchain

Blockchain is a distributed digital ledger technology that enables secure, transparent, and tamper-proof recording of transactions. Unlike traditional centralized systems, such as banks or payment platforms, blockchain eliminates the need for intermediaries by allowing P2P transactions. Every transaction is verified by a network of computers (nodes) and recorded in blocks that are cryptographically linked, forming an immutable chain of data. Originally developed for Bitcoin, blockchain technology now has applications far beyond cryptocurrencies.

6.10.3.1 How blockchain works

Each block in a blockchain contains a timestamp, transaction data, and a cryptographic hash of the previous block, ensuring that data cannot be altered once recorded without affecting the entire chain. All nodes in the network maintain a copy of the blockchain, and any change would require simultaneous consensus across the entire network, a virtually impossible task. This distributed and secure nature makes blockchain resistant to hacking and fraud. Transactions are often structured in Merkle trees for efficient verification and storage.

6.10.4 Key features and applications blockchain is known for four main features

Transparency, immutability, decentralization, and permissionless access (in public blockchains). These qualities allow parties who may not trust each other to collaborate securely. It offers significant benefits across various sectors, including finance, supply chain, healthcare, real estate, and government. Because it can automate and secure the flow of value and data, blockchain is seen as a revolutionary step in digital transformation, often called "the next generation of the Internet."

Advantages of blockchain technology:

- Record precision: Records are irrevocable and cannot be changed; therefore they are authentic and effective in finance, education, and healthcare.
- Cost reduction: gets rid of middlemen such as banks and lowers transaction costs.
- Decentralization: It has a network of global computers, which means it has no central control.
- Quick transactions: Transactions are fast with no delays by third parties. Privacy protection: The personal data is encrypted and is available only by means of the use of private keys.
- Security: Advanced cryptography eliminates tampering and supports trust.
- Challenges and limitations: Expensive setup and maintenance: The cost of implementation may be high, particularly for small businesses.

- Slowness: Public blockchains are slow because of consensus on a network level.
- Possibility of abuse: An anonymous character can be abused to do illegal activities such as laundering money.
- Absence of regulation: Regulation of the use of the blockchain or cryptocurrency is lacking in many governments.
- Scalability problems: As users and transactions increase, the system may be slowed down, and efficiency may be diminished.

6.10.4.1 Why combine artificial intelligence with blockchain?

When AI and blockchain work together, they create a secure, intelligent, and transparent system that can transform multiple industries. While AI brings intelligence and learning, blockchain provides security and trust, as shown in Table 6.5.

6.10.5 Data

Power AI can sift through massive amounts of data to find patterns, predict outcomes, and improve decision-making.

- Fraud detection: AI identifies anomalies in data; blockchain makes sure the data hasn't been tampered with.
- Smart verifications: AI can build classifiers (models that detect or predict things), which blockchain can verify and store securely.

Table 6.5 Integration of AI and blockchain across key applications.

Feature	Description
Improved Business Models	AI analyzes trends while blockchain ensures data reliability.
Global Verification	Blockchain allows cross-border validation of data/models created by AI.
Smarter Compliance and Audits	Audits become real-time and automated using AI, with traceable records stored on the blockchain.
Smarter Finance	AI-powered financial advice and fraud detection; blockchain ensures trust in data.
Predictive Analysis	AI predicts future trends; blockchain ensures the data backing it is genuine.
Intelligent Retail	Personalization through AI using trusted blockchain-sourced customer data.
Digital IP Rights	Creative content (music, art, etc.) can be protected and tracked using AI and stored on a blockchain.
Transparent Governance	Automated and fair governance systems with public blockchain records and AI monitoring.

(*AI*, artificial intelligence).

- Marketing automation: In customer-facing industries, AI uses data collected via blockchain to deliver targeted and automated marketing.

6.10.6 Conclusion

The integration of AI in robotics is not just a technological advancement that represents a transformative shift in how machines learn, think, and collaborate with humans. As AI continues to evolve, its synergy with robotics will lead to smarter, more autonomous systems capable of operating effectively across an ever-expanding range of real-world applications. Integration into robotic systems, especially through vision and adaptive learning, has revolutionized tasks that require precision, adaptability, and efficiency. From aerospace to packaging and logistics, AI is unlocking new levels of autonomous functionality, reshaping the capabilities of modern industrial robots.

AI disclosure

During the preparation of this work, the author(s) used Napkin AI in order to generate images. After using this tool/service, the author(s) reviewed and edited the content as needed and take(s) full responsibility for the content of the publication.

References

Ahmad S, Arya SK, Gupta S, Singh P, Dwivedi SK. 2023 Study of cryptographic techniques adopted in blockchain. In: 4th international conference on intelligent engineering and management, ICIEM. 9798350341126. India: Institute of Electrical and Electronics Engineers Inc; 2023. Available from: http://ieeexplore.ieee.org/xpl/mostRecentIssue.jsp?punumber=10164834, https://doi.org/10.1109/ICIEM59379.2023.10166591.

Antonopoulos AM. Mastering bitcoin: unlocking digital cryptocurrencies; 2017.

Buterin V. A next-generation smart contract and decentralized application platform; 2014. Available from: https://ethereum.org/en/whitepaper.

Gupta S, Sadoghi M. Blockchain transaction processing. Encyclopedia of Big Data Technologies 2019:366–76. Available from https://doi.org/10.1007/978-3-319-77525-8_333.

MIT's bitcoin-inspired enigma lets computers mine encrypted data; 2015.

Naidu G, Mishra R. Blockchain technology architecture and key characteristics. International Journal of Advance Research and Innovative Ideas in Education 2018.

Paliwal M, Saraswat P. An overview on blockchain. International Journal of Innovative Research in Computer Science & Technology 2021:127–30. Available from https://doi.org/10.55524/ijircst.2021.9.6.29.

Patel S, Bathla R, Gupta D. 2024 deploying a cryptocurrency technique using hashing and web integration. In: 11th International conference on reliability, infocom technologies and optimization (trends and future directions), ICRITO. 9798350350357. India: Institute of Electrical and Electronics Engineers Inc; 2024. Available from: http://ieeexplore.ieee.org/xpl/mostRecentIssue.jsp?punumber=10522096, https://doi.org/10.1109/ICRITO61523.2024.10522142.

Quantum secured blockchain; 2022. Available from: https://www.quantumblockchains.io/wp-content/uploads/2023/06/QBCK_WhitePaper.pdf.

Singh R, Agarwal AK, Naaz R, Kumar R, Vijay R. 2022 A study of cryptographic primitives in the context of blockchain's data integrity and privacy. In: Proceedings of the 2022 11th international conference on system modeling and advancement in research trends, SMART. 9781665487320. 1589-1595. India: Institute of Electrical and Electronics Engineers Inc; 2022. Available from: http://ieeexplore.ieee.org/xpl/mostRecentIssue.jsp?punumber=10046659, https://doi.org/10.1109/SMART55829.2022.10046712.

Song C, Chen L, Wu X, Li Y. A secure data sharing model utilizing attribute-based signcryption in blockchain technology. Sensors 2025;25(1):160. Available from https://doi.org/10.3390/s25010160.

Tomorrow's apps will come from brilliant (and risky) bitcoin code; 2014.

Trustworthy blockchain data sharing model based on multi-authority ABE and zk-SNARK; 2022.

CHAPTER 7

Image-guided surgery and artificial intelligence algorithms

Kande Archana[1], V. Kamakshi Prasad[2], and Laxmaiah Mettu[1]
[1]Department of CSE-Data Science, CMR Engineering College, Hyderabad, Telangana, India,
[2]Department of CSE, Jawaharlal Nehru Technological University, Hyderabad, Telangana, India

7.1 Introduction

Medical robotics and image-guided surgery (IGS) have advanced significantly during the last 20 years, transforming contemporary surgical practice. These developments have greatly improved patient outcomes, reduced risks, and increased surgical precision. More dexterity, stability, and access to hard-to-reach anatomical areas are now possible with robotically-assisted operations, making them less invasive and more effective than traditional techniques. Surgeons may now plan and perform procedures with previously unheard-of accuracy because of image-guided technologies that enable real-time visualization of internal structures. These technologies' incorporation of artificial intelligence (AI) accelerates the revolution in surgical care. Large-scale datasets can be analyzed by AI-driven algorithms to aid in intraoperative decision-making, forecast surgical results, and aid in diagnosis. Medical image interpretation, robotic instrument guidance, and dynamic surgical plan adaptation based on real-time data are all being accomplished with machine learning (ML) and deep learning models. By using data-informed precision treatments, these advancements are not only increasing surgical safety but also allowing for more individualized patient care. Even with these successes, there are still a number of obstacles to overcome. Widespread adoption is hampered by problems such as the high cost of robotic systems, the requirement for specialized training, data privacy issues, and the incorporation of AI into the current healthcare infrastructure. Furthermore, great thought must be given to the ethical issues and legislative barriers surrounding autonomous surgical systems. The cutting-edge of medical robotics and AI-assisted surgery, identifying present operational and technological obstacles, and suggesting strategic avenues for further progress are the objectives of this proposal. Enhancing system interoperability, increasing affordability, and guaranteeing ethical deployment will be prioritized. By tackling these objectives, the study hopes to aid in the creation of cutting-edge surgical technologies that are more accessible, intelligent, and safe in a variety of clinical contexts.

AI-Powered Developments in Medical Robotics
DOI: https://doi.org/10.1016/B978-0-443-27783-2.00004-1

7.1.1 Background of medical robotics and image-guided surgery

In the modern healthcare industry, medical robotics and IGS have become game-changing technologies that are revolutionizing surgical techniques and greatly improving clinical results. These developments minimize invasiveness while increasing surgical precision, safety, and efficiency, which results in shorter recovery periods, a lower chance of problems, and happier patients. With the unmatched precision and accuracy that robotic-assisted surgeries (RASs) provide, intricate treatments can be carried out with few incisions and little damage to surrounding tissue. At the same time, IGS makes use of cutting-edge imaging technologies, such as magnetic resonance imaging (MRI), computed tomography (CT), and ultrasound, to visualize interior anatomy in real-time, improving the surgeon's ability to maneuver and perform surgery in delicate or difficult-to-reach places. The da Vinci Surgical System, which is still in widespread use in fields including general surgery, gynecology, and urology, marked a turning point in the development of surgical robotics. This device pushes the limits of minimally invasive surgery (MIS) by providing improved dexterity, tremor filtration, and three-dimensional (3D) imaging. By enabling focused interventions and enhancing the precision of tumor resections and implant placements, image-guided navigation systems have also improved the precision of procedures in neurosurgery, orthopedics, and cancer. Notwithstanding these encouraging advancements, a number of obstacles prevent these technologies from being widely used. Significant obstacles include the high cost of purchase and upkeep, restricted availability in environments with limited resources, and the challenging learning curve involved in becoming proficient with robotic systems. Furthermore, incorporating new technologies into standard surgical procedures necessitates specialized training and a strong infrastructure (Lanfranco et al., 2004). Research is still being done to improve user-friendly interfaces, system integration, and affordability in order to overcome these problems. As advancements continue, it is anticipated that the combination of robotics, AI, and imaging will result in surgical systems that are even more intelligent, flexible, and easily accessible, hence revolutionizing patient care in the future.

7.1.2 The role of artificial intelligence in surgical procedures

A new era of intelligent, data-driven surgical care has been ushered in by AI, which has profoundly changed the field of surgical robotics and image-guided treatments. AI technology is significantly improving surgical outcomes and patient safety by improving decision-making, streamlining workflows, and reducing human error. In order to identify intricate patterns and produce predictive analytics that support surgical planning and intraoperative decision-making, sophisticated AI systems can evaluate enormous volumes of diverse medical data, including imaging scans and patient histories. Deep learning-based image processing, which makes it possible to segment organs, tissues, and anatomical structures in real-time, is one of the most significant uses of AI. This makes it possible to precisely target and navigate during surgery, particularly when working on delicate or important areas. By helping surgeons with tasks like automatic suturing, tremor reduction, motion stabilization, and adaptive control, AI-powered robots expand these capabilities. By reducing unintentional tissue injury, these features not only increase accuracy but also surgical safety. Moreover, AI-enhanced haptic feedback devices offer tactile sensations that improve surgeons' situational awareness and control. AI is also transforming surgical training and education. ML-powered simulation platforms allow for immersive, scenario-based learning settings where students may rehearse techniques and get immediate performance feedback. These platforms aid in speeding up the learning curve for intricate robotic procedures and provide objective criteria for competence evaluation. Predictive analytics powered by AI can also provide intraoperative advice, foresee possible difficulties, and tailor treatment plans according to patient profiles. However, there are important issues that need to be resolved if AI is to be successfully incorporated into surgical practice. These include adhering to strict regulatory requirements, guaranteeing the interpretability and transparency of AI systems, and resolving ethical issues with algorithmic bias and data privacy. AI will continue to have a revolutionary impact on surgical treatment when these problems are fixed.

7.1.3 Objectives of the proposal

Medical robotics and IGS have transformed modern healthcare, increasing precision, reducing surgical risks, and improving patient outcomes. AI has played a critical part in this shift, enabling better decision-making, improved surgical navigation, and real-time data analytics. This proposal proposes to advance the industry by solving critical problems and employing AI-based solutions to improve RAS. The major goal of this project is to provide an advanced AI-powered platform for medical robotics and IGS. This system will combine ML algorithms, deep learning-based image processing, and real-time surgical data analytics to help surgeons improve accuracy and efficiency during procedures. AI in surgical robotics is predicted to improve preoperative planning, intraoperative guidance, and postoperative assessments, resulting in better patient care. This project focuses on improving real-time image processing techniques in order to increase the precision of image-guided surgeries. Current imaging modalities, such as MRI, CT scans, and ultrasound, provide critical information to surgeons, but combining these with AI algorithms can dramatically improve precision. AI-based segmentation techniques will be investigated to improve anatomical mapping and enable improved imaging of complicated structures during surgery (Litjens et al., 2017). Furthermore, the study will use AI-driven predictive analytics to foresee probable difficulties and recommend corrective actions during procedures (Rajpurkar et al.,

2018). Another goal of this idea is to create an adaptive robotic control mechanism powered by AI. Traditional robotic surgical systems use preprogrammed instructions, whereas AI allows for real-time modifications depending on patient-specific variances. This adaptability will allow for more tailored surgical procedures while reducing the likelihood of complications. The incorporation of AI-driven feedback systems in robotic surgery will also allow for enhanced haptic feedback and motion stabilization, making robotic-assisted procedures more efficient.

Furthermore, the project seeks to examine the function of AI in automated surgical skill assessment and training. Deep learning models will allow for the analysis of surgical performance and the provision of objective feedback to surgeons in training. This AI-based evaluation methodology will aid in evaluating critical characteristics such as precision, speed, and error rates, resulting in better surgical education and competency development. Furthermore, this study will look into the ethical and regulatory implications of AI in medical robotics. AI-powered surgical devices must follow strict safety and ethical requirements to ensure patient safety. The study will look at existing regulatory frameworks and make recommendations for the safe integration of AI into surgical practices (Lanfranco et al., 2004). The long-term goal of this proposal is to help progress AI-driven medical robots by nurturing breakthroughs that may be applied in clinical practice. The study's findings will seek to bridge the gap between AI technology and real-world surgical applications, guaranteeing that AI-powered robotic systems reach the greatest levels of safety, precision, and reliability. By achieving these objectives, this proposal will play a significant role in shaping the future of AI in medical robotics and IGS.

7.2 Overview of image-guided surgery

The rapidly developing discipline of IGS combines cutting-edge imaging technology with surgical navigation systems to improve the safety, accuracy, and precision of medical treatments. Driven by the increasing need for better patient outcomes and lower procedural risks, IGS was first created in the late 20th century for neurosurgery and has subsequently spread to orthopedics, ENT, cardiology, and minimally invasive operations. Fundamentally, IGS gives surgeons a dynamic and in-depth picture of the patient's anatomy by combining real-time imaging data with navigational aids. This combination makes it possible to precisely localize essential structures and surgical targets, which is especially important for delicate or difficult operations. Preoperative imaging (such as CT or MRI scans), intraoperative imaging (such as fluoroscopy, ultrasound, or intraoperative CT), tracking systems (such as optical or electromagnetic), and a computer-based navigation platform that combines and shows real-time data are the main parts of an IGS system. Different imaging modalities meet distinct surgical objectives. For example, fluoroscopy allows for continuous monitoring during interventional operations and ultrasound provides real-time soft-tissue visualization, while

CT and MRI provide high-resolution anatomical detail. Together, these instruments improve the surgeon's spatial awareness, reduce tissue injury, and facilitate MIS. IGS has a significant effect on surgical accuracy. It lowers the risk of problems, speeds up recovery, and lessens intraoperative uncertainty. For example, IGS guarantees maximum clearance during tumor resections while preserving healthy tissue. Integration with robotics and AI holds promise for significantly improving IGS capabilities as technology develops, leading to safer and more efficient surgeries. To sum up, IGS is a revolutionary development in surgical practice that offers a potent combination of imaging and operating strategy that will continue to influence medical practice in the future.

7.2.1 Definition and historical development of image-guided surgery

The use of real-time imaging technologies to help surgeons plan, navigate, and perform surgical procedures with greater accuracy and precision is known as IGS. When X-ray-based guiding was initially used to assist surgical procedures in the early 20th century, that is when IGS got its start. However, the field saw significant breakthroughs in the late 20th and early 21st centuries, particularly with the advent of advanced imaging modalities such as CT, MRI, and the integration of robotic-assisted technologies. The use of stereotactic frames in neurosurgery was one of the first significant advancements in IGS. These tools made it possible for surgeons to precisely locate and target deep-seated brain lesions, which was a significant advancement in the development of guided surgical techniques. The foundation for increasingly intricate processes requiring precise anatomical guidance was established by this invention. The advent of computer-assisted navigation devices transformed the surgical field by the 1990s. By combining preoperative and intraoperative imaging data, these devices made it possible to provide real-time guidance throughout surgeries. IGS consequently emerged as a key component of MIS in a variety of specialties, such as cancer, cardiovascular surgery, and orthopedics (Lee, n.d.). A variety of imaging modalities, including CT, MRI, fluoroscopy, and ultrasound, are combined with optical or electromagnetic tracking systems in contemporary IGS platforms. These instruments greatly enhance spatial orientation and lower intraoperative risks by giving surgeons a dynamic, 3D image of the operating field. With the incorporation of robotics, augmented reality (AR), and AI, IGS is still developing today, increasing surgical accuracy and improving patient outcomes. IGS is positioned to play an even bigger role in precision surgery in the future as these technologies advance.

7.2.2 Key components of image-guided surgery systems

The foundation of IGS systems is a collection of advanced technologies that cooperate to increase surgical accuracy and boost therapeutic results. In order to provide precise anatomical guidance throughout the surgical procedure, these systems usually comprise a number of essential components (Santiago et al., 2021). Preoperative imaging, which entails obtaining high-resolution scans such as CT, MRI, or positron emission tomography (PET) images prior to surgery, is one of the fundamental components of an IGS system. These scans offer comprehensive anatomical maps that are used as a guide for intraoperative navigation and surgical planning. Surgeons can define surgical aims, identify important structures, and foresee possible difficulties with the use of preoperative imaging. Another essential element that offers real-time updates during the process is intraoperative imaging. Technologies such as intraoperative CT or MRI, fluoroscopy, and ultrasound aid in tracking the surgical procedure's progress and modifying the surgical strategy as necessary. In dynamic surgical settings where tissue changes or patient positioning may impact accuracy, this real-time feedback is extremely helpful. Both optical and electromagnetic tracking devices are used to continuously check the patient's anatomy and the location of the surgical tools. By enabling exact localization and orientation, these technologies guarantee that surgical instruments are positioned precisely in relation to the patient's body. Advanced navigation software integrates and visualizes all of this data, giving a 3D, real-time picture of the operating field. This interface helps surgeons plan their motions, steer clear of important structures, and perform surgeries with greater assurance. These elements work in concert to produce a system that gives surgeons more situational awareness and decision-making skills. The integration and improvement of these elements will raise the precision, security, and effectiveness of image-guided surgical procedures even more as technology develops.

7.2.3 Imaging modalities

IGS uses a variety of imaging modalities, each of which makes a distinct contribution to the precision and efficacy of surgical operations. Preoperative planning, intraoperative navigation, and postoperative evaluation are all supported by the vital anatomical and functional data that these technologies offer (Patel, n.d.). MRI and CT are two of the most often used preoperative imaging modalities. Planning operations involving the skull, spine, and joints is made easier using CT's high-resolution, cross-sectional pictures of bone and thick tissues. On the other hand, MRI offers remarkable contrast for soft-tissues, making it possible to see the brain, spinal cord, muscles, and internal organs with precision. The basis for surgical mapping and navigation is these comprehensive preoperative images. The ability of ultrasound to visualize soft-tissue structures in real-time and dynamically makes it a popular tool for intraoperative imaging. Being portable, noninvasive, and radiation-free, it is a useful tool for directing operations, including vascular surgery, tumor resections, and biopsies. When rapid, in-the-moment evaluations are required during surgery, ultrasound is quite helpful. Fluoroscopy is a technique that uses continuous X-ray imaging and is especially helpful for procedures that include implantation, catheter placement, or other device placement. Its real-time features provide precise device positioning and minimize procedural errors by enabling surgeons to instantly change their procedures based on monitoring instrument movement.

PET is frequently used in conjunction with CT (PET/CT) for applications requiring functional imaging, such as tumor localization or metabolic activity monitoring. PET is a valuable tool for surgical planning and assessing the course of a disease, despite its less frequent intraoperative application. The foundation of IGS is made up of three imaging modalities, which together allow for safer, more accurate, and more successful surgical procedures in a variety of medical specialties.

7.2.4 Navigation systems

A key element of IGS is navigation systems, which offer real-time tracking and visualization to improve surgical control and precision. The location and orientation of surgical instruments with respect to the patient's anatomy are continually monitored by these systems, which make use of cutting-edge technology like electromagnetic, optical, or robotic tracking. Low-frequency magnetic fields produced by a transmitter positioned close to the surgical site are used in electromagnetic tracking. These fields are detected by tiny sensors built into surgical instruments, which then send the navigation system their spatial coordinates. This technique is particularly helpful in limited-line-of-sight situations and in minimally invasive operations. Conversely, optical tracking uses reflecting markers and infrared cameras that are affixed to patient-fixed reference frames and instruments. These markers' movements are tracked by the cameras in three dimensions, enabling extremely precise localization. The extreme precision of optical systems makes them popular, but they need a clear line of sight between the tracking cameras and the markers. The most sophisticated type of surgical navigation is represented by robotic tracking devices. These systems provide extremely precise, repeatable, and stable movements by integrating robotic arms with real-time image and tracking data. The robotic tools can be used directly by surgeons or guided by navigation algorithms, improving their capacity to carry out intricate procedures with little human error. Navigation software that superimposes real-time data on preoperative 3D models of the patient's anatomy is combined with these tracking technologies. Surgeons benefit from improved spatial awareness, decreased intraoperative uncertainty, and the ability to carefully carry out surgical plans thanks to this combination of actual and virtual worlds. Navigation systems are growing more precise, intuitive, and able to serve a variety of surgical specialties, including neurosurgery, orthopedics, and more, as IGS technology advances.

7.2.5 Real-time feedback and visualization

In IGS, real-time feedback and visualization are crucial components that greatly enhance surgical precision and patient safety. Surgeons can make timely corrections during difficult surgeries because of these features, which allow them to continuously evaluate anatomical structures, instrument locations, and procedural progress. Advanced technologies like virtual reality (VR) and AR are being included in IGS systems more frequently in an effort to improve this process (Roberts, 2022). Through the use of AR, digital data, including surgical plans, 3D anatomical models, and vital alarms, can be superimposed directly into the surgeon's field of vision. By combining virtual data in real-time with the actual operating field, surgeons may look at underlying tissues without making big incisions, which supports MIS. By improving spatial awareness and assisting in the identification of vital structures such as tumors or blood arteries, AR lowers the possibility of unintentional injury. In contrast, VR produces a completely immersive digital environment that enables surgeons to work with a 3D model of the patient's anatomy. VR is being investigated more and more for intraoperative guiding, even though its primary applications are preoperative planning, surgical rehearsal, and teaching. Before going into the operating room, surgeons can evaluate surgical routes, model procedures, and gain a better understanding of intricate anatomical linkages. AR and VR work together to make surgery more user-friendly, engaging, and knowledgeable. In addition to enhancing accuracy and confidence during treatments, these technologies help shorten operating times and lower the risk of problems. By making treatments safer, more effective, and more customized, the incorporation of extended reality into IGS has the potential to completely transform surgical practice as hardware and software capabilities continue to develop.

7.2.6 Types of image-guided surgery

In several medical disciplines, IGS has emerged as a crucial tool, providing customized solutions that improve accuracy, safety, and results in intricate procedures. Because of its adaptability, it may be tailored to the particular anatomical and procedural needs of many surgical specialties, each of which uses the optimal imaging modalities and navigation systems. IGS is essential in neurosurgery for treating abnormalities of the brain and spine. Surgeons can carefully identify and target brain tumors, vascular abnormalities, or functional areas while causing the least amount of damage to healthy tissues by utilizing high-resolution imaging methods like MRI and CT scans in conjunction with stereotactic navigation systems. There is a much lower chance of neurological problems when one can maneuver across sensitive neuronal structures. IGS is frequently utilized in orthopedic surgery for fracture healing, spinal fusion, and joint replacement. Restoring joint function and lowering revision rates require precise implant location and alignment guidance, which CT-based navigation devices assist with. 3D imaging and intraoperative fluoroscopy help with real-time adjustments during hardware implantation and bone restoration. IGS helps cardiovascular surgery by using imaging methods, including CT angiography, ultrasound, and fluoroscopy, to direct procedures such as bypass grafting, valve repair, and stent implantation. Accurate catheter navigation and reduced procedural risks are ensured by real-time imaging of blood flow and vascular structure. IGS facilitates the accurate location and excision of malignancies during oncological surgery while preserving healthy tissue. Tumor boundaries are frequently determined using PET/CT and MRI, and image-guided biopsy or resection techniques aid in obtaining distinct margins, which are essential for the effectiveness of treatment. All things considered, IGS is a game-changing technology that keeps influencing contemporary surgical practice in a variety of specialties by providing patient-specific, image-enhanced guidance for better clinical results.

7.2.7 Intraoperative navigation

Tracking surgical tools in real-time and continuously in relation to the patient's anatomy while the procedure is being performed is known as intraoperative navigation. Surgeons may execute delicate procedures with greater assurance and safety thanks to this technology, which is crucial for giving them precise spatial awareness. The risk of harming vital tissues and structures is decreased by intraoperative navigation, which continuously updates the position of instruments inside the surgical area. This approach is very helpful for doing surgery on delicate and complicated regions like the spine and brain. Serious neurological injury or loss of function can arise from even little variations in these processes. An interactive 3D map of the patient's anatomy is produced by intraoperative navigation systems by combining current tracking data with preoperative imaging data, such as MRI or CT scans. In order to avoid vital locations such as blood arteries, nerves, and functional brain regions, surgeons utilize this map to precisely guide instruments along predetermined trajectories. For instance, intraoperative navigation in neurosurgery facilitates deep brain stimulation, tumor resection, and biopsy operations by offering precise, real-time images of the intricate structure of the brain. Similarly to this, it helps with precise pedicle screw placement and neural element decompression during spine surgery, both of which are essential for preserving neurological function and spinal stability. Intraoperative navigation offers advantages beyond increased precision. Additionally, it helps to enhance patient outcomes, cut down on radiation exposure (when fluoroscopy is substituted), and shorten surgical timeframes. The precision and safety of surgery are being further improved by these navigation systems, which are growing increasingly complex, intuitive, and integrated with other technologies like robotics and AR as technology develops. A key component of contemporary surgical practice, intraoperative navigation allows for accurate, well-informed procedures that preserve critical tissues and maximize therapeutic outcomes.

7.2.8 Preoperative planning

In IGS, preoperative planning is a crucial stage that uses cutting-edge imaging technologies to produce an extensive and thorough surgical procedure roadmap. High-resolution scans, such as CT, MRI, or PET, are used in this planning step to precisely record anatomical information unique to each patient. Surgeons can better comprehend the many spatial linkages and complexities inside the patient's anatomy by carefully examining these photographs (Kim, 2023a). Surgeons can determine the optimal surgical strategy for each patient based on the data acquired during preoperative planning. The best places for incisions, routes to the target area, and placements of vital structures that need to be maintained are all determined by this customized planning. Surgeons can then improve their methods, eliminating needless tissue damage and making the treatment less intrusive. Additionally, preoperative planning makes a substantial contribution to better operation results. Surgeons can prevent intraoperative surprises, shorten operating hours, and improve intervention accuracy by foreseeing possible difficulties and anatomical variances in advance. This careful planning is particularly crucial for intricate surgeries where accuracy is crucial, such as neurosurgery, orthopedic reconstructions, and oncological resections. By providing surgical teams with a common visual reference, preoperative planning facilitates interdisciplinary collaboration in addition to surgical guidance. By making it possible to clearly communicate the intended surgery and its hazards, it also helps with patient education and informed consent. In summary, the incorporation of advanced imaging into IGS's preoperative planning process converts unprocessed anatomical data into useful information. In the end, this improves patient outcomes and speeds up recovery by enabling surgeons to provide safer, more efficient, and customized surgical treatment.

7.2.9 Postoperative monitoring

In order to evaluate IGS effectiveness and promote the best possible patient recovery, postoperative surveillance is essential. Continuous monitoring following surgery with cutting-edge imaging technologies enables physicians to assess healing progress, identify issues early, and direct additional therapies as needed. Postoperative follow-up frequently uses high-resolution imaging techniques such as CT, MRI, and ultrasound. Physicians can verify that anatomical repairs or resections have been correctly accomplished thanks to these procedures, which offer a detailed visualization of the surgical site. In neurosurgery, for instance, postoperative MRI scans are used to confirm that tumors have been completely removed while maintaining the integrity of nearby vital tissues. Imaging is a crucial tool for early detection of any problems, in addition to verifying surgical success. Before they show up clinically, postoperative surveillance can identify problems, including infections, hematomas, edema, or incorrect implant placement. Timely therapies that may avert more serious consequences or the need for revision procedures are made possible by early detection. Additionally, postoperative imaging aids in rehabilitation planning by offering information on the condition of the healing bones and tissues, which enables recovery programs to be customized to the patient's unique circumstances. Periodic imaging, for instance, evaluates implant integration or bone fusion in orthopedic surgery, helping to inform decisions on physical therapy and weight-bearing.

By encouraging proactive management and individualized follow-up, the incorporation of image-based postoperative monitoring into patient care improves the overall standard of care. Additionally, it supports the expanding discipline of outcome-based medicine, which uses data-driven evaluations to guide ongoing advancements in surgical methods and technology. In conclusion, one of the main pillars of comprehensive IGS care is postoperative surveillance with sophisticated imaging. It guarantees the longevity of surgical outcomes, promotes early problem-solving, and eventually aids in safer, quicker, and more fruitful patient recoveries.

7.2.10 Impact of image-guided surgery on surgical precision

Modern surgical practice has been profoundly changed by IGS, which improves accuracy, lowers risks, and makes it possible to carry out intricate treatments successfully. By combining cutting-edge navigation and imaging technology, IGS reduces human error, prevents needless tissue injury, and enhances overall surgical results (Johnson, n.d.). Providing accurate instrument tracking and real-time visualization is one of IGS's main benefits. Because of this accuracy, surgeons may work with more assurance, particularly in high-risk and delicate areas such as the brain, spine, and cardiovascular system. Patients heal more quickly and avoid complications when they can see and move around important structures. Additionally, IGS encourages minimally invasive techniques that lessen infection risk, shorten hospital stays, and cause less damage to surrounding tissues. In operations when standard surgery could be too risky or unfeasible, this is quite helpful. Patients have better long-term results, faster healing, and less postoperative pain as a result. Looking ahead, IGS is set to undergo yet another transformation with the incorporation of AI. AI-powered navigation and imaging systems are being created to help with decision-making, anticipate surgical results, and automate image interpretation. Large datasets can be analyzed by these intelligent algorithms to find trends, improve anatomical mapping, and offer suggestions in real-time while doing surgery.

Future developments could include ML-powered enhanced AR environments, predictive analytics for customized surgical planning, and autonomous robotic systems that collaborate with doctors. These developments will raise the bar for what may be accomplished with surgery while simultaneously increasing accuracy and efficiency. In conclusion, by improving accuracy and lowering risks, IGS has already changed the surgical environment. It is anticipated that as AI-enhanced technologies advance quickly, their capabilities will increase even further, bringing in a new era of intelligent, precisely directed surgical care.

7.3 Artificial intelligence algorithms in medical robotics

Medical robotics is undergoing a revolution thanks to AI, especially in surgical applications. Healthcare practitioners are reaching previously unheard-of levels of accuracy, productivity, and patient safety by incorporating AI-powered algorithms into surgical systems. From preoperative planning and intraoperative guidance to post-operative monitoring, AI is revolutionizing every aspect of surgery, ultimately resulting in better clinical outcomes and quicker patient recovery. These days, autonomous and robotically-assisted surgical systems use a variety of AI approaches. Large volumes of patient data are analyzed by ML algorithms to enhance procedural methods, forecast problems, and facilitate individualized surgical planning. These algorithms improve real-time decision-making by continuously improving their outputs and learning from past situations. Complex medical pictures like MRI, CT, and ultrasound scans are well-processed by deep learning, a kind of ML. It makes it possible to accurately and automatically identify surgical targets, tumor boundaries, and anatomical features. When it comes to tasks such as tissue segmentation or surgical anomaly detection that call for intricate pattern recognition, deep learning models are especially useful. Another essential element of AI in surgery is computer vision. It enables real-time visual data interpretation from cameras and imaging devices by robotic systems. With these capabilities, surgical robots can identify tools, track the movement of tissue, and adjust as needed during an operation. This results in improved outcomes for MIS, more stable movements, and improved hand–eye coordination.

AI is also making it possible for autonomous surgical systems to carry out particular operations, such as cutting, suturing, or biopsy sampling, under supervision. In addition to improving procedural consistency, these technologies lessen the physical strain on surgeons. Altogether, AI is revolutionizing surgical robotics, allowing for safer, more intelligent, and more efficient surgical treatment. AI technology promises to further expand the realm of what is feasible in contemporary surgery as it develops.

7.3.1 Introduction to artificial intelligence in surgery

Modern healthcare is changing as a result of the use of AI in surgical practice. AI is paving the way for a new era of precision-driven, patient-centered surgery by automating difficult activities, lowering the possibility of human mistakes, and improving surgical efficiency. Large volumes of medical data, such as imaging scans, patient histories, and intraoperative parameters, can be analyzed by AI-powered systems to help surgeons make better, data-driven decisions. Clinicians can customize operations to each patient's unique anatomy and risk variables thanks to these technologies' ability to recognize patterns and forecast surgical outcomes. The combination of AI and RAS is among the most significant advancements. By analyzing real-time surgical data and providing actionable insights, AI algorithms in these devices give real-time advice. AI, for instance, can assist a robotic system in modifying the pressure or position of an instrument in response to visual cues or tissue resistance, increasing precision and lowering the risk of issues. Additionally, AI makes it possible for surgeons to receive real-time feedback during procedures, warning them of any irregularities or departures from the surgical plan. This degree of situational awareness lowers the chance of mistakes and enhances intraoperative decision-making. Additionally, by examining patterns in tissue reactions or vital signs, AI can foresee potential issues and enable prompt intervention. These developments have led to a notable improvement in patient outcomes. AI-powered surgeries are typically speedier, more reliable, and less intrusive, which frequently results in shorter hospital stays, quicker recovery periods, and fewer postoperative complications. In summary, the use of AI in surgery marks a significant advancement in surgical innovation, particularly when combined with robotic technologies. It changes the future of surgical care by providing surgeons with improved instruments for accuracy, safety, and flexibility.

7.3.2 Types of artificial intelligence algorithms used in surgery

A wide variety of algorithms power AI in surgery, allowing devices to learn from enormous medical datasets and assist with vital surgical procedures. These algorithms, which include neural networks, supervised learning, unsupervised learning, and reinforcement learning (RL), are the foundation of intelligent systems that provide real-time intraoperative guidance, surgical planning, and diagnosis. In AI-based surgical applications, supervised learning is frequently utilized for tasks including anatomical structure identification, risk prediction, and illness classification. These algorithms are very useful for supporting diagnosis and making surgical decisions since they are trained on labeled datasets, which teach them to link particular inputs, such as patient histories or medical pictures, with known results. Unlabeled data may include hidden patterns or relationships that can be found through unsupervised learning. Finding novel disease subtypes, grouping similar patient characteristics, or spotting surgical defects that might not be immediately obvious to doctors are all made easier using this method. It strengthens comprehension of intricate medical issues and facilitates individualized treatment planning. By using rewards or penalties to drive trial-and- error, RL allows systems to learn the best course of action. In surgical robotics, RL can teach autonomous systems to carry out certain tasks, such as suturing or navigating intricate anatomical paths, by continuously enhancing performance in response to environmental feedback.

For analyzing high-dimensional, complicated data, such as real-time video feeds, MRIs, and CT scans, neural networks, particularly deep learning models, are essential. These models give surgeons dynamic and precise insights by identifying patterns, segmenting tissues, and even simulating surgical outcomes. When combined, these AI algorithms enable surgical systems to carry out intricate analyses, improve surgical accuracy, and support in-the-moment decision-making. It is anticipated that these technologies will become ever more essential to safe, effective, and individualized surgical care as they develop.

7.3.3 Machine learning

Modern surgery is being revolutionized by ML, which allows robotic devices to learn from enormous volumes of medical data and gradually get better at what they do. ML models promote individualized decision-making, improve surgical precision, and improve patient outcomes through ongoing learning and pattern recognition. Through the examination of data from thousands of prior surgical procedures, including imaging scans, procedural videos, sensor inputs, and postoperative outcomes, ML algorithms are able to spot patterns and connections that would not be apparent to human observers. With this information, surgical robots can improve their actions, adjust to complicated settings, and react more cleverly to changes that occur in real-time while performing procedures. The capacity of ML to predict patient-specific outcomes is one of its main advantages in surgical robotics. ML models can forecast the risk of problems, estimate recovery timeframes, and direct the selection of the best surgical techniques based on a patient's medical history, anatomy, and disease features. Surgeons are able to arrange more focused and efficient interventions because of this individualized strategy. Additionally, ML is essential to risk assessment and mitigation. ML systems can warn surgeons about possible dangers during an operation by examining trends in prior data that are connected to unfavorable outcomes or surgical blunders. In high-stakes situations, this preemptive advice promotes safer decision-making and improves situational awareness. Furthermore, ML models get more precise and effective over time as they are exposed to more surgical data. Robotic-assisted surgical systems are guaranteed to stay current with changing best practices and procedural innovations because of their capacity for continual learning. In conclusion, surgical robots can now learn, adapt, and help with extremely difficult procedures thanks to ML. It makes data-driven insights possible, enhances safety, and opens the door to more intelligent, individualized surgical care.

7.3.4 Deep learning

Artificial neural networks are used in deep learning, a potent type of ML, to comprehend and react to intricate surgical situations. Deep learning has emerged as a key approach for improving the capabilities of AI-assisted surgical systems by simulating the human brain's capacity to identify patterns and learn from data (Patel, n.d.). Convolutional neural networks, or CNNs, are among the deep learning architectures that have the most influence on surgery. CNNs are particularly well-suited for evaluating medical pictures, including real-time video feeds from endoscopic cameras, MRIs, CT scans, and X-rays. They are crucial for intraoperative guidance and surgical planning because of their capacity to automatically extract and learn pertinent elements from raw picture data. CNNs' excellent accuracy in identifying abnormalities in medical imaging is one of their main advantages. CNNs, for example, can detect vascular irregularities, tumors, or lesions that may be hard for the human eye to see, allowing for more accurate surgical targeting and an earlier diagnosis. This degree of specificity reduces the possibility of missing important structures and enables better-informed decision-making. Deep learning models help with real-time tissue classification in the operating room, which enables surgical robots to discriminate between normal and aberrant tissues as treatments are being performed. This feature improves intraoperative safety by allowing surgeons to successfully remove sick tissue while protecting important structures.

Deep learning also helps with precise robotic movement guidance. AI systems may modify tool trajectories, improve cutting pathways, and preserve spatial awareness by analyzing dynamic visual inputs and continuously learning from previous processes. This ensures that delicate movements are performed with the least amount of error possible. In conclusion, deep learning, especially via CNNs, is revolutionizing surgery by improving robotic system responsiveness and picture interpretation accuracy. Its use keeps pushing the boundaries of data-driven, patient-specific, minimally invasive surgical care.

7.3.5 Computer vision

A key component of AI-driven surgery is computer vision, which gives robotic systems the ability to precisely interpret and process visual information from the operating room. Computer vision enables surgical robots and support systems to see, comprehend, and react in real-time to intricate anatomical and procedural cues by simulating human vision and augmenting it with computational power (Kim, 2023b). Improving navigation is one of computer vision's primary functions in surgery. AI-powered systems are able to follow surgical tools, monitor patient anatomy, and precisely direct the actions of the robot by evaluating live video feeds and medical imaging data. This feature increases the accuracy of minimally invasive operations and lessens the need for manual control. Anatomical recognition is another important use. In the surgical area, sophisticated algorithms are able to recognize and label a variety of tissues, organs, arteries, and anomalies. This automatic detection enables more confident dissection or resection and helps avoid unintentional damage to important structures. This function is crucial for making decisions in real-time during complex surgeries, such as neurosurgery or oncological procedures. Additionally, AR overlays in which 3D models or digital annotations are projected onto the surgeon's field of vision during the procedure are made possible by computer vision. These overlays successfully combine actual and virtual information to improve situational awareness by displaying surgical plans, highlighting target locations, or revealing hidden structures beneath the surface. Additionally, ML is enabling computer vision systems to continuously improve, enabling them to adjust to different operating situations, lighting conditions, and anatomical variances. This flexibility guarantees strong performance for a variety of treatments and patient types. In conclusion, computer vision greatly enhances the potential of AI-assisted surgery by offering visual intelligence, enhancing surgical precision, and facilitating safer, more successful results through sophisticated navigation and real-time augmented guidance.

7.3.6 Artificial intelligence applications in surgical robotics

By providing intelligent support across the whole surgical workflow, preoperative planning, intraoperative assistance, and postoperative monitoring, AI is essential to improving the capabilities of surgical robots. By increasing precision, effectiveness, and patient safety, these AI-powered technologies are revolutionizing surgical practice (Martinez, 2022). AI systems examine patient-specific information, including lab results, radiology scans, and medical histories, to produce comprehensive surgical maps during preoperative planning. This makes it possible to precisely plan according to the anatomy and condition of the patient. AI lowers risks and improves readiness by assisting surgeons in simulating surgeries, anticipating potential issues, and selecting the best surgical course. AI offers real-time assistance throughout the intraoperative phase by means of robotically-assisted MIS. AI systems use sensor data and real-time video streams to precisely direct robotic arms. In order to preserve stability and lower the possibility of human mistakes, these systems can modify instrument motion in response to tissue resistance or visual cues. Intelligent suturing, in which robotic systems learn to carry out delicate suturing operations autonomously or semiautonomously, represents a significant advancement in this field. These devices increase surgical efficiency, decrease errors caused by tiredness, and improve consistency. During surgery, AI also helps to optimize workflow. It can monitor the operation's progress, forecast the course of action, control surgical instruments, and guarantee efficient team collaboration. Time spent in the operating room is decreased by this process simplification, and it minimizes unnecessary disruptions.

AI continues to contribute to postoperative surveillance following surgery by evaluating imaging and recovery data to identify potential problems early. AI-powered solutions have the ability to notify physicians of any problems and suggest prompt actions, which can lead to better results and quicker recuperation periods. In conclusion, AI is pushing surgical robotics to new heights by intelligently assisting each stage of surgery, ultimately leading to safer, quicker, and more customized operations.

7.3.7 Robotic-assisted surgery

By fusing robotic technology with AI, RAS systems like the popular da Vinci medical System have revolutionized the medical field by improving the accuracy, control, and safety of intricate procedures (Sanders, 2023). These technologies provide surgeons with instruments that are more accurate, stable, and dexterous than those of humans. In RAS systems, AI is essential to enhancing surgical performance. AI improves instrument control using sophisticated algorithms, allowing robotic arms to perform extremely complex movements that would be challenging or impossible for human hands to perform alone. In MIS, where access to internal organs is restricted, and high-precision is essential, these movements are extremely helpful. Improved visualization is a major benefit of AI-integrated RAS. Surgeons may observe anatomical features in great detail using high-definition 3D imaging and AI-powered image processing, which enhances decision-making and lowers the chance of error. AI can also help with tumor identification, highlighting important anatomical features, and superimposing real-time instructions based on preoperative data. AI helps with haptic feedback simulation in addition to visual upgrades, giving the surgeon a tactile experience through robotic controls. Important characteristics that are frequently overlooked in conventional laparoscopic surgery, such as tissue texture, resistance, and tension, can be evaluated with the aid of this simulated tactile response. By providing stable and repeatable robotic motions, AI also aids in the removal of human constraints, such as hand tremors or fatigue. This is especially helpful for lengthy, intricate processes that demand sustained precision. All things considered, AI-powered robotic-assisted surgical technologies are changing the landscape of surgery. They enable surgeons to carry out complex procedures more skillfully by providing increased dexterity, better visualization, and insightful feedback, which improves patient outcomes, reduces recovery periods, and minimizes problems.

7.3.8 Autonomous surgical systems

By making it possible to create completely autonomous surgical devices, AI is quickly expanding the frontier of surgical innovation. These cutting-edge platforms promise previously unheard-of levels of accuracy, effectiveness, and safety by performing surgical procedures with little to no human involvement. Deep reinforcement learning (DRL), an AI technique that enables machines to learn optimal strategies through trial and error in simulated or real-world contexts, is at the core of these systems. By continuously improving their actions based on feedback and performance results, DRL helps robotic systems in autonomous surgery to master complex tasks, including tissue manipulation, dissection, and suturing. Real-time data processing is also essential to autonomous surgical systems. To dynamically understand the surgical environment, they combine input from multiple sources, including force sensors, high-resolution cameras, and medical imaging. Without direct surgeon control, this feature enables the robot to make snap decisions, adjust to unforeseen anatomical variances, and carry out accurate maneuvers. Large datasets of annotated surgical procedures are being used to train these computers to replicate expert surgical practices. They eventually have the ability to spot trends, foresee difficulties, and choose the best surgical strategy for every patient. This improves the consistency of intricate processes and lowers output variability. Common human limits in surgery, such as weariness, tremors, and making decisions under pressure, are also addressed by the move toward autonomy. AI-driven autonomous systems can drastically lower surgical errors, increase procedural efficiency, and shorten operating durations by reducing these variables. Even though clinical use of fully autonomous surgery is still in its infancy, continued research and development are hastening the process of wider adoption. It is anticipated that these intelligent systems will soon be essential instruments for carrying out risky and delicate surgeries with unparalleled accuracy and dependability.

7.3.9 Artificial intelligence for preoperative planning and decision support

In order to generate highly customized surgical plans, AI algorithms are increasingly using patient-specific data, such as radiology scans, medical histories, and clinical reports. Surgeons may precisely customize operations to each patient's unique anatomy and condition by integrating and evaluating these disparate databases, which improves safety and efficacy (Jackson, 2023). CT scans, MRIs, and ultrasounds are examples of radiological pictures that offer comprehensive anatomical and pathological details. Advanced image processing and pattern recognition are used by AI-powered systems to extract important information, such as tumor size, location, or vascular architecture, which serve as the basis for a personalized surgical roadmap. Surgeons are better able to foresee difficulties and choose the best entry locations and methods thanks to this accurate mapping. AI evaluates overall risk variables by taking into account comorbidities, test data, and patient medical history in addition to imaging. These intricate datasets are analyzed by ML models to find any side effects or surgical technique contraindications. This thorough knowledge enables medical professionals to decide whether open or MIS, robotic support, or other therapies are best. Surgical planning is further improved by AI-driven decision-support systems, which learn from enormous databases of previous cases. These systems are able to forecast probable outcomes, such as recovery time, infection risk, or the chance of success for different techniques, by comparing the profile of the current patient with prior data. Surgeons can choose techniques that optimize favorable outcomes while lowering risks thanks to this foresight.

AI can continue to help throughout surgery by modifying the plan in real-time in response to intraoperative observations, offering dynamic guidance that is in line with the patient's changing state. In conclusion, surgery planning becomes a customized, data-driven procedure thanks to AI's capacity to incorporate and analyze patient-specific data. Better patient outcomes, more educated surgical techniques, and a new benchmark for precision medicine in the operating theater are the outcomes of this.

7.3.10 Real-time artificial intelligence assistance during surgery

In order to greatly increase patient safety and surgical accuracy, AI is essential for giving real-time help during surgical procedures. AI tools help surgeons make well-informed decisions in dynamic and frequently challenging situations by continually monitoring intraoperative data and providing timely alerts (Smith and Brown, 2022). AI's ability to monitor vital signs, including heart rate, blood pressure, oxygen saturation, and respiration rate in real-time during surgery, is one of its primary uses. These physiological markers are continuously analyzed by sophisticated algorithms that look for minute variations that could indicate new problems, such as bleeding, hypoxia, or heart distress. AI allows for quick action by rapidly informing the surgical team, potentially averting negative results. AI-powered robotic equipment actively adjusts to the changing intraoperative environment in addition to monitoring patients. These systems use visual information and sensor inputs to precisely modify force application and instrument movements. For instance, the robot can autonomously adjust its movements to prevent injury, preserve stability, and improve the surgical technique if it detects unexpected tissue resistance or movement. Through real-time analysis of previous and present data patterns, AI also predicts likely difficulties. AI can spot trends or risk factors that may not be immediately apparent to the human eye by using prediction models that have been trained on large databases of surgical cases. This predictive ability improves readiness and response by enabling surgeons to foresee problems before they fully materialize.

Furthermore, robotic aid during surgery can be continuously improved thanks to AI-driven feedback loops. In order to ensure increasingly improved performance, the system learns from each movement and adjusts to surgeon preferences, patient anatomy, and procedural differences. In conclusion, the incorporation of AI into the surgical setting offers crucial real-time assistance via predictive analytics, adaptive robotic control, and attentive monitoring. This collaboration promotes more accurate surgical performance, lowers mistakes, and eventually leads to safer, better patient outcomes.

7.4 Data-driven techniques for surgical efficiency

Surgery has seen a significant transformation with the adoption of data-driven approaches, which have increased accuracy, optimized processes, and enhanced patient outcomes. These methods are transforming the planning, execution, and assessment of surgeries by utilizing big data, ML, and real-time analytics. Big data, which includes enormous volumes of patient data gathered from imaging investigations, wearable technology, electronic health records (EHRs), and surgical equipment, is at the heart of this revolution. Clinicians can find previously unattainable patterns and insights by analyzing this abundance of data. Large datasets, for example, make it possible to identify risk factors for problems and provide information for patient-specific evidence-based surgical methods. Big data is expanded upon by ML, which creates prediction models to aid with surgical decision-making. ML algorithms examine past surgical cases to forecast results, evaluate risks, and improve treatment strategies. These models improve safety and efficiency by assisting surgeons in choosing the best methods, foreseeing issues, and customizing care paths. A further crucial element made possible by data-driven systems is real-time decision-making. In order to give prompt alarms and advice during surgery, AI-powered platforms continuously process live data streams, including intraoperative imaging, instrument placements, and physiological signals. This real-time data helps surgeons adjust to unforeseen circumstances, reduce mistakes, and guarantee accurate performance of intricate procedures.

Additionally, by managing surgery schedules, coordinating interdisciplinary teams, and automating repetitive operations, data-driven approaches help to optimize workflow. Better resource use, lower expenses, and shorter operating times result from this. In conclusion, the use of ML, big data analytics, and real-time AI is transforming surgery into a more accurate, flexible, and patient-focused field. These developments hold out the prospect of sustained improvements in surgical results, effectiveness, and general healthcare quality.

7.4.1 The role of big data in surgery

Surgery has undergone a transformation since the advent of data-driven techniques, which have improved patient outcomes, workflow efficiency, and precision. Together, technologies like real-time decision-making, ML, and big data analytics have significantly increased surgical safety and efficiency (Johnson, 2021). Big data, which refers to the enormous amounts of structured and unstructured medical data produced every day, is in the vanguard of this change. This information is gathered from a variety of sources, such as intraoperative monitoring systems, imaging repositories, wearable medical technology, surgical logs, and EHRs. Researchers and physicians can uncover previously unnoticed but significant patterns and trends by examining these huge datasets. Healthcare practitioners can identify relationships between patient characteristics, surgical methods, and postoperative outcomes, for instance, by mining data from EHRs and surgical logs. These understandings aid in improving surgical procedures, foreseeing possible issues, and creating individualized treatment programs for each patient. These evidence-based modifications have a direct impact on safer and more successful procedures. Big data facilitates predictive analytics in addition to retrospective analysis. Risks such as infection, hemorrhage, or bad reactions can be predicted using ML algorithms that have been trained on prior surgical data. By lowering surprises and enhancing readiness, this foresight helps surgeons with preoperative planning and intraoperative decision-making.

Furthermore, dynamic modifications to the operation are made possible via real-time data integration during surgery. AI systems can give prompt feedback and alerts by continuously monitoring physiological data and surgical tool placements, which improves accuracy and patient safety even further. In conclusion, big data has transformed raw data into knowledge that can be put to use, thereby revolutionizing surgical care. More individualized, effective, and successful surgical procedures are being promoted by this data-driven development, which eventually helps patients and doctors alike.

7.4.2 Machine learning and predictive analytics

By efficiently utilizing a variety of data types, such as structured records and unstructured clinical notes, medical professionals can greatly enhance preoperative and postoperative treatment (Lee and Chen, 2023). A deeper comprehension of patient circumstances is made possible by this thorough data utilization, which supports more informed clinical decisions and individualized care routes. Predictive analytics that predict surgical results and possible problems are made possible by ML algorithms, which are essential to this procedure. Large volumes of historical surgical data are analyzed by ML models, which reveal trends and risk variables that conventional analysis could miss. By anticipating issues such as infection, bleeding, or delayed healing, this predictive power enables doctors to put preventive measures into place and modify care plans as necessary (Patel and Singh, 2020). Neural networks and deep learning are two ML algorithms that have demonstrated exceptional success in deciphering intricate medical data. These models can accurately anticipate patient reactions and surgical risks and are excellent at processing a variety of inputs, including imaging, lab findings, and patient demographics. Deep learning systems, for instance, can evaluate preoperative images to evaluate tissue viability or spot minute indicators that point to increased risk after operation. AI-powered models support customized treatment plans, which helps clinicians even more. These tools eliminate uncertainty during surgery by combining patient-specific data and predictive insights, assisting surgeons in making the best choices in real-time. By ensuring that procedures are matched to each patient's needs, this customized method improves surgical precision and results.

Furthermore, AI systems have the capacity to continuously learn from fresh data, improving their forecasts and suggestions over time. Both surgical planning and postoperative care continue to advance as a result of this dynamic learning. In conclusion, integrating thorough data use with cutting-edge AI techniques improves surgical care throughout the spectrum, from preoperative evaluation to postoperative recuperation, resulting in safer, more efficient, and more individualized treatment outcomes.

7.4.3 Patient outcome prediction

By estimating anticipated surgical outcomes using historical patient data, including demographics, medical history, and intraoperative variables, predictive models have emerged as a crucial tool in contemporary surgical treatment. These models give physicians valuable insights that enhance patient counseling, maximize treatment plans, and lower adverse events by utilizing this abundance of data. In order to anticipate significant outcomes, including complication rates, duration of hospital stay, recovery times, and readmission chances, predictive models fundamentally examine trends from previous surgeries. Clinical information, such as lab results, imaging findings, and intraoperative factors like blood loss or anesthesia duration, is linked with patient demographics, such as age, sex, and comorbidities. The algorithms may produce customized risk evaluations based on each patient's distinct profile, thanks to this thorough input. The potential of predictive modeling to improve patient counseling is one of its main advantages. Surgeons can help patients make more informed decisions about their care by giving them more precise information about possible risks and expected results. Treatment plans are in line with patient preferences and objectives because of this transparency, which also builds confidence. Additionally, by identifying patients who are more likely to experience difficulties or require a longer recovery time, predictive analytics aid in individualized treatment planning. Surgical methods, perioperative care, and postoperative monitoring can all be customized by clinicians. For example, in order to increase resilience prior to surgery, patients who have been identified as high-risk may get more rigorous monitoring or prehabilitation programs.

The decrease in hospital readmissions and surgical complications is another significant effect of predictive models. Healthcare teams can better manage resources, take preventative action, and intervene early by proactively recognizing risk factors. This results in less stress on healthcare systems, safer surgeries, and faster recovery times. In conclusion, by guiding customized therapies, improving outcome estimation, and ultimately improving patient safety and healthcare efficiency, predictive models that use patient-specific data greatly improve surgical care.

7.4.4 Surgical risk assessment

By using clinical and real-time patient data to forecast possible complications and adverse events, risk assessment models have emerged as essential tools in contemporary surgical practice. By providing surgeons with practical insights, these models facilitate better decision-making and prompt actions to enhance patient outcomes and safety. Fundamentally, risk assessment models use a variety of data sources, including intraoperative measurements like vital signs and surgical parameters, as well as preoperative clinical data like patient history, comorbidities, and test findings. AI-powered apps can spot early warning indicators and minute patterns in this data by processing it continuously, something that conventional monitoring techniques could overlook. AI's capacity to quickly evaluate thousands of patient cases is one of its main benefits in risk assessment. ML algorithms use massive datasets to discover intricate connections between surgical results and patient factors. Because of this thorough training, the system can precisely classify individuals based on their unique risk profiles, identifying those who might be vulnerable to issues such as infections, bleeding, or organ failure. By giving surgeons real-time risk assessments, these AI-driven risk models improve surgical practice by assisting them in making intraoperative decisions about surgical technique adjustments, anesthetic plan modifications, and emergency preparedness. Additionally, by identifying patients who need more specialized treatment or close monitoring to avoid readmissions, the models aid in postoperative management.

AI-powered solutions lower ambiguity and assist surgeons in implementing individualized intervention methods by providing evidence-based risk estimations. In addition to increasing surgical accuracy, this individualized strategy also helps to enhance resource allocation and overall healthcare efficiency. To sum up, AI-enabled risk assessment models revolutionize surgical treatment by utilizing a wealth of clinical and real-time data to proactively anticipate issues, hence facilitating safer, more intelligent, and more successful surgical procedures.

7.4.5 Real-time data integration and decision-making

Intraoperative decision-making has been transformed by the combination of real-time data from surgical tools, patient vital signs, and cutting-edge imaging technology. AI-driven technologies provide surgeons with quick, actionable insights that improve accuracy and safety during intricate surgeries by continuously gathering and evaluating these many data streams (Kim and Robinson, 2020). Metrics, including blood pressure, heart rate, oxygen saturation, force feedback, instrument placement, and live image from modalities such as ultrasound or endoscopy, are all included in real-time data capture. Together, these data provide a thorough picture of the patient's current state and the surgical setting, enabling dynamic modifications to be made during the procedure. Surgeons may quickly monitor important parameters thanks to AI-powered dashboards that combine this abundance of data into understandable, user-friendly statistics and visuals. These platforms can point out departures from typical patterns, identify possible issues before they become serious, and provide evidence-based suggestions for remedial measures. When millimeter-level accuracy is essential during complex or MIS, this kind of support is especially helpful. Additionally, AI systems integrate real-time intraoperative data as well as preoperative data, including imaging, patient history, and predictive models. The creation of customized alerts and instructions based on each patient's distinct anatomical and physiological profile is made possible by this integration. For instance, the technology can instantly alert the surgeon to reevaluate or alter the strategy if a sharp drop in blood pressure occurs in tandem with a certain instrument movement.

By automating ongoing data processing, the ability to receive real-time alarms and recommendations not only increases surgical accuracy but also lessens the cognitive load on the surgical team. This enables doctors to concentrate on important choices and technological implementation, while AI manages intricate data interpretation. In conclusion, using AI-driven dashboards to leverage real-time data greatly improves surgical safety and precision by offering prompt, contextualized insights and assistance, which eventually improves patient outcomes in intricate surgical procedures.

7.4.6 Artificial intelligence-driven instrument tracking

Surgical instrument monitoring has been greatly enhanced by AI and computer vision technologies, which are essential for increasing patient safety and operational effectiveness. By enabling automated, accurate surgical tool monitoring during procedures, these technologies guarantee that the appropriate instruments are available when needed. Manual instrument tracking, which is frequently used in traditional surgical workflows, is prone to human error and can result in missing or misplaced instruments. These mistakes can lengthen operating hours, postpone procedures, and, in extreme situations, endanger patient safety. By continuously detecting and tracking instruments using video feeds from operating room cameras or sensors built within the instruments, AI-driven computer vision systems help to alleviate these difficulties. These systems can identify and distinguish between different surgical tools in real-time by using deep learning models that have been trained on large datasets of surgical scenes and equipment. If an instrument is lost, improperly prepared, or about to be utilized at an unsuitable moment, this feature allows the operating room team to get automated alerts. By predicting the surgeon's subsequent instrument requirements based on the procedural phase, automated tracking also optimizes the surgical workflow. The system may alert robotic assistants and scrub nurses to quickly prepare and supply the necessary tools, reducing downtime and enabling the surgical team to stay focused on important duties.

These AI-based tracking solutions also help with better inventory control and documentation. They improve adherence to safety procedures and lower expenses related to instrument loss or damage by producing precise records of instrument usage and sterilization cycles. In conclusion, AI and computer vision technologies revolutionize surgical instrument management by improving workflow efficiency, lowering the chance of errors, and increasing accuracy. As a result, surgical procedures run more smoothly, there are fewer disruptions, and patients receive safer, better treatment.

7.4.7 Real-time visualization of surgical sites

By offering improved, real-time images of anatomical components during procedures, AR and AI-based imaging technologies are revolutionizing surgical visualization. These developments have a special effect on MIS, where visibility and access issues are major obstacles (Khan, 2023). AR integrates live intraoperative data with preoperative scans, such as CT or MRI, by superimposing digital pictures onto the surgeon's field of vision. A thorough, 3D map of the patient's anatomy is produced by this fusion and can be seen directly on the surgery site or on specialized displays, such as AR glasses or monitors. Through improved spatial awareness and depth perception, AR enables surgeons to more precisely navigate intricate anatomy without making big incisions. Through real-time processing and interpretation of imaging data, AI plays a crucial role. Intraoperative images are analyzed by deep learning algorithms to identify important features, including tumors, blood arteries, and nerves. This automated identification increases surgical precision and lowers the possibility of unintentional damage to important tissues. Combining AR and AI-based imaging improves the surgeon's capacity to make quick, well-informed decisions and modify procedures in response to dynamic input. For instance, these technologies enable more accurate tissue manipulation and tool placement during laparoscopic or robotic procedures by precisely localizing concealed lesions or borders.

By avoiding tissue damage, cutting down on operating time, and lowering postoperative complications, AR and AI not only enhance technical performance but also improve patient outcomes. Shorter hospital stays and a quicker recovery are encouraged by the improved visualization, which supports less invasive techniques. Furthermore, AR systems increase access to specialized treatment by facilitating surgical training and collaboration by allowing distant experts to view the operational field and offer real-time guidance. In conclusion, the use of AI-based imaging and AR in MIS greatly improves surgical accuracy, safety, and efficiency, which eventually results in better patient outcomes and a wider acceptance of innovative surgical methods.

7.4.8 Optimization of surgical workflows

By avoiding tissue damage, cutting down on operating time, and lowering postoperative complications, AR and AI not only enhance technical performance but also improve patient outcomes. Shorter hospital stays and a quicker recovery are encouraged by the improved visualization, which supports less invasive techniques. Furthermore, AR systems increase access to specialized treatment by facilitating surgical training and collaboration by allowing distant experts to view the operational field and offer real-time guidance. In conclusion, the use of AI-based imaging and AR in MIS greatly improves surgical accuracy, safety, and efficiency, which eventually results in better patient outcomes and a wider acceptance of innovative surgical methods. AI-driven analytics generate flexible timetables that optimize OR use by leveraging previous data on surgery durations, team performance, patient flow, and resource availability. By taking into account variables unique to each patient, case complexity, and surgeon experience, ML algorithms are able to forecast surgery timings more precisely. This accuracy makes it possible to allocate surgical blocks dynamically, which lowers the likelihood of schedule overruns and gaps. AI aids in workflow optimization beyond scheduling by examining trends in staff movements, instrument usage, and task sequences. By identifying bottlenecks and streamlining procedures such as instrument sterilization or patient transfer, insights from this data can reduce the amount of time that passes between surgeries. In order to guarantee that the appropriate resources are available when needed, AI-powered solutions may also predict staff availability and supply requirements.

Healthcare institutions can reduce expenses related to overtime, cancellations, and underused OR capacity by increasing scheduling accuracy and operational flow. Additionally, shorter wait times for patients result from more effective OR management, enhancing access to prompt surgical care. To sum up, AI analytics-powered data-driven scheduling and workflow optimization improve operating rooms' productivity and economy. Greater patient satisfaction, more efficient surgical workflows, and greater resource utilization result from this.

7.4.9 Time reduction in surgical procedures

By evaluating enormous volumes of data from prior operations, ML models are progressively revolutionizing surgical workflows by suggesting the best steps to take throughout the procedure. By eliminating pointless movements and interventions, these intelligent devices aid in streamlining procedures and enhancing surgical accuracy and efficiency (Lee and Chen, 2023). Historically, the optimum course of action in surgical treatments is mostly determined by the experience and intuition of the surgeon. To find best practices and ideal sequence patterns, ML models may now analyze historical surgical data, including operative recordings, instrument trajectories, and patient outcomes. These models can suggest the best course of action in real-time by identifying the most effective pathways, guaranteeing more streamlined and uniform processes. These suggestions are supplemented in the operating room by AI-guided robotic assistance, which physically performs or assists with tasks determined by the analysis of the ML model. To lessen the cognitive strain on the surgeon and surgical team, robotic systems can, for example, autonomously position equipment, hold tissues in place, or modify camera views to provide the optimal viewpoints. By removing pointless operations and streamlining instrument transitions, these AI-powered solutions reduce redundancy. They therefore help to drastically reduce surgical times without sacrificing precision or security. A shorter anesthetic duration also helps patients recover more quickly and reduces dangers.

Furthermore, AI systems improve their recommendations over time by continuously learning from every process they watch or help with. By customizing recommendations to certain surgeons' methods and patient profiles, this adaptive learning capability enables them to support precise and individualized surgery. In conclusion, robotic help guided by AI and ML models is transforming the way surgeries are conducted. They decrease surgery time, enhance productivity, and raise the standard of patient care by suggesting effective process stages and streamlining procedures.

7.4.10 Minimization of human error

By reducing human error, standardizing processes, and enhancing patient safety, automation and AI are revolutionizing modern surgery. AI greatly improves the accuracy and dependability of operations by incorporating intelligent alert systems, real-time data analysis, and ML algorithms into surgical workflows (Patel and Singh, 2020). The capacity of AI to support precise navigation is one of its main benefits in surgery. By aligning surgical instruments with patient anatomy, sophisticated imaging systems and AI algorithms provide real-time guidance. By avoiding important structures, this accuracy lowers the chance of problems and guarantees that the best surgical practices are followed. AI is also essential for standardizing processes. AI algorithms are able to identify best practices and ideal task sequences by evaluating vast amounts of surgical data. Standardized protocols that decrease variability across processes and increase overall consistency are developed as a result of these discoveries. AI-guided robotic systems can do delicate or repetitive operations with high accuracy, thus lowering the possibility of human error. Furthermore, AI-powered automatic alert systems offer real-time notifications of possible hazards, including irregular vital signs, improper use of instruments, or deviation from the intended surgical course. By acting quickly in response to these proactive alarms, surgical teams can stop problems before they become more serious.

Surgical operations continue to be revolutionized by the incorporation of data-driven techniques, which combine AI, ML, and real-time analytics. In addition to improving intraoperative performance, these tools facilitate decision-making at every stage of the surgical process, from preoperative planning to postoperative care. In conclusion, automation and AI mark a substantial advancement in surgical innovation. These tools are changing the future of surgery and increasing patient outcomes in quantifiable ways by decreasing human error, streamlining processes, and guaranteeing safe, effective treatments.

7.5 Case studies and applications

Robotic and AI-assisted operations are among the most revolutionary developments in contemporary medical technology. For patients undergoing a variety of surgical procedures, these devices provide unmatched precision, decreased invasiveness, and enhanced results. AI and robotics are changing the surgical care landscape, from intricate neurosurgeries to minimally invasive orthopedic and cardiac procedures. Every stage of surgery is improved by AI-driven algorithms, including postoperative monitoring, planning, and execution. ML models evaluate patient-specific data prior to surgery in order to develop customized surgical plans. Robotic devices, such as the da Vinci Surgical System, give surgeons improved dexterity, reduced tremor, and 3D visualization during surgery. These features enable extremely accurate motions, especially in confined or sensitive anatomical areas. Smaller incisions, less blood loss, quicker healing times, and a decreased chance of problems are thus common outcomes of treatments. AI-powered real-time imaging and navigation technology helps surgeons even more by providing submillimeter accuracy, guiding tools, and dynamic viewing of patient anatomy. In neurosurgery, where avoiding vital tissues is crucial, these developments are very helpful. Despite their potential, a number of obstacles still exist. Widespread adoption is hampered by the high prices of robotic systems and AI integration, especially in environments with limited resources.

Significant training for surgeons and medical personnel is also necessary for successful adoption, and persistent issues with data privacy, autonomy, and the moral application of AI must be resolved. However, robotics and AI in surgery have a bright future. It is anticipated that further advancements in real-time data processing, deep learning, and robotic tool downsizing would increase system performance and reduce costs. A new age of safer, more intelligent, and more individualized surgical treatment is anticipated as these technologies become more developed and widely available, increasing their incorporation into standard clinical practice.

7.5.1 Artificial intelligence-assisted image-guided neurosurgery

The accuracy, security, and effectiveness of brain and spinal procedures have greatly improved because of AI-assisted image-guided neurosurgery. Because of the intricacy of the central nervous system and the close proximity of vital structures, these procedures frequently call for the highest level of accuracy. Neurosurgeons can now carry out highly targeted procedures with the least amount of harm to nearby healthy tissues by fusing AI with real-time imaging techniques, such as MRI and CT (Marescaux and Soler, 2004). By evaluating massive datasets, such as MRI and CT scans, ML algorithms are essential for preoperative planning because they may detect anomalies like cancers, vascular malformations, or spinal lesions. AI is particularly good at segmenting tumors, correctly differentiating between healthy and diseased tissues, and mapping important brain regions like those involved in speech, motor function, and cognition. This guarantees that surgical strategies are designed to remove as much tumor as possible while maintaining neurological function. AI systems combined with real-time intraoperative imaging during surgery offer dynamic updates on the location of surgical instruments in relation to the patient's anatomy. By improving spatial awareness, these tools enable neurosurgeons to respond to anatomical shifts brought on by fluid loss or brain enlargement, a phenomenon known as "brain shift."

Real-time adaptation of AI systems to these changes updates surgical maps and lowers the possibility of unintentional injury. Additionally, minimally invasive techniques are supported by AI-enhanced imaging guidance, which allows neurosurgeons to use smaller incisions to target deep-seated or difficult-to-access tumors. Patients recover more quickly, have shorter hospital stays, and have less postoperative discomfort as a result. To sum up, AI-assisted image-guided neurosurgery is a major advancement in medical technology. Surgeons can improve patient outcomes for patients with complex neurological diseases by using ML and real-time imaging to increase precision and safety.

7.5.2 Robotic-assisted minimally invasive surgery

Robotic-assisted MIS has dramatically improved patient outcomes by shortening recovery times, lowering surgical trauma, and boosting precision. The da Vinci Surgical System, one of the most popular robotic-assisted platforms, enables surgeons to conduct difficult surgeries with greater dexterity and magnified 3D visualization. Robotic-assisted operations have shown superior results in urological surgeries, notably radical prostatectomy, with less blood loss, shorter hospital stays, and lower complication rates. Similarly, in gynecological treatments, robotic platforms improve precision in hysterectomy and myomectomy, lowering postoperative complications in comparison to traditional laparoscopic approaches (Marescaux and Soler, 2004). ML algorithms built into robotic surgical systems help to improve real-time decision-making. AI-powered robotic helpers can examine intraoperative data and provide predictive analytics to help surgeons perform the procedure. Challenges in robotic-assisted MIS include the high cost of robotic platforms, the requirement for significant training, and worries about the learning curve associated with these technologies. However, further breakthroughs in AI and robotics are projected to refine and broaden the applications of robotic-assisted MIS.

7.5.3 Applications in orthopedic surgery and spine procedures

AI and robotics have had a dramatic impact on orthopedic surgery, particularly joint replacement and spinal surgeries. AI-powered preoperative planning tools assist surgeons in determining the appropriate implant placement, resulting in better patient outcomes (Marescaux and Soler, 2004). Robotic-assisted orthopedic treatments, such as those performed with the Mako robotic-arm system, offer unprecedented precision in complete knee and hip replacement. These technologies use preoperative CT scans to generate a detailed 3D model of the patient's anatomy, allowing surgeons to receive real-time feedback during bone preparation and implant positioning. Robotic-assisted navigation devices, such as Mazor X, are being used in spinal fusion and deformity correction surgeries. These technologies improve screw placement precision, lowering the likelihood of problems such as nerve injury and implant misalignment. Furthermore, AI systems use postoperative recovery data to tailor rehabilitation regimens for orthopedic patients. Using data-driven insights, healthcare professionals can optimize postsurgical treatment and improve long-term mobility outcomes. Despite these advances, robotic orthopedic surgery has significant hurdles, including high prices and limited accessibility, particularly in underdeveloped countries (Marescaux and Soler, 2004). Ongoing research and development aim to make these technologies more affordable and accessible in the future.

7.5.4 Cardiac surgery with image-guided robotics

The use of AI and robots in cardiac surgery has improved precision and reduced the invasiveness of treatments such as coronary artery bypass grafting and mitral valve replacement. AI-powered imaging approaches, such as real-time intraoperative imaging and predictive analytics, empower surgeons to make data-driven decisions while lowering surgical risks. Robotic-assisted cardiac surgeries, such as those conducted with the da Vinci System, have been demonstrated to reduce surgical trauma and hospital stays when compared to traditional open-heart procedures (Marescaux and Soler, 2004). These robotic devices offer increased maneuverability and stability, allowing surgeons to conduct difficult cardiac surgeries more accurately.

Furthermore, AI-enhanced imaging aids in preoperative planning by creating 3D models of the heart and vascular system, allowing for patient-specific procedure customization. ML models can also help anticipate postsurgical problems, allowing for early intervention techniques that enhance patient outcomes. Despite the advantages, limitations in robotic-assisted cardiac surgery include high equipment costs, the requirement for specialized training, and the difficulty of incorporating AI-driven decision-making tools into surgical workflows (Marescaux and Soler, 2004). Future advancements aim to solve these difficulties by strengthening AI models, lowering prices, and increasing access to advanced robotic surgical techniques.

7.6 Challenges and limitations of image-guided surgery and artificial intelligence algorithms

A new era of precision and MIS has been brought about by the combination of AI and IGS, but in order to guarantee their safe, moral, and successful application in clinical practice, a number of significant issues need to be resolved. These obstacles, which have logistical, ethical, and technical components, make it difficult for AI-powered surgical devices to be widely used. The quality and variety of the data are among the most significant technological constraints. Large, annotated datasets are essential for training and validating AI algorithms. Model accuracy and generalizability may be hampered by inconsistent imaging quality, a lack of standardized procedures, and restricted access to a variety of patient data. Inaccurate projections brought on by faulty or biased data raise the possibility of surgical mishaps. Another significant obstacle is system integration. AI must work in unison with current imaging platforms, surgical instruments, and hospital information systems. Robust engineering and validation are necessary to guarantee real-time performance, interoperability, and system stability under high-stakes surgical situations, but they can be time-consuming and resource-intensive. Adoption is made more difficult by moral and legal considerations. There are still unanswered questions about algorithm openness, accountability, informed consent, and data protection. For example, it's not clear who is legally liable if an AI system makes poor advice that results in a surgical complication: the surgeon, the hospital, or the manufacturer. The creation of precise legal and ethical frameworks is necessary to address these issues. Future developments in AI-powered surgery will hinge on a number of factors, including enhancing AI explainability and transparency, creating patient-specific AI models, and putting in place regulatory guidelines that protect patient safety without limiting creativity. In conclusion, even if IGS and AI have revolutionary potential, achieving their full potential will require overcoming present obstacles. To create a safer, more intelligent surgical future, physicians, engineers, ethicists, and legislators must work together strategically.

7.6.1 Technical challenges

AI-powered IGS has the potential to revolutionize surgery, but in order to guarantee accuracy, security, and effectiveness, a number of technological issues need to be resolved. The need for real-time processing is one of the main obstacles. Instantaneous data interpretation is essential for surgical procedures, particularly when high-resolution imaging modalities like intraoperative ultrasonography, CT, and MRI are used. Massive datasets are produced by these imaging technologies, which puts a heavy computational burden on processors and necessitates high-bandwidth, low-latency data transport systems.

In AI-based navigation systems, latency presents another difficulty. Surgical accuracy can be compromised by even slight lags in image processing or system reactions, which could lead to inaccurate instrument motions or delayed decision-making at crucial times. Sustaining the safety and efficacy of surgery requires regular, real-time feedback. Limitations in hardware also add to the complexity. Many operating rooms lack the hardware, such as powerful servers, AR-compatible screens, and high-performance GPUs, necessary to enable AI-integrated platforms. Additionally, misalignment problems may arise with AR-based guidance systems that superimpose virtual anatomical maps on the patient during surgery. Inaccurate AR predictions raise the possibility of surgical complications by causing incorrect instrument placement (Wong et al., n.d.). To fully realize the potential of AI-enhanced IGS in clinical practice, these technical constraints must be resolved.

7.6.2 Data accuracy and quality

The caliber, variety, and precision of training data have a significant impact on how well AI performs in surgical applications. Large, well-annotated datasets are necessary for AI algorithms to learn efficiently. But inadequately labeled data, skewed sampling, or a lack of visual diversity can seriously impair model performance and produce inaccurate predictions that could jeopardize surgical judgment. For instance, when used in a variety of therapeutic contexts, AI systems that were primarily trained on data from a certain demographic group may find it difficult to generalize to larger groups, raising the possibility of incorrect diagnoses or surgical complications. The existence of image artifacts is a significant obstacle in AI-assisted medical imaging. Low signal-to-noise ratios, patient movement, or irregular scanner calibration can all cause these distortions, which can impair the quality and accuracy of images. Such artifacts have the potential to interfere with AI-based picture segmentation and classification procedures during surgery, which could result in incorrect identification of disease regions or anatomical structures (Park and Lee, n.d.). Patient safety may be at risk due to improper surgical planning or navigation brought on by this deception. Improving dataset quality, encouraging demographic variety in training data, and putting strong artifact detection and repair methods in imaging systems are all crucial for increasing AI reliability in surgical settings.

7.6.3 System integration issues

There are a number of infrastructure and practical obstacles to overcome when incorporating AI algorithms into traditional surgical procedures. The incompatibility of AI systems with the outdated hardware that many healthcare facilities currently utilize is a significant obstacle. The smooth adoption of cutting-edge AI-driven platforms is hampered by the lack of computational power or connectivity required for real-time AI processing in older imaging equipment, surgical instruments, and hospital information systems (Naik et al., 2022). These systems can be expensive and time-consuming to upgrade, which creates logistical and financial obstacles to deployment. Furthermore, interoperability may be hampered by software-hardware compatibility problems across robotic platforms, AI-powered navigation systems, and current surgical infrastructure. In addition to delaying integration, these discrepancies may also reduce workflow effectiveness and increase the risk of mistakes during crucial processes.

The absence of uniform standards for incorporating AI into surgical settings is another major barrier. Hospitals must deal with uncertainties about compliance, data management, and liability in the absence of defined regulatory or procedural frameworks. Investment in AI technologies may be deterred by this ambiguity. Surgeons and medical personnel must also receive extensive training for optimal adoption. Both technical expertise and a change in surgical mindset are necessary to become proficient in using AI-assisted instruments. The efficiency of AI technologies in the operating room may be limited by this steep learning curve, which may also hinder adoption (Malczewska et al., 2021). To fully realize AI's promise in surgery, these issues must be resolved.

7.6.4 Ethical and legal considerations

AI in surgery raises difficult moral and legal issues, especially with relation to patient autonomy, informed consent, and responsibility. It is crucial that AI systems' procedures continue to be open and understandable as they help with diagnostic, surgery planning, and intraoperative decisions more and more. To make educated decisions regarding care, patients and healthcare professionals need to be able to comprehend how AI generates its suggestions. In the absence of this clarity, cooperative decision-making and patient trust could be jeopardized. Making sure patients maintain their autonomy when AI is used to make decisions is a significant ethical challenge. Information regarding the function, constraints, and possible hazards of AI must be incorporated into consent procedures. Not providing patients with enough information could jeopardize their rights and legal protections. Liability for AI-related errors, such as misdiagnosis or surgical consequences, is still unclear from a legal perspective. There are concerns regarding who is responsible for the healthcare facility, the surgeon, or the AI creator. These circumstances are not adequately covered by current medical laws, which leads to uncertainty in malpractice and litigation instances. Establishing precise legal and ethical frameworks will be essential as AI in surgery develops further in order to guarantee responsibility and safeguard patient rights.

7.6.5 Accountability in artificial intelligence-driven surgery

Significant accountability issues are introduced by AI-assisted surgical operations, particularly in the event of mishaps. It becomes difficult to assign precise blame for unfavorable results when AI systems offer intraoperative assistance or recommendations. Although assessing and confirming AI-generated insights is ultimately the job of surgeons, the increasing dependence on automated decision-making may make this difficult to do (Wong et al., n.d.). Determining who is at fault, the software developer, the AI algorithm, or the surgeon's judgment, becomes a controversial topic if a surgical error results from flawed AI analysis or incorrect predictions. Legal actions and malpractice claims are complicated by this uncertainty, which may leave patients without obvious options for redress. Regulatory agencies must create thorough and precise rules for liability in AI-assisted surgery in order to solve these issues. All parties involved, including medical institutions, AI developers, and healthcare practitioners, should have their roles and duties clearly defined in these recommendations. Furthermore, to guarantee that AI tools are dependable, understandable, and applied correctly in clinical contexts, stringent safety, testing, and performance criteria must be implemented. By establishing these rules, surgical practice will be able to safely and ethically incorporate AI technology while also increasing trust in these technologies and defining accountability.

7.6.6 Data privacy and security

Strict safeguards are required to protect data privacy and regulatory compliance when using medical data to train AI models, especially in light of regulations such as the General Data Protection Regulation and the Health Insurance Portability and Accountability Act. The transmission and storage of patient imaging and health records within AI-driven surgical systems expose these extremely sensitive data points to serious cybersecurity concerns, such as hacking, data breaches, and unauthorized access (Park and Lee, n.d.). These systems must include strong encryption protocols, safe access controls, and efficient data anonymization methods that eliminate personally identifiable information from datasets in order to protect patient confidentiality. While preserving the usage of medical pictures for AI training and surgical guidance, such approaches aid in preventing illicit data use. Concern over AI companies' commercial exploitation of private surgical data is also growing. Sensitive patient data may be abused for financial gain unrelated to medical treatment in the absence of adequate control, betraying patient confidence and ethical norms (Naik et al., 2022). Therefore to ensure that AI developers and healthcare providers adhere to strict privacy regulations, improved regulatory frameworks and audits are crucial for monitoring data-handling methods. Encouraging public trust and the responsible development of AI-assisted surgery depend heavily on maintaining the confidentiality and integrity of medical data.

7.6.7 Limitations in current artificial intelligence capabilities

Even with major improvements, context-based reasoning and real-time decision-making in surgical settings remain substantial problems for present AI models. The majority of AI systems rely largely on statistical correlations and pattern recognition from training data, which limits their capacity to handle unforeseen complications or unique surgical settings that don't fit into previously acquired patterns (Malczewska et al., 2021). This restriction is crucial in surgery because quick decisions based on context are necessary due to unforeseen circumstances. The reliance of AI systems on sizable, meticulously annotated datasets for training is another significant barrier. The scarcity of instances and the intricacy of annotations in highly specialized surgical domains can make it challenging to acquire such substantial, high-quality data. This scarcity reduces the AI's flexibility by limiting its exposure to a variety of surgical changes. Furthermore, a lot of AI models use supervised learning, which leaves them susceptible to out-of-distribution cases, or examples that differ greatly from the training data. These conditions may cause AI forecasts to lose their accuracy, which could be dangerous in intricate or uncommon surgical situations (Malczewska et al., 2019). Future advancements must concentrate on strengthening contextual comprehension, integrating unsupervised and RL, and expanding data diversity and quality in order to raise AI's credibility and safety in surgery.

7.6.8 Understanding contextual decision-making

The incapacity of AI in surgery to completely understand the contextual subtleties that significantly impact surgical decision-making is one of its main limitations. Even while AI is very good at processing large volumes of clinical and imaging data, it does not have the same intuitive judgment, expertise, or flexibility that experienced surgeons do in the operating room. These human traits are crucial, particularly in settings that are dynamic and uncertain. For example, AI systems may find it difficult to adapt to changing situations in trauma surgery, where conditions can change quickly and necessitate quick, on-the-spot alterations. Their efficacy in highly changeable situations is limited by their dependence on prelearned patterns and algorithms, which makes real-time adaptation difficult (Wong et al., n.d.). This drawback emphasizes how important human skill is to AI-assisted surgical procedures. AI can improve planning, accuracy, and efficiency, but it cannot take the place of surgeons' rapid thinking and sophisticated grasp of intricate procedures. As a result, the best surgical results require a team effort, in which AI complements but does not replace the crucial decision-making function of experienced surgeons.

7.6.9 Patient-specific variability

Anatomical variances, physiological reactions, and genetic factors unique to each patient that have a significant impact on surgery results are frequently difficult for AI models to account for. Because existing AI systems are mostly trained on big, generalized datasets that might not accurately represent individual patient variations, personalized medicine continues to pose a substantial problem. When applied to a variety of patient populations, this constraint lowers the precision and dependability of AI-driven suggestions. For example, differences in bone density, vascular architecture, or tissue properties can have a major influence on surgical planning and execution in specialties such as orthopedic and cardiac surgery. AI models that ignore these subtleties might offer less-than-ideal advice, which could have an impact on patient safety and the efficacy of treatment. AI systems must develop by incorporating patient-specific biomarkers and utilizing predictive analytics catered to each person's unique genetic and physiological characteristics in order to meet these obstacles. More accurate surgical planning and decision-making would be made possible by incorporating individualized data, which would enhance results and lower complications (Park and Lee, n.d.). The future of precision surgery depends on creating AI systems that can adjust to these individualized characteristics. This will guarantee that AI tools advance beyond broad forecasts and toward really personalized patient care.

7.7 Future directions

The field of IGS is about to undergo a radical transformation because of advancements in robotic automation, AI, VR, AR, and predictive analytics. The precision and personalization of surgical treatments will be greatly increased by these state-of-the-art technologies, which will eventually improve patient outcomes. Surgeons will be able to plan and perform procedures with previously unheard-of confidence and detail because of AI's ability to interpret complex imaging data more accurately. An immersive, real-time visualization of anatomical features will be made possible by VR and AR technologies, which will improve the surgeon's spatial awareness and allow for more accurate navigation during surgeries. Large volumes of patient data will be used by predictive analytics to forecast surgical risks and results, enabling better-informed decision-making based on the unique characteristics of each patient. By permitting precise, repeatable movements that are beyond human capability, robotic automation will significantly improve surgical accuracy while lowering the risk of complications and minimizing tissue damage. When combined, these developments will enhance access to sophisticated surgical care in a variety of healthcare settings, encourage individualized treatment plans, and improve surgical precision. The future of IGS offers safer, more effective procedures with better results and wider availability as these technologies easily integrate into clinical processes, ushering in a new era of surgical innovation.

7.7.1 Advanced artificial intelligence models for personalized surgery

AI models are developing quickly, becoming more and more competent to handle data in real-time and facilitate highly customized surgical planning. To forecast surgical results and spot possible hazards, ML algorithms examine patient-specific data, including imaging, medical history, and physiological readings. This enhances the overall efficacy of treatment by enabling surgeons to customize operations to each patient's particular anatomy and condition. It is anticipated that, in the future, AI-powered systems will provide advanced decision-support capabilities that can adapt to shifting intraoperative circumstances. These devices could give prompt feedback, recommend changes, and assist in predicting problems before they occur by continuously monitoring real-time data throughout operation. This kind of flexibility would improve surgical procedures' accuracy and safety.

AI's capacity to streamline procedural operations can also shorten operating room hours, minimize human error, and enhance resource use. From preoperative planning to postoperative care, these developments will help surgeons make better, data-driven decisions at every stage of operation. All things considered, the incorporation of adaptive AI and real-time data processing into surgical practice holds the potential to revolutionize operative procedures by making them safer, more effective, and more individualized for each patient (Naik et al., 2022).

7.7.2 Integration of virtual reality and augmented reality with image-guided surgery

Real-time intraoperative navigation and surgical training are being transformed by VR and AR technology. By projecting vital information straight onto the surgical field, AR overlays improve surgeons' anatomical visualization and allow for more accurate and assured execution of intricate procedures. By improving spatial awareness and emphasizing important structures, this real-time augmentation lowers errors and enhances surgical results. At the same time, VR-based simulations have emerged as crucial resources for medical training and education. Surgeons can repeatedly practice a variety of procedures in these immersive environments without endangering patients in a secure, regulated environment. Through skill development, technique improvement, and preparation for uncommon or difficult patients, this practical experience helps surgeons become more prepared for real operations. By enhancing preoperative preparation through comprehensive virtual walkthroughs and providing improved intraoperative guidance through real-time AR support, the continuous development of immersive technology promises to further elevate surgical practice. These instruments will continue to change the surgical landscape as they advance in sophistication and become more commonly used, improving safety, accuracy, and customization of operations to the unique anatomy and requirements of each patient (Malczewska et al., 2021).

7.7.3 Predictive analytics and outcome-based surgery

Big data and AI-driven predictive analytics are revolutionizing surgical treatment by making it possible to anticipate surgical outcomes and evaluate patient-specific risks. These algorithms can find trends and correlations about complications, recovery times, and overall operative success rates by examining large datasets from prior procedures. Surgeons can better grasp the possible difficulties and adjust their strategies with the aid of this abundance of information. More individualized surgical planning is made possible by these data-driven insights, which also help physicians predict patient-specific risks and improve perioperative management techniques. By giving more precise estimates for recovery times and potential consequences, predictive analytics also aids in well-informed patient counseling.

In the future, more proactive operating room decision-making will be made possible by developments in predictive modeling. Dynamic risk assessment during surgery will be made possible by real-time integration of patient data with past results, enabling surgical teams to quickly modify their methods and interventions to reduce errors. It is anticipated that these enhancements will lessen surgical complications, decrease hospital stays, and enhance patients' long-term health results. In conclusion, predictive analytics is a major step toward more customized, outcome-driven treatment by utilizing AI and big data to improve surgical precision and safety (Malczewska et al., 2019).

7.7.4 Evolution of robotic surgery: from assistance to autonomy

From a simple tool aid to permitting partially autonomous procedures, robot-assisted surgery has undergone substantial development. In order to improve dexterity and precision during MIS, early robotic systems mainly functioned as extensions of the surgeon's hands. But the advent of AI has brought about a new era in which, with human supervision, robotic equipment can now carry out intricate surgical procedures with a greater degree of autonomy. AI-enabled robotic devices that can perform autonomous surgical procedures by utilizing real-time sensor data, sophisticated imaging, and ML algorithms are the subject of the next stage of development. These technologies perform interventions more reliably and precisely than manual approaches alone by fusing AI-powered decision-making with the unmatched precision of robotic equipment. Such robots may dynamically modify their activities by continuously evaluating intraoperative data, limiting human error, and lowering the variability that comes with traditional surgery. Shorter recovery periods, fewer complications, and better surgical outcomes are the effects of this. A greater number of MIS are anticipated to be carried out by these AI-driven robotic systems as they develop, providing surgeons with improved support and opening up new options for intricate operations requiring extraordinary precision. In the end, the combination of robotics and AI holds the potential to transform surgical procedures by providing safer, more effective, and incredibly accurate procedures (Naik et al., 2022).

7.7.5 Potential for global healthcare improvements and access

Cutting-edge surgical technologies have enormous potential to reduce healthcare inequalities worldwide. Expert surgical consultation and planning can be done remotely thanks to AI-powered telemedicine technologies, which offer specialized expertise to underdeveloped areas where there may not be enough qualified surgeons. These technologies, when combined with remote robotic surgery equipment, enable skilled surgeons to carry out intricate treatments remotely, removing geographic restrictions that restrict access to high-quality care. AI-driven surgical instruments and telecommunications infrastructure are becoming more and more affordable, making their use in rural and low-resource environments more and more possible. By granting patients around the world access to cutting-edge operations that were previously only available in major medical institutions, this democratization of advanced surgical capabilities has the potential to drastically reduce health inequities.

Additionally, AI-assisted monitoring and real-time data sharing during remote surgeries improve safety and accuracy, guaranteeing results that are on par with in-person procedures. Regardless of a patient's location, the widespread adoption of these technologies promises to enhance recovery rates, lower complications, and raise surgical standards globally. In conclusion, the combination of robots, telemedicine, and AI is revolutionizing healthcare delivery by increasing access to, affordability of, and equity in sophisticated surgical care globally. This development is an important step in reducing the disparity in surgical outcomes between underprivileged and affluent groups (Park and Lee, n.d.).

7.8 Conclusion

The application of AI has led to impressive developments in IGS, greatly improving surgical accuracy, productivity, and patient outcomes. Modern surgery is changing as a result of the integration of AI-driven technologies, including robotic-assisted systems, VR, AR, and ML. With the use of these instruments, surgeons may more clearly perceive intricate anatomical features, improve surgical planning, and perform treatments with previously unheard-of accuracy. Large volumes of patient data are analyzed by ML algorithms, which enhance intraoperative decision-making by anticipating possible issues and recommending customized strategies. With the use of immersive, real-time representations made possible by VR and AR technologies, surgeons may navigate fragile tissues with less risk and invasiveness. In the meantime, AI-enabled robotic-assisted platforms improve control and dexterity, enabling more accurate movements than those of human capabilities. Looking ahead, completely autonomous robotic technologies and sophisticated predictive analytics will likely shape IGS's future. DRL and ongoing data feedback will enable autonomous systems to carry out complex surgical procedures with less human involvement, lowering errors and increasing consistency. Large datasets will be used by predictive analytics to forecast patient-specific results and optimize customized surgical approaches. Furthermore, data-centric, AI-driven strategies will be essential to increasing access to high-tech surgical care worldwide, particularly in underprivileged areas. These developments will lower healthcare expenses while simultaneously improving patient safety by increasing procedural precision and decreasing complications. A new era of innovation in medical robotics and patient care will be ushered in as AI and IGS technologies continue to advance and work together to redefine surgical norms and workflows. Through safer, more effective, and highly customized procedures that benefit patients in a variety of healthcare settings, this integration promises to raise the bar for surgery globally.

References

Jackson M. Real-time. AI Assistance in Surgery 2023;14:221–38. Available from https://doi.org/10.1016/j.sairev.2023.0145.

Johnson T. Artificial intelligence in modern surgery. AI in Surgery Journal n.d; 10–24, https://doi.org/10.1007/s56789-022-01567.

Johnson P. Enhancing surgical outcomes with predictive analytics. AI in Healthcare Journal 2021;12(2):200–14. Available from https://doi.org/10.1109/AIHC.2021.202345.

Khan MM. The role of AI in neurosurgery. Clinical Case Reports and Clinical Study 2023;10(1):1–4. Available from https://doi.org/10.61148/2766-8614/jccrcs/158.

Kim J. Advanced imaging in postoperative monitoring. Postoperative Care Journal 2023;10:55–72. Available from https://doi.org/10.1089/pcj.2023.0010.

Kim J. AI Applications in Robotic-Assisted Surgeries 2023;12:55–72. Available from https://doi.org/10.1055/s-0043-1789456.

Kim S, Robinson J. Time reduction in surgeries through robotics and AI. Medical Robotics Journal 2020;15(3):185–99. Available from https://doi.org/10.1089/mrj.2020.0015.

Lanfranco AR, Castellanos AE, Desai JP, Meyers WC. Robotic surgery: a current perspective. Annals of Surgery 2004;239(1):14–21. Available from https://doi.org/10.1097/01.sla.0000103020.19595.7d.

Lee M. Surgical Technology International n.d. https://doi.org/10.3109/08921990.2021.1875432.

Lee K, Chen M. Machine learning in surgical risk assessment. International Journal of Surgery 2023;58(1):88–102. Available from https://doi.org/10.1016/j.ijs.2023.01.009.

Litjens G, Kooi T, Bejnordi BE, Setio AAA, Ciompi F, Ghafoorian M, et al. A survey on deep learning in medical image analysis. Medical Image Analysis 2017;42:60–88. Available from https://doi.org/10.1016/j.media.2017.07.005 Available from: http://www.elsevier.com/inca/publications/store/6/2/0/9/8/3/index.htt.

Malczewska A, Frampton AE, Mato Prado M, Ameri S, Dabrowska AF, Zagorac S, et al. Circulating MicroRNAs in Small-bowel Neuroendocrine Tumors. Annals of Surgery 2019;274:e1–9. Available from 10.1097/sla.0000000000003502.

Malczewska A, Frampton AE, Mato Prado M, Ameri S, Dabrowska AF, Zagorac S, et al. Circulating MicroRNAs in small-bowel neuroendocrine tumors: a potential tool for diagnosis and assessment of effectiveness of surgical resection. Annals of Surgery 2021;274(1) E1, Available from https://doi.org/10.1097/SLA.0000000000003502, http://journals.lww.com/annalsofsurgery/pages/default.aspx.

Marescaux J, Soler L. Image-guided robotic surgery. Surgical Innovation 2004;11(2):113–22. Available from https://doi.org/10.1177/107155170401100208.

Martinez C. The Future of Robotic-Assisted Surgery 2022;7:189–204. Available from https://doi.org/10.1007/s00345-022-01890-4.

Naik N, Hameed BMZ, Shetty DK, Swain D, Shah M, Paul R, et al. Legal and ethical consideration in artificial intelligence in healthcare: who takes responsibility? Frontiers in Surgery 2022;9. Available from https://doi.org/10.3389/fsurg.2022.862322.

Park SH, Lee, JM. n.d. Data privacy in healthcare: in the era of artificial intelligence. Healthcare Informatics Research 77–84, https://doi.org/10.4258/hir.2023.29.2.77.

Patel R. n.d. Advancements in Computer Vision for Surgical AI. https://doi.org/10.1016/j.cvsa.2022.110987.

Patel S. n.d. Journal of Surgical Innovation Navigation Systems for Minimally Invasive Procedures. 75–89, https://doi.org/10.1093/jsi/122032.

Patel R, Singh T. Real-time data in operating rooms: a game changer? Surgical Innovations 2020;33(4):312–25. Available from https://doi.org/10.1177/1553350620904321.

Rajpurkar P, Irvin J, Ball RL, Zhu K, Yang B, Mehta H, et al. Deep learning for chest radiograph diagnosis: a retrospective comparison of the CheXNeXt algorithm to practicing radiologists. PLoS Medicine 2018;15(11) Available from https://doi.org/10.1371/journal.pmed.1002686, http://medicine.plosjournals.org/perlserv/?request=index-html&issn=1549-1676.

Roberts D. Applications of IGS in various specialties. World Journal of Surgery 2022;46:983–97. Available from https://doi.org/10.1007/s00268-021-06345-8.

Sanders. The rise of AI-powered autonomous surgical systems 2023;5:33–48. Available from https://doi.org/10.1089/auts.2023.0045.

Santiago I, Barata M-J, Figueiredo N, Parés O, Matos C. Early conformational changes at tumour bed and long term response after neoadjuvant therapy in locally-advanced rectal cancer. European Journal of Radiology 2021;140:109742. Available from https://doi.org/10.1016/j.ejrad.2021.109742.

Smith J, Brown L. Big data in surgery: a new frontier. Journal of Medical Informatics 2022;45(3):145–59. Available from https://doi.org/10.1016/j.jmi.2022.03.004.

Wong JE, Smith, AL, Gupta, RK. n.d. Artificial intelligence in orthopaedic surgery. Journal of Orthopaedic Surgery and Research 1–9. https://doi.org/10.1186/s13018-023-03456.

CHAPTER 8

Neurosurgery with artificial intelligence-assisted robotics

Manjeet Rege[1] and Dharani Kumar[2]
[1]Department of Software Engineering and Data Science, University of St. Thomas, St. Paul, MN, United States,
[2]School of Business, Woxsen University, Hyderabad, Telangana, India

8.1 Introduction

8.1.1 Historical context of neurosurgery

Neurosurgery has been an extreme field in medicine since ancient times, when the surgeons tried to go to the end of what could be done surgically in the human body. Even humans have been trying to correct their brain problems as far back as the discovery of early ancient skulls, which have shown remnants of trepanation, a crude form of surgery on the head, in the treatment of headaches, psychopathological conditions, or spiritual misalignment. These were the primitive attempts done without any anesthesia or antisepsis, and success was mainly attributed to luck and basic knowledge of brain mechanics. The 19th and 20th centuries represented a break in the field. Modern neurosurgery was born at a time when anesthesia was discovered, when antiseptic measures were discovered, and when the anatomy of the nervous system was developed. The invention of technologies such as the operating microscope, stereotactic frames, and first-generation CT and MRI scanners changed the possibilities of localizing and reaching the lesion precisely. Neurosurgeons would now be able to access distinct brain structures with greater precision while minimizing collateral damage to critical cortical and brainstem regions (Char et al., 2018).

8.1.2 Contemporary challenges in neurosurgical practice

In spite of these developments, neurosurgery is full of problems. These actions usually involve considerable dexterity on a manual level, awareness of situations, and immediate decision-making. The high cost of learning, as well as the vast physical and cognitive requirements of the surgeon, is capable of affecting a procedure. During the surgical activity, the diagnosis is still based on preoperative static images and surgical expertise, which is not enough to handle complex dynamic alterations in anatomical and physiological states. Moreover, there are inequalities that exist in the world in terms of exposure to well-trained neurosurgeons and the advanced surgical infrastructures, and this results in the untreating and undertreating of patients.

AI-Powered Developments in Medical Robotics
DOI: https://doi.org/10.1016/B978-0-443-27783-2.00015-6

8.1.3 Technological paradigm shift: the role of robotics

Robotic surgery came as an option to overcome the restrictions of human hands and stamina. Robotic systems bring more stability, accuracy, and scalability to the movement. There are even higher stakes in the case of neurosurgery because the mind and the spinal cord are unforgiving, and any deficiency of even a millimeter will result in irretrievable deficiencies. Submillimeter accuracy is obtainable with robotic platforms and eliminates tremor or enhances tool trajectory and also opens areas previously inaccessible. Artificial intelligence (AI) extends such robot systems to smartly automate the operating room. Surgical planning, intraoperative strategies, anomaly detection in real-time, and the analysis of massive quantities of data are possible due to algorithms. The joint efforts of AI and robotics are the basis of a new era of surgical interventions, which become personalized, based on data, and extremely accurate.

8.1.4 Why artificial intelligence is integral to modern neurosurgery

This is because AI is in a unique position to help with most of the information-related loads that neurosurgeons have to deal with. It is already either tested in clinical trials or applied to pilot projects around the world to interpret radiologic images automatically, segment tumors, and predict the intraoperative complications. Machine learning models can give evidence-based advice or issue high-risk trajectory alerts based on thousands of neurosurgical procedures. Moreover, surgery with the help of reinforcement learning systems is capable of fitting the surgery maneuvers according to the real-time feedback and learning effective strategies in multivariable situations. In addition to the operating room, AI is used in surgical planning and selection of the patient, as well as postoperative recovery processes. The advent of prediction models is assisting neurosurgeons in recognizing the best methods to use with each individual patient, decreasing the paradigm of complications and maximizing outcomes. Robotic systems can incorporate these tools to constitute a feedback loop of planning, taking action, and adjusting.

8.1.5 The convergence of robotics and artificial intelligence: a new surgical ecosystem

The synthesis of AI and robotic systems is especially effective in the work of the neurosurgeon. Robotic arms perform elaborate movements clumsily and inaccurately, but under the control of AI, which analyzes and interprets real-time conditions in order to direct and modify their movements. In other words, AI can process the intraoperative imaging of the tumor resection and make modifications so that the surgeon is able to avoid damage to any vital structures, whilst the robot will do all the modifications that were just analyzed by the AI with minimal delay. This allows context-sensitive and dynamic interventions that respond better to the anatomy of the individual patient and interpatient variability during surgery. The coming up of such technologies is leading to the progression of mixed-surgery environments, with human surgeons, robotic systems, and AIs collaborating. Such ecosystems are built on the strengths of each of the elements, including the creativity and experience of the surgeon, the mechanical stability of the robots, and the computing abilities of the AI. The combination of the two will facilitate the promotion of safer, more effective, and more accessible neurosurgical care.

8.1.6 Structure of this chapter

The chapter critically examines the principles and concepts of AI-enhanced robotics in neurosurgery on a systematic basis. It offers a historical landscape, an overview of current technologies, and conducts an appraisal of clinical applications in major neurosurgical procedures. The discussion also consists of an analysis of issues, such as data bias, regulatory barriers, and training issues. It also addresses future perspectives, such as autonomous neurosurgical systems, tele-robotics, and the application of virtual reality and augmented reality (AR) in operating rooms. The focus is to provide an in-depth insight into the ways in which AI and robotics are transforming neurosurgery now and predict a future when such devices become part of the regular practice in various medical facilities. It is against this background that this paper hereby takes a concentrated examination of the role of robotics in neurosurgery.

8.2 Role of robotics in neurosurgery

In neurosurgery, a robot provides an outlook on one of the most important advances to achieve submillimeter accuracy, intraoperative safety, and the robustness of surgical outcomes. Since the field of neurosurgical intervention is getting increasingly complicated, the robotic systems should not be seen as a replacement for the surgeon but as a powerful tool that enables humans to do the impossible. It talks about the operative possibilities, technical capabilities, and how robotics has revolutionized the sphere of neurosurgery in many other aspects.

8.2.1 Evolution of robotic neurosurgery

The first use of surgical robotics in neurosurgery would be traced back to the 1980s, when the crude mechanical arms were incorporated in the stereotactic application of biopsy needles. Today, robotic systems have been developed since then and have gone beyond being mere passive mechanical assistants to semiactive and actively controlled platforms able to possess integrated imaging and navigation skills. The switch toward

AI-enhanced systems has raised autonomy, accuracy, and responsiveness even more. Current neurosurgery robotic platforms are developed to become a part of imaging tools, navigation options, and data-rich interfaces so that instrument insertion can be enabled through precise placement in previously unknown locations. All these systems not only minimize the fatigue of the surgeons and trembling of their hands but also allow new manners of surgery where it was considered inoperable.

8.2.2 Submillimeter accuracy and stability

Maintaining a trajectory of a tool with submillimeter accuracy is one of the signature benefits of robotics in neurosurgery. This is especially vital in hard-seated lesions of eloquent parts of the brain, where there is a risk of having disastrous results even with a small diversion. The robotic arm presents reliable, repeatable movements that do not depend on tremor, fatigue, or nonuniform manual manipulation. Together with intraoperative real-time imaging, such systems improve the reliability of targeting procedures that include deep brain stimulation (DBS), stereotactic biopsies, and laser interstitial thermal therapy.

8.2.3 Instrument stabilization and dexterity

Complex and lengthy procedures are stabilized through the use of advanced robot platforms on the instrumentation. In contrast to the meager methods where minor variations could be experienced, the robotic systems offer consistency of tool orientation as well as depth. Multidegree-of-freedom joints and end-effectors are added to some systems, enabling access through narrow surgical corridors that are characteristically minimally invasive, for example, skull base or intraventricular surgery. Motion scaling can also be applied on the robotic interface such that coarse movement of the hand can be scaled to a finely resolved microsurgical movement.

8.2.4 Enhanced visualization and imaging integration

New-age robotic systems are constructed to work easily with sophisticated intraoperative imaging modalities, including MRI, CT, and fluorescence microscopy. These systems enable the surgeons to see the anatomical structures continuously and make modifications to their techniques according to real-time developments. Such novel technologies as AR overlay, which projects 3D models of neuroanatomy onto the operating scene right into the operating field, enhance depth perception and spatial awareness. Stable platforms supporting endoscopes and cameras. The reliable platforms provided by robotic arms succeed in improving surgical viewing in narrow or deep fields. The steady viewpoint and possible fine-grained positioning of these robotic holders enhance the safety and efficiency of the procedure.

8.2.5 Applications in complex procedures

Robotics has been an essential tool in taking risks in procedures that require large amounts of precision and repeatability. Robotic pedicle screw insertion in spinal neurosurgery has shown improvement of complications (malpositioning and neurological damage) to the extent that malpositioning and neurological damage are almost nonexistent. In the case of skull base tumors and vascular malformations, the robotic systems allow the use of a method that involves little brain retraction and exposure, minimizing postoperative morbidity. Further, minimally invasive methods like ablation of a tumor using a needle, evacuation of an intracerebral hemorrhage, and biopsy of deep lesions have been possible due to the robotic-assisted stereotactic devices. Such processes, which invariably tend to be risky, can be conducted today with reduced chances of complications following the improved control and safety allowances ensured by the robotic platforms.

8.2.6 Human–robot collaboration

Against the assumption that robotics is already interested in substituting human surgeons, the new generation of robot systems is developed to cooperate with them. The robotic arm is only a mechanical assistant to the surgeon, who keeps control over the operative decisions, and the work of the robotic arm is limited to such operations as alignment of the trajectory, dampening of the movements, and conduction of tools. This cooperation model enables better decisions to be made as the intuition of a surgeon is coupled with the consistency of the machine. There is also hands-on cooperative control available, where the surgeon controls the robotic arm in real-time, and the system adds resistance or correction through trajectories. This improves accuracy and also assists in an educational environment due to the added safety limits provided to the trainee surgeon.

8.2.7 Training, learning curves, and standardization

Despite the fact that the process of robotics involvement within neurosurgical practice implies a certain learning curve, such training programs as simulations and specific educational plans demonstrate the possibility to minimize the learning period. The robotic systems are frequently coupled with a virtual reality simulator and haptic feedback environment, by which residents and trainees can build fine manual dexterity and confidence in the procedure steps in a safe manner. In addition, robotics ensures the transfer of techniques among surgical teams and institutions as they become standardized. Robotic-assisted methods enhance repeatability and allow more objective monitoring of outcomes due to the coding of procedural steps and the reduction of operator variability. It is a vital standardization in multicenter trials and allows the creation of evidence-based surgery guidelines.

8.2.8 Cost considerations and economic impact

Although the initial expenses of robotic systems can be high, a number of studies indicate that the technologies can save money invested in some healthcare services by lowering complication rates, shortening hospitalizations, and avoiding revision surgeries. Robotically assisted high-volume centers can afford the payment of the initial investment in the long run due to the procedural efficiency and better outcomes they deliver. Besides, the portability of robotic platforms in various fields of surgery, that is, neurosurgery, orthopedics, and otolaryngology, additionally increases their value offering. The cross-specialty ROI is quite common in cases when institutions approach robotics as a strategic tool and are supported by the streamlined workflow and increased patient satisfaction.

8.2.9 Limitations and future potential

Although they hold great potential, existing robotic systems are also limited by the provision of haptic feedback in a few directions, the need for imaging, which has to be precise, and their complex installation. Further developments will probably be in mind with the applications engaging tactile sensing, autonomous correction, and easier user interfaces. Even further potential of robotics used in neurosurgery is related to the concepts of growing miniaturization, wireless control, and AI-powered flexibility of its application. Simultaneously, the democratization of access to robotic platforms-related efforts is also going across the world to create systems that are cost-effective and modular to be used in low-resource settings. Such activities will play a key role in addressing healthcare disparities and making sure that the advantages of the use of robotic neurosurgery are fairly distributed globally (Fig. 8.1).

8.3 Neurosurgery with artificial intelligence-assisted robotics

8.3.1 Artificial intelligence integration in neurosurgical robotics

The cognitive element of next-generation neurosurgical robotics would be AI, which would allow machines to not only perform precise operations but also the ability to analyze, adjust, and learn in real-time (Kashif et al., 2024). This part has elaborated further on the fundamental application of AI in neurosurgery due to its establishment in a preoperative plan, intraoperative implementation, and postoperative analysis (Mansour, 2025). Through analyzing important AI techniques, clinical application examples, and recent breakthroughs, we can give a unified idea of how AI enhances the functionality of robotic neurosurgical tools.

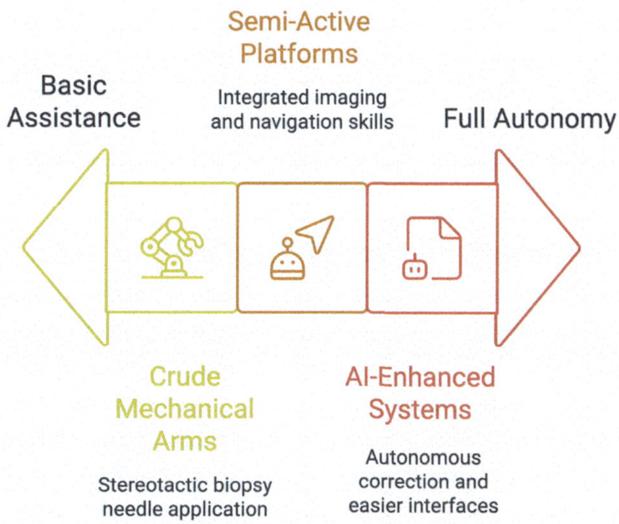

Figure 8.1 From basic assistance to full autonomy.

8.3.2 Fundamentals of artificial intelligence in surgery

Neurosurgery applications of AI may be divided more generally into supervised learning, unsupervised learning, reinforcement learning, and deep learning methods (Mansour, 2025). The training of these computational models consists of huge sets of data containing radiological images, electrophysiological signals, genomic profiles, and clinical outcomes (Kashif et al., 2024). They are able to learn a pattern, to predict, and can even provide indications on the surgical plans that traditional human clinicians might not see intuitively (Wah, 2025). Supervised learning is especially relevant in classification problems, for example, tumor boundaries, whereas unsupervised learning can infer hidden structure in unlabeled data, for example, discovering new subtypes of a disease (Kashif et al., 2024). Reinforcement learning, based on behavioral psychology, enables the surgical systems to be continuously improved through interaction with the environment, whereas deep learning has an advantage in tasks involving high-dimensional pattern recognition, for example, analysis of real-time video feeds in surgery (Wah, 2025).

8.3.3 Image-guided navigation and tumor segmentation

Image processing is one of the most influential applications of AI in neurosurgery (Awuah et al., 2024). It is possible to segment tumor margins, ventricular systems, and eloquent cortex automatically using convolutional neural networks trained with thousands of brain MRIs (Liu et al., 2024). All this serves to speed up surgical planning, lessen interobserver variability, and provide the required level of patient-specific customization (Mansour, 2025). Superior AI such as nnU-Net has hit benchmark segmentation of glioma, meningioma, and brain metastasis (Liu et al., 2024). These systems naturally fit into robotic systems to allow real-time updating of the anatomical maps in response to intraoperative variables, such as brain shift, to improve the accuracy and safety of navigation (Mansour, 2025).

8.3.4 Real-time intraoperative decision support

Neurosurgical operations are dynamic and require real-time adapting systems (Wah, 2025). With AI-based decision support machinery, such intraoperative decision-making data as stereotactic coordinates, brain shift measurements, and neurophysiological signals are analyzed, and their results provide real-time corrections of robotic trajectories (Liu et al., 2024). Considering the example of DBS procedures, there are patient-customizable electrode positioning choices that could utilize AI-generated optimization based on the character of the patient response and electrophysiological feedback (Liu et al., 2024). During tumor replications, reinforcement learning models can propose the most effective resectioning lines without damaging functional zones, and they could be adapting themselves on a constant basis, given the acquisition of new images and sensor data (Wah, 2025).

8.3.5 Predictive modeling and outcome forecasting

Surgical decision-making is being changed by predictive analytics through modeling of risks and outcomes (Liu et al., 2024). Based on the longitudinal training sets containing machine learning algorithms, AI systems can be used to forecast postoperative complications, including hematomas, infections, or even cognitive impairment (Schmidgall et al., 2025). These models have been able to take into consideration a wide range of variables, such as patient demographics, genetic markers, intraoperative measurements, and hospital resource integration (Liu et al., 2024). The findings will enable these forecasting tools to give the surgeons more opportune decision-making in the treatment processes by individualizing the treatment pathways and being proactive in addressing risks that can be identified (Schmidgall et al., 2025). These predictions are used to change workflows on robotic platforms, thereby selecting less invasive means in patients with a higher risk of wound infection or enhanced recovery protocols in the predicted delayed healing (Liu et al., 2024).

8.3.6 Artificial intelligence for robotic motion planning and control

The high-end robots need motion planning with intelligence to complete complicated trajectories within the thick anatomy and variable brain and spinal structure (Schmidgall et al., 2025). AI helps during robotic path planning by ideally calculating insertion depths, tool angles, and traveling vectors (Schmidgall et al., 2025). This is in the context of minimally invasive surgery, where the operating space is confined; therefore, AI is used so that robotic arms do not damage important neurovascular structures and the exact targets of operations are reached (Schmidgall et al., 2025). Also, AI control systems control feedback of robotic motion correction based on tissue resistance, tool deflection, or unexpected movement, with unprecedented surgical precision (Liu et al., 2024).

8.3.7 Artificial intelligence-enhanced visualization and augmented reality

The add-on to a surgical perception that computer vision models augment is that critical structures are overlaid in the live videos of the endoscopes and microscopes (Kashif et al., 2024). They are models in which real-time interpretations of both 2D and 3D visual data take place, which increases the identification of anatomical characteristics that can be concealed by bleeding or abnormal pathology (Mansour, 2025). The potential of AR supported by AI is the ability to combine preoperative planning data with the intraoperative view and have representations of intuitive spatial orientation of complex procedures (Mansour, 2025). According to the analysis data processed by AI, surgeons have the potential to visualize tumor margins, blood vessels, or functional networks in order to operate on the actual operative field (Kashif et al., 2024).

8.3.8 Integration with genomic and omics data

Neurosurgery requires integration at levels other than imaging and electrophysiology in brain surgery (Awuah et al., 2024). The presence of genomic, transcriptomic, and proteomic information has now been integrated into AI models to make surgical approaches personalized (Awuah et al., 2024). As an example, the molecular subtype of the tumor that can be deduced with the help of the AI-driven genomic analysis is now becoming an important factor in selecting tumor resection strategies (Wah, 2025). These multimodal AI systems not only construct complete patient profiles that inform not only the extent of tissue to remove but also which tissue regions to biopsy to send to molecular testing or which pathways to attack in pre or postalias treatment (Wah, 2025).

8.3.9 Autonomous surgical subsystems

Although complete autonomy of neurosurgery is a vision, a few semiautonomous systems will be connected (Schmidgall et al., 2025). These consist of automated drill tips during cranial access, robotic laser ablation instruments that vary according to the thermal feedback acquired, and automatic suture placement modules (Liu et al., 2024). These subsystems are controlled by AI-based on the real-time input to achieve safety and precision (Schmidgall et al., 2025). Their algorithms have fail-safes that signal the surgeon and remove automation when something unusual comes up, and human control remains without sacrificing speed (Liu et al., 2024).

8.3.10 Federated learning and privacy-preserving artificial intelligence

The lack of diversity in annotated data is one of the major drawbacks in training AI models in the field of neurosurgery because of privacy issues (Guan et al., 2024). Federated learning enables collaborative training of models without exchanging patient data among institutions (Lee et al., 2024). On the contrary, the transmission and aggregation of model updates, rather than raw data, provide data security and increase the capacity of learning (Guan et al., 2024). The methodology allows the formation of AI models with global robustness so that they generalize across demographics, types of imaging modalities, and procedural methods (Lee et al., 2024).

8.3.11 Expert predictions and current research

For neurosurgery, AI–robotic constructs are already being tested in such centers of excellence as MIT, Stanford, and the Fraunhofer Institute (Ghimire et al., 2025). Certain projects like the CSAIL intelligent surgical platform at MIT have even understood closed-loop systems in which AI not only guides but also performs surgical steps with human oversight (Ghimire et al., 2025). It would seem to be agreed among experts that, in as short as 10 years, subtasks such as cortical mapping, navigation, and resection, even in some cases, would be the common AI–robotic functioning (Wah, 2025). The market leaders are directed at explainable AI models, the enhancement of the sensor integration, and interoperable platforms to increase adoption and confidence (Wah, 2025). To conveniently summarize, the problem of neurosurgical robotics where AI is integrated is not a one-dimensional development but rather a multifaceted change encompassing perception in real-time, predictive intelligence, precision control, ethical data utilization, etc. (Kashif et al., 2024). It alters the way neurosurgeons interact with data, medical equipment, and, in general, they make decisions during surgery (Mansour, 2025).

8.4 Clinical applications

The advent of the use of AI-aided robotics within the neurosurgical practice field marked the beginning of a new space of precision, safety, and efficiency in the practice. These technologies are being used in many fields, such as functional neurosurgery, tumor resection, spine surgery, and vascular interventions. In both of these spheres, AI is involved in improving decision-making, whereas robotics takes care of gradual, micro precise performance. This section goes further about the most impactful clinical applications and talks about such ongoing research, pilot programs, and other expert opinions that predict their future uptake.

8.4.1 Deep brain stimulation

Functional neurosurgery, especially DBS, is one of the most innovative applications of AI-augmented robotics. DBS refers to the implantation of electrodes in deep structures of the brain, such as the subthalamic nucleus or globus pallidus, in order to treat Parkinson's disease, dystonia, and essential tremor. Conventionally, this surgery was associated with a lot of surgical expertise and was time-consuming. AI can be used to optimize this process using preoperative planning based on brain imaging and patient-oriented connectomic data. MRI and tractography data are analyzed by algorithms to allow the identification of optimal electrode trajectories that subject eloquent regions to minimal risk. In surgery, robotics can follow through on these plans to a precision of submillimeters, and AI incorporates real-time microelectrode recordings and patient feedback to optimize final placement. Reduced operative time and enhanced clinical outcomes have been reported in a few centers that made use of such systems, including pilot programs in Germany and Japan. In addition, the closed-loop DBS systems are currently being investigated. They have AI algorithms to monitor continuously the brain signal of patients and automatically adjust stimulation parameters to produce a smart neuromodulation platform that responds to the demands of the moment.

8.4.2 Brain tumor resection

AI and robotics have transformed the possibilities of neuro-oncology by investigating what tumor surgery can be and could be by improving safety and the completeness of tumor removal. The eventual effort should always be to have maximum resection without damaging the brain tissue around it. This is enabled by the AI in the form of image segmentation, volumetric analysis, and real-time tissue classification. As an example, if preoperative MRIs are available for thousands of patients, the boundaries of the tumor can be accurately determined using machine learning models. These boundaries are usually input into robotic guidance systems, which direct resection tools to ensure accurate and safe tumor removal. During the surgery, AI contributes to the decision-making process by updating the images in a short period, and combined with with fluorescence imaging or Raman spectroscopy, tissue differentiation becomes possible. The pilots in the United States and Europe revealed that AI–robotic systems can lower the necessity of revision surgeries and deficits that take place postoperatively. Stanford and University College London scientists are working on deep learning programs that can identify tumor tissue in real-time, possibly enabling surgeons to change their tactics during surgery.

8.4.3 Spinal neurosurgery

Robotic systems used in spine surgery help in their precision and stability, especially during pedicle screw insertion, correction of deformities, and fusion. The main complications of such procedures traditionally were connected to high variability in screw placement and, as a result, common nerve injury or implant failure. AI helps in planning the trajectories according to vertebral anatomy and CT scans carried out prior to the operation. These plans are carried out by robotic arms more consistently and with less radiation exposure compared to the manual methods or the use of fluoroscopy. New systems such as Mazor X and ExcelsiusGPS already include AI modules that make constant changes in positioning using intraoperative imaging. Mayo Clinic and Cleveland Clinic have shown increased accuracy, less blood loss, and decreased operation time using these platforms. Furthermore, AI can forecast the potential postoperative problem, for example, adjacent segment disease or hardware fracture, and intervene in the surgery plan.

8.4.4 Neurovascular interventions

Less commonly, neurovascular cases (aneurysm coiling and arteriovenous malformation [AVM] embolization) are Android invading AI–robotic systems. Such procedures involve the use of microcatheters in entering the tortuous vasculature toward the vulnerable targets. AI simulates the activity of blood circulation and vessel wall tension, which allows interventionalists to select the best actions. Robotics allows steady control of the catheter, and the rupture and embolic stroke are reduced. Emerging systems are on trial at a few centers in Canada and Singapore, and their results appear good.

8.4.5 Epilepsy surgery

In defining epileptogenic zones to be resected or ablated, stereoelectroencephalography is gaining increased assistance in AI and robotics. Machine learning examines patterns of EEGs, imaging, and genes of patients to propose points of focus. Then, they are possibly inserted with extreme precision by robotic arms into electrodes that cover various routes. The technologies minimize the possibility of hemorrhage and increase accuracy in the localization of seizures; as a result, the interventions are more efficient and less invasive. Predictive analytics also assist in predicting which patients are most likely to respond to surgery, thus enhancing patient selection procedures and obtaining consent from the patients.

8.4.6 Pediatric neurosurgery

Pediatric AI-augmented robotics is becoming more popular in complicated cranial reconstruction and tumor surgery. The possibility to create projections of cranial growth and predict the outcomes of surgeries, in particular, is of great use to developing brains. The precision of the robot minimizes the damage to sensitive tissues, and AI assists in optimizing resection objectives with functional preservation. AI-based cranial morphometry is being tested in institutions such as Boston Children's Hospital to maximize the results in disorders like craniosynostosis. These quantitative tools guarantee that surgical interventions will match with long-term developmental objectives.

8.4.7 Emerging areas and multidisciplinary collaborations

On top of the established processes, interdisciplinary projects are investigating the use of AI–robots in pain surgery, the installation of a brain–computer interface, and endoscopic pituitary surgery. As an example, motor-impaired patients use AI algorithms to place electrodes with the help of robotic arms when conducting BCI research. Innovation is moving faster due to collaborative programs between neurosurgeons, biomedical engineers, and data scientists. Massive initiatives to put robotic devices and advanced neural mapping technologies together are financed by the European Human Brain Project and the NIH BRAIN Initiative.

8.4.8 Summary

AI-guided robotics in clinical work has become a reality and is no longer a theoretical construct in clinical practice. They are transforming the process of planning, performing, and perfecting by the neurosurgeons. Better precision, safety of the patient, and customization are visible and seen within the functional, oncological, spinal, and vascular fronts. With these systems, their range will be vast, eventually redefining what neurosurgical care can do.

8.5 Benefits and outcomes

The introduction of robotics with AI support to the process of neurosurgery has already yielded quantifiable gains in the level of clinical practice, clinical outcomes, and patient safety. The steady merger of smart technologies with precision-therapeutic equipment has not only leveled up the technical skills of a surgeon but also introduced new standards of procedural uniformity and even postoperative rehabilitation. In this section, multidimensional advantages of these technologies are discussed on the basis of clinical trials, pilot projects, and the opinion of experts.

8.5.1 Precision and consistency

One of the greatest opportunities of robotic systems is that, potentially, such systems can perform the surgical plans with the accuracy of submillimeters in place of human beings, resulting in minimized risk of error on the part of human beings. Combined with AI-based planning applications, surgeons bring surgical outcomes within striking distance of consistency across operators and organizations. Analysis of robotic-assisted pedicle screw insertion, to give one example, demonstrates an accuracy rate of above 95%, compared to 8590% with manual methods. Such predictability is especially useful during high-risk studies, for example, brainstem tumor biopsies or DBS, where even an infinitesimal deviation of the optimal trajectory can leave the patient with permanent deficits.

8.5.2 Reduced operative time and fatigue

Preoperative planning is facilitated by the use of AI, where the plan makes use of several algorithms to automate imaging segmentation, trajectory design, and risk assessment. Robotic platforms can perform tedious or complex procedures during an operation, and thus the operative time is spent substantially lower, including drilling, bipolar electrode insertion, and tool orientation. Less exposure to anesthesia not only decreases the risk to the patients, but it also enables surgical teams to do more procedures in a certain time period. What is also important is that the ergonomic benefit offered by the robot minimizes fatigue of the surgeon, who is the main contributor to intraoperative errors, particularly in long or microsurgical procedures.

8.5.3 Enhanced patient safety and faster recovery

Less tissue abuse that results from AI–robotic systems causes minimally invasive procedures, which lowers blood volume in the operating room, the likelihood of infection, and postsurgery challenges. In spinal surgeries using robotic systems, percutaneous methodologies are used, keeping the musculature intact and reducing the size of the incision. This helps in reducing the length of hospitalization, faster functional recovery, and patient satisfaction. The pilot studies of North American and European centers point to a decrease in the average length of stay at the hospital of patients who have a neurosurgical operation performed by a robotic assistant by 25–40%.

Additionally, feedback loops and real-time monitoring generated with AI guarantee that any deviations from the surgical plan are identified and fixed in a timely manner. Such responsiveness increases intraoperative safety, especially when operating within a dynamic environment where any brain shift, swelling, or changes in the vasculature might undermine outcomes.

8.5.4 Improved decision-making and cognitive relief

AI systems are cognitive enhancers to the surgical team and handle and process the huge intra-operative data. Such systems eliminate noise, point out anomalies, recommend the best responses, and leave the surgeon time to make higher-order decisions. As another example, during tumor resections, AI can be used to amalgamate the histopathological prediction, volume data, and the patient-specific neurofunctional map to propose boundaries of resection to optimize both oncological and functional outcomes. This mental recording also lowers the mental load in the high-stakes procedures. AI capabilities that are essential to identify patterns across many simultaneous inputs can be used to supply the surgical team with valuable insights, which would otherwise be impossible to process manually in real-time: neurophysiological signals, stereotactic coordinates, and imaging data.

8.5.5 Evidence-based personalization

The AI models that have been trained using big data can state how an individual patient is likely to respond, and hence a tailor-made treatment can be prescribed. Such forecasts are used in surgical planning, in choosing and selecting devices, and in postsurgery care strategies regarding the genetic profile of a patient and his/her anatomical peculiarities and comorbidities. Examples include being used in epilepsy surgery, where AI can analyze aspects of the EEG, images, and clinical data to predict the postsurgery results of epilepsy and, therefore, assist in determining the most suitable cases who should undergo the surgery.

Such predictive insights, in the longer term, make possible proactive care pathways. When patients at increased risk of complications or slow recovery are identified, the care teams will be able to redistribute resources more effectively and introduce prevention strategies that will make the long-term neurological outcomes better.

8.5.6 Quality assurance and training

The documents of AI-based documentation on all stages of a robotic intervention enable a closer look at technical performance, the instruments used during surgery, and surgical flow. The sets of data are invaluable to quality assurance, audit of complications, and surgical benchmarking. This information can assist institutions in recognizing inefficiencies in the processes of work and setting milestones in training as well as in their ability to ensure the meeting of safety standards. Robotic platforms with AI simulators and performance feedback systems provide a well-organized and fast-learning environment for residents and fellows. These tools give the prospective neurosurgeons an opportunity to train their skills using objective assessment so that they can be competent at the time of operating on live patients.

8.5.7 Scalability and global impact

The upfront cost may be expensive to make the initial investment in robotic platforms, but the savings in the revision rates, fewer complications, and shorter hospital stays have long-term consequences resulting in saving many costs. Institutional value is also enhanced by the scalability of robotic systems in all specialties, including neurosurgery, orthopedics, and ENT.

In addition, the inclusion of AI enables their adjustment to the monitoring of events remotely, supervision of surgical procedures tele-surgically, and the possibility of eventual complete tele-robotic surgery. Such scalability makes AI-assisted robotics a potential method to close the gap in the neurosurgical care of the most underserved patients across the world, allowing the equivalent interventions in low-resource areas without access to highly skilled neurosurgeons.

8.5.8 Summary

The advantages of AI-assisted robotics in neurosurgery cut across the technical classification, operation, and patient-oriented care aspects. The systems boost the accuracy of procedures, minimize complications, accelerate recovery, and improve the efficiency and adequacy of surgical delivery. They are gaining increased usage, indicating a general orientation to use data to optimize neurosurgical care at a system level, and they create fresh standards of safety, individualization, and availability worldwide.

8.6 Challenges

8.6.1 Data quality and bias

AI model performance relies on the diversity, accuracy, and scope of training data. These datasets may be biased, and biases in these datasets may give rise to skewed model behavior that negatively affects clinical equity and safety (Char et al., 2018).

8.6.2 Ethical and legal considerations

The problems of liability, informed consent, and algorithmic transparency have not been addressed yet. Clinical accountability and medico-legal examination have serious problems with respect to the interpretability of black-box AI systems.

8.6.3 Implementation and training

High-performance AI–robotic projects would require investment in infrastructure, engaging new working methods together with interdisciplinary collaboration, and putting data science skills in neurosurgical training programs.

8.6.4 Regulatory frameworks

One of the reasons why clinical integration is not the case is because there are no harmonized global regulations in reference to AI in surgical robotics. Contentious issues are the criteria used by national and international health authorities to validate software, approve devices, and design clinical trials to bring innovation; these differences in criteria may cause delays and legal ambiguities.

8.6.5 Regional variations in adoption

The unequal availability of the technology's infrastructure, health funding, and clinical experience promotes uneven use of AI-aided robotics in neurosurgery. The faster implementation may work well in high-income countries, and resource-constrained ones may have a long road to travel because of the cost, access, and policies. Bias and Data QualityAIs actions depend on the width, correctness, and variety of the training sets. Due to underlying biases in such datasets, they may result in bias in the model behavior, negatively affecting clinical equity and safety (Char et al., 2018).

8.6.6 Ethical and legal considerations

Such matters as liability, the ability to make informed consent, and transparency of algorithms have not been resolved. Black-box AI systems thus give rise to vital issues of clinical accountability and medico-legal interrogation due to the lack of interpretability of their workflow and decision rules.

8.6.7 Implementation and training

Interdisciplinary cooperation and a combination of infrastructural investments are required to roll out AI–robotic systems successfully, as well as to integrate data science abilities in neurosurgical education.

8.7 Future directions

8.7.1 Autonomous closed-loop systems

Autonomous control systems may be included in future robotic platforms to adjust surgical parameters according to real-time physiologic and anatomical feedback to further eliminate the need for direct control.

8.7.2 Tele-neurosurgery

Tele-robotic systems, which become possible thanks to the availability of fast data transmission means, can allow access to high-level care to rural areas or remote locations by allowing expert neurosurgeons to work remotely.

8.7.3 Integrative precision neurosurgery

Combining genomics, pictures, and intraoperative information, AI-based systems are capable of developing a personalized surgery plan that is maximized in the short-term performance and long-term neurological performance.

8.7.4 Ultra-minimally invasive robotics

Advancements in nano and microrobotic tools have the potential to expand neurosurgical surgery to areas that are inaccessible until now and reduce the morbidity of the patient even further.

8.8 Conclusion

AI, as applied to the system of robotic surgery, is one of the paradigms that will change neurosurgical practice. These technologies are not add-ons but the fuel to transform brain and spine surgery to precision, personalization, and predictive optimization. Despite some technical and ethical issues, the trend of innovation leads to the fact that normal surgical practice will soon be redefined.

Key takeaways:

- On AI-assisted robots, neurosurgical operations involve more precision, consistency, and safety.
- Predictive analytics and real-time decision support facilitate personalized care of the patients.
- The adoption differs between countries because of infrastructural and regulatory differences.
- The current research and pilot programs indicate the trend toward semiautonomous and tele-neurosurgical solutions.

Actionable insights:

- **For Neurosurgeons:** Make investments in the interdisciplinary training that will cover AI literacy and proficiency in the field of robotic systems. Take part in clinical trials and play a role in outcome registries that prove new technologies.
- **For Healthcare Policymakers:** Ensure the creation of adaptive regulatory mechanisms that strike a balance between achieving innovation and the safety of

the patient. Provide funding where there is an underprovision of resources so as to ensure equity of access.

The next generation of neurosurgery will be synonymous with a seamless blend between human knowledge and machine intelligence toward leading a next generation of safe, smarter, and more effective patient care. AI in robotized surgical tools is a revolutionary paradigm in neurosurgical practices. Digital technologies are not add-ons but drivers that embark on a transformation of precision, personalization, and predictive optimization of brain and spine surgery. There are still some technical and ethical issues, but the trend of innovation indicates that new norms in the sphere of surgery are expected soon. The future of neurosurgery is facing the perfect combination of both human and machine intelligence to solve the challenge of safe, smart, and more efficient care of patients at large.

References

Char DS, Shah NH, Magnus D. Implementing machine learning in health care: addressing ethical challenges. New England Journal of Medicine 2018;378(11):981–3. Available from https://doi.org/10.1056/NEJMp1714229, http://www.nejm.org/medical-index.

Ghimire, S., Shrestha, S., Shrestha, D., & Awal, S. (2025). Robotics in neurosurgery: The past, present and the future. Egyptian Journal of Neurosurgery, 40, 96. https://doi.org/10.1186/s41984-025-00395-9.

Guan H, Yap P-T, Bozoki A, Liu M. Federated learning for medical image analysis: a survey. Pattern Recognition 2024;151:110424. Available from https://doi.org/10.1016/j.patcog.2024.110424.

Kashif M, Muthana A, Al-Qudah AM, Hoz SS. The influence of artificial intelligence on neurological surgery and patient outcome. Surgical Neurology International 2024;15 Available from https://doi.org/10.25259/SNI_321_2024, https://surgicalneurologyint.com/surgicalint-articles/the-influence-of-artificial-intelligence-on-neurological-surgery-and-patient-outcome/.

Liu Y, Wu X, Sang Y, Zhao C, Wang Y, Shi B, et al. Evolution of surgical robot systems enhanced by artificial intelligence: a review. Advanced Intelligent Systems 2024;6(5). Available from https://doi.org/10.1002/aisy.202300268.

Mansour MA. How AI is redefining patient care and surgical excellence. ScienceDirect 2025.

Schmidgall S, Opfermann JD, Kim JW, Krieger A. Will your next surgeon be a robot? Autonomy and AI in robotic surgery. Science Robotics 2025;10(104). Available from https://doi.org/10.1126/scirobotics.adt0187, https://www.science.org/doi/10.1126/scirobotics.adt0187.

Wah JNK. Journal of Robotic Surgery 2025.

CHAPTER 9

Cardiovascular robotics and artificial intelligence integration

Ravi Teja Kothuru
Shiley-Marcos School of Engineering, Applied Artificial Intelligence, University of San Diego, San Diego, CA, United States

9.1 Introduction

9.1.1 The rise of artificial intelligence in medical robotics

The advances made in surgical treatment with advancements in artificial intelligence (AI) and robot-assisted surgery represent a new era of innovation and efficiency in the field of surgery. The development and refinement of robotic surgical systems have advanced rapidly in the last 10 years, allowing a higher level of complex surgeries to be performed successfully and with a greater degree of efficiency in terms of the time taken, the range of motion of the instruments, and the accuracy of the surgery performed (Apare et al., 2024). The relative importance of these two factors—flexibility and accuracy—is particularly relevant for complex and potentially life-threatening surgery, where nursing staff and other healthcare providers can greatly enhance the successful management of patients through appropriate levels of surgical skill, dependability of robotics, and adaptability of robotic surgical technologies to meet the needs of patients on an individual basis. The growth of AI-based solutions for robotic surgery is driven primarily by a desire for more efficient and effective implementation of robotic surgical processes, improved quality of surgical results through increased efficiency of the surgical process, and the need for increased levels of patient-specific treatment. AI systems allow robotic surgical systems to analyze a tremendous amount of heterogeneous data from a variety of sources, detect patterns, simulate the results of many different surgical parameters, and provide dynamic decision support during the surgical procedure (Knudsen et al., 2024). In this way, we are starting to see the beginnings of hybrid intelligent systems that combine machine learning with robotic precision, as described in new surgical platforms reviewed by Picozzi et al. (2024). These hybrid systems are not only improving the accuracy of procedures but also lowering complications and recovery, thus changing the landscape of modern surgical practice. AI works with robotic systems to enhance their predictive capabilities and adaptive control functions by enabling robotic systems to predict and anticipate the likelihood of surgical complications developing and automatically adjust their adherence to standard instrument movement rules to change the procedural course to avoid these complications.

AI-Powered Developments in Medical Robotics
DOI: https://doi.org/10.1016/B978-0-443-27783-2.00003-X

Additionally, through ongoing education based upon previous surgical results, AI integration will create a dynamic resource of learning and knowledge that can better prepare future surgical practices. In this respect, hybrid intelligent systems—which utilize both machine learning and robotic technology with precision—will establish a new benchmark for surgical care, resulting in better patient outcomes and increased risk reduction during the procedure (Picozzi et al., 2024).

9.1.2 Cardiovascular surgery: a high-stakes domain for robotics

Cardiovascular surgery is marked by its complexity compared to the other surgical specialties, as shown in Fig. 9.1 (Rajput, 2024), and is often complicated by the complicated anatomy of the heart, the need to maintain hemodynamic stability, and the high litigation risk involved with a surgical mistake; thus it is a prime candidate for advanced robotic technology (Chitwood, 2022; Saeed et al., 2023). With the heart continuously moving, surgeons require extreme dexterity when performing surgical interventions on the soft, delicate tissues of the heart, which can make it difficult for even highly skilled surgeons to perform manual surgical interventions. Therefore new techniques like automated transcatheter tricuspid valve replacement (TTVR) with hybrid-assisted intelligence are beginning to show the potential to transform surgical performance. Improved systems are producing improved precision, more consistent workflows, shorter procedure times, and possibly fewer complications for patients.

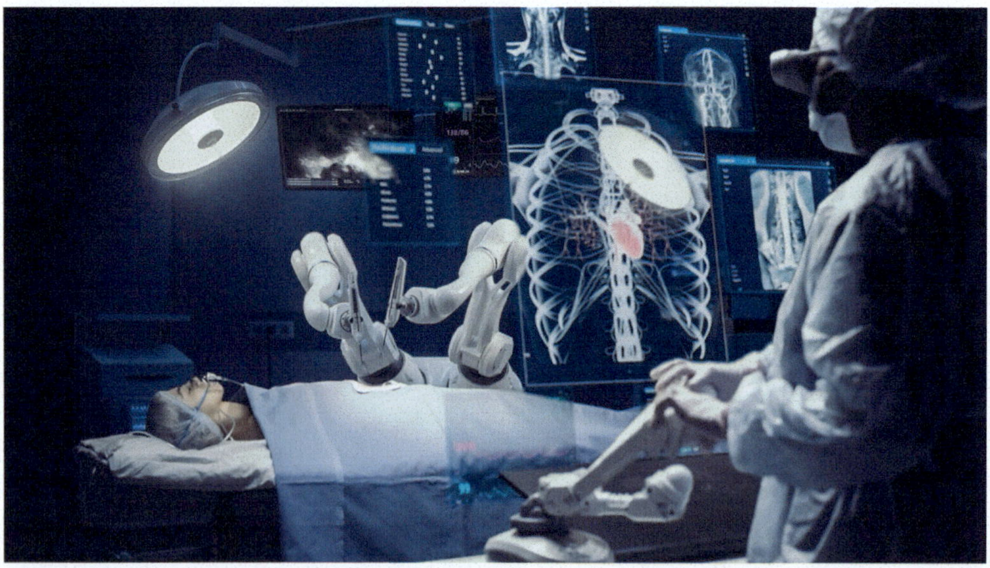

Figure 9.1 An example image of cardiovascular surgery with robotics (Rajput, 2024). This image is an example of cardiovascular surgery utilizing robotics technology.

The use of AI and robotics in cardiovascular surgery enables us to provide safer, more effective, and more personalized care for cardiac patients (Wang et al., 2024).

9.1.3 Objectives and scope of the chapter

The objective of this chapter is to review the use of AI in the field of cardiovascular robotic surgery, including its current applications, opportunities, and directions for future development. Areas of interest in this chapter are autonomous robotic systems, localized sensor-based technology linked to the Internet of Things (IoT), and deliberation around the ethics of implementing these technologies in clinical practice. The chapter will emphasize the need for multidisciplinary teams and standard protocols for clinical applications, with reference to recent literature and clinical inquiry (Arjomandi Rad et al., 2025; Knudsen et al., 2024). The overriding purpose is to explore, in-depth, a comprehensive understanding of the value of AI-driven robotic systems in innovating the practice of cardiovascular surgery—improved procedural safety and efficiency and intelligent connected operating ecosystems (Li et al., 2024; Saeed et al., 2023). The chapter will provide context on the technological enablers and translational hurdles to support a roadmap for stakeholders seeking to implement, regulate, and enhance their AI-powered medical device solutions for cardiovascular care.

9.2 Evolution of cardiovascular robotics systems

9.2.1 Milestones and technological progress

- **1960–70s.** Cardiopulmonary bypass allowed open-heart surgery and valve replacement (Bouchard et al., 2019).
- **Late 1990s.** Robotic telemanipulation provided the accuracy of surgical manipulation through minimally invasive surgery. Early systems such as AESOP, then ZEUS and da Vinci, provided the ability to repair a mitral valve robotically (Chitwood, 2022).
- **2000s.** Robotic telemanipulation was applied to coronary artery bypass grafting (CABG), septal defect repair, and tumor removal. With better 3D views and wristed instruments with camera control, the outcomes improved: less trauma, better blood handling, and shorter stays (Bouchard et al., 2019).
- **2010–20s.** AI, haptics, and imaging facilitated the shift to employing semiautonomous systems to provide intraoperative support.
- **Current and future scenario.** With AI systems coming online, robotic systems are beginning to merge precision, data, and autonomy, where IoT will also be integrated, as shown in Fig. 9.2.

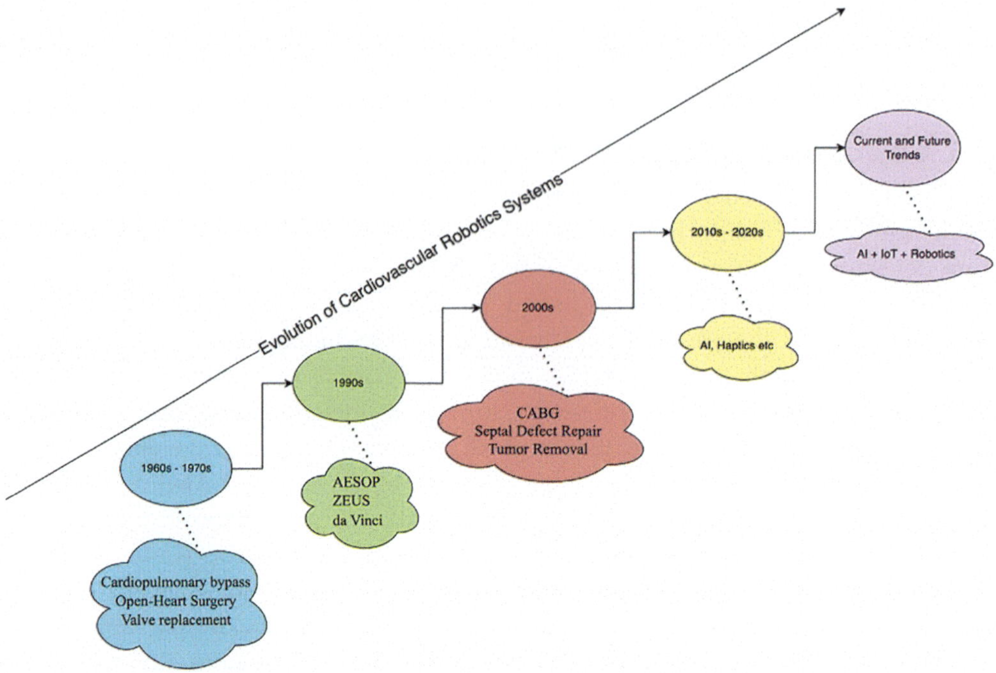

Figure 9.2 Evolution of cardiovascular robotic systems.

9.2.2 Key platforms: da Vinci, CorPath, and beyond

Fig. 9.3 shows the da Vinci Surgical System (Intuitive Surgical), which is the benchmark for robotic-assisted cardiac surgeries (Gumbs et al., 2020). The features of multiarm endoscopy, 3D vision, and tremor cancellation have opened complex procedures such as mitral valve repairs, intracardiac tumor resections, and coronary bypass grafting (Chitwood, 2022). The da Vinci System also helped standardize robotic approaches not only across cardiac centers but also across peripheral centers.

After this, the CorPath GRX (Corindus/Siemens Healthineers) allowed for robotic percutaneous coronary intervention (PCI) and peripheral vascular interventions to be done through a cockpit that is shielded from radiation (Picozzi et al., 2024). The modern systems now offer real-time AI-assisted imaging and further robotic control for more complex tasks such as robotic transcatheter tricuspid valve replacement (TTVR) (Wang et al., 2024). Development of these new platforms offers haptic feedback, performance monitoring, and automation that enhances variable elements, traditionally killed by fixed-function tools, into responsive, sensory-enhanced robotic systems (Knudsen et al., 2024; Picozzi et al., 2024).

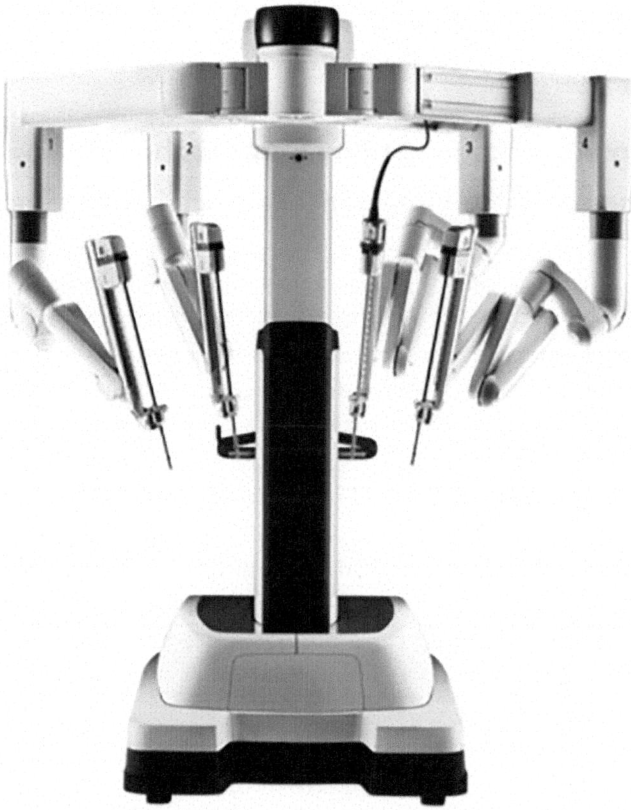

Figure 9.3 da Vinci Surgical System (Gumbs et al., 2020).

9.2.3 Unique clinical and technical challenges

- Cardiac anatomy and motion provide specific challenges for robotic systems, where even small errors can be catastrophic.
- The absence of real-time haptic feedback detracts from surgeon confidence, notwithstanding ongoing investigations in force-sensing technologies (Knudsen et al., 2024).
- Steep learning curves necessitate interdisciplinary collaboration, namely amongst cardiology, radiology, and engineering teams.
- High system costs and the re-education of workflow make adoption of cardiovascular robotics especially limiting for hospitals with fewer resources (Bouchard et al., 2019).
- AI integration introduces additional regulatory challenges related to transparency, reliability, veracity, quality, and liability for adaptive intraoperative systems (Mohammadi et al., 2024).

Together, these barriers limit the more widespread implementation of cardiovascular robotic platforms and their potential advantages.

9.3 Artificial intelligence in cardiovascular robotics

9.3.1 From automation to intelligence: artificial intelligence's expanding role

The field of robotics applied to cardiovascular surgery has progressed beyond the use of mechanized implements, with the introduction of increased avenues for autonomous, AI-supported operating procedures and methods. In particular, the use of robotic-assisted laparoscopy has evolved from the initial use of robotic platforms, for example, the Da Vinci Surgical robot, to provide surgeons with a precision tool with limited independence (Chitwood, 2022). Today, surgery can be performed completely autonomously, thanks to machine learning technology, computer vision advancements, and sophisticated sensor feedback systems that provide nearly real-time adaptive control for many surgical procedures. AI is now functioning as an operating partner, helping to continuously assess data collected from the operating room, augment surgical imaging capabilities, direct instrument movements with precision, and alert the operating team in real-time about potential complications (Knudsen et al., 2024).

In addition to moving beyond strictly programmed tasks, the combination of predictive intelligence with the operating environment anticipates challenging anatomical situations, optimizes surgical trajectories, and reduces the likelihood of surgical errors. Hybrid systems such as robotic-assisted TTVR procedures using a robotic arm serve as examples of this emerging trend, which combines the expertise of surgeons with that of AI to ultimately enhance procedural outcomes, reduce procedural risk, and ultimately increase surgical efficiency (Wang et al., 2024). The development of intelligent AI-based robotic systems enables shifting robotic surgical systems from passive tools to intelligent operating companions through their ability to engage in adaptive decision-making during complex cardiovascular surgical procedures.

9.3.2 Artificial intelligence in surgical planning, navigation, and real-time decision support

The way AI is revolutionizing cardiovascular operations using personalized 3D models prepared before surgery, overcoming both standard 2D imaging and subjective evaluation of images by doctors, has become the industry standard. By employing both deep learning and physics-based models for the simulation of blood flow within the body, present within 3D models developed prior to surgery and utilized during aortic stenosis (AS) procedures (Ozturk et al., 2024). By capturing minute differences in cardiac anatomy and predicting how blood would move through the various arteries and veins in the body, this technology has enabled more precise approaches to performing surgical intervention. The 3D model that has been generated can then be used by a surgeon during surgery in conjunction with the reconstructed preoperative images and the real-time sensory input provided by robotic devices used in the operating room (video, force, and position), to combine them into one image to allow the surgeon accurate navigation during the procedure. This significantly decreases the workload required to navigate during surgery and reduces the chance of error due to the reliance on real-time visual evaluations of unusual conditions during complex cardiac procedures. Additionally, predictive algorithms of AI can identify potential issues for the surgeon before they occur, thus giving prior notice before the issue occurs and having the potential of providing valuable information to the surgeon during surgery, thereby improving patient safety and surgical results (Knudsen et al., 2024).

9.3.3 Skill assessment and feedback systems

AI is revolutionizing how we train and evaluate surgical proficiency through two main mechanisms: First, AI-driven feedback platforms provide new forms of objective, precise, and repeatable assessment that replace or augment traditional means (observation by a "subject matter expert") when evaluating surgeon competency by measuring robotic kinematics, tool trajectories, applied force, and surgical efficiency (Knudsen et al., 2024). In addition, collected data through AI systems can be utilized to enhance residency programs, inform curriculum development, and establish standardized benchmarks for surgical competence, thereby enabling continual skill advancement over time. Second, AI-based systems produce a real-time learning environment by offering immediate clues, predicting potential errors, and presenting corrective action step-by-step. AI systems also provide "closed-loop" feedback whereby they can respond to measured surgeon competency by automatically changing the amount of robotic assistance offered. Moreover, predictive AI models utilized in complex congenital heart surgery are capable of predicting patient outcomes, developing appropriate procedural plans, and gathering the historical performance of both the individual surgeon and surgical team in order to maximize both patient safety and enhanced surgical team efficiency (Mohammadi et al., 2024).

9.4 Artificial intelligence-enhanced robotic interventions: case highlights

Following the foundational progress that has taken place in cardiovascular robotics, there have also been new advancements with hybrid intelligence systems, which improve the accuracy and adaptability of procedures such as valve replacements.

9.4.1 Hybrid intelligence in tricuspid valve replacement

To better appreciate hybrid intelligence, it is first important to understand the basic robotic procedure for a TTVR.

9.4.1.1 Robotic procedure overview

Robotic cardiac surgery has come a long way from its early days of mitigating mitral valve repair to more complex interventions, like interventions to the tricuspid valve (Chitwood, 2022). Known as the "forgotten valve," the tricuspid valve was limited by its anatomic positioning and more functional complexity with regard to locations and procedures. The advantages of utilizing robotics, specifically in TTVR, provided a way to enhance dexterity, endoscopic visualization stability, and filtering of surgeon tremor to navigate these procedural barriers (Bouchard et al., 2019). Robotic arms give surgeons the ability to operate through reduced access points with better control, less trauma, and improved recovery.

9.4.1.2 Hybrid artificial intelligence: concept and implementation

A hybrid intelligence framework for TTVR surgery combines both human expertise and ML-based systems, as presented in the published literature (Wang et al., 2024), and integrates the use of robotic technology into this surgical procedure. The framework contains hybrid perception systems that utilize context-based trajectory planning, computer vision, and learning in real-time during the performance of the procedure. The AI algorithm can enhance the navigation accuracy, valve alignment, and precise deployment of implanted devices using virtual guidance and adaptive path planning through the synthesis of multimodal procedural information, including fluoroscopy, ultrasound, and robotic kinematic signals, among others (see Fig. 9.4).

One important aspect of this system is that it has been created so that humans and AI can work together by placing the human input into the decision-making cycle of the AI. The AI has been trained on clinical heuristics that were developed from expert surgical workflows, and the AI's default behavior is based upon these heuristics. The AI also receives direct feedback from the surgeon in real-time while they are performing critical steps of the procedure. The AI receives corrective actions from the surgeon after making mistakes, which is a process used in reinforcement learning, and the information provided to the AI allows it to learn over time how to improve its trajectory planning and control policies. The adaptability of the AI during an unpredictable anatomic variance or unpredictable procedural complexity is vastly improved, while at the same time, the development of augmented autonomy mitigates many of the risks associated with complete automation and permits the preservation of human authority over life-threatening decisions (Wang et al., 2024). In essence, this system provides a new framework for how AI plays a role in robotic procedures, allowing AI to not only assist

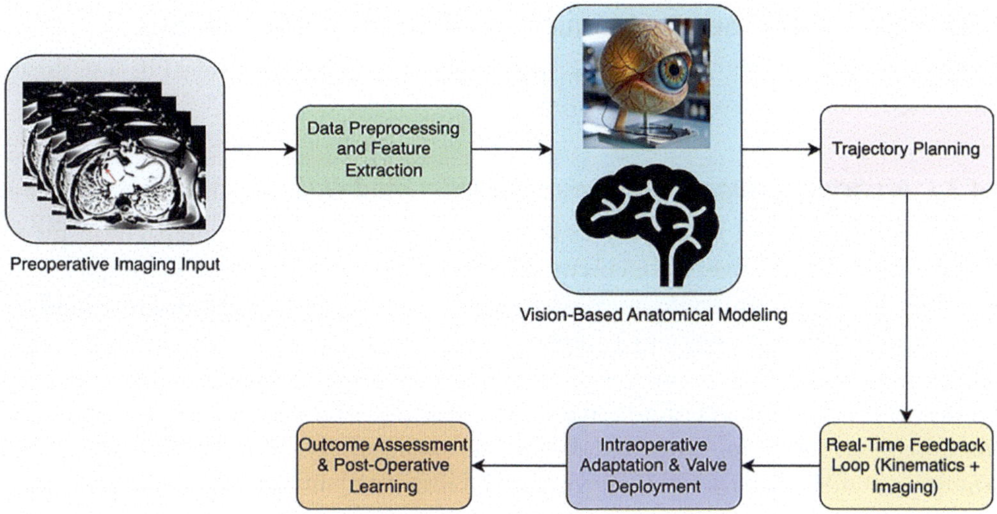

Figure 9.4 Hybrid AI workflow in robotic tricuspid valve replacement.

in the precision of the task but also to supplement surgical judgment and create a greater situational awareness, as well as a higher level of procedural safety.

9.4.1.3 First in vivo findings and future prospects

In the initial in vivo study, the feasibility and safety of the system were demonstrated in pigs, with accurate positioning and anchoring of the prosthetic tricuspid valve while the device was guided by limited amounts of human guidance (Wang et al., 2024). The robotic platform exhibited intelligent adaptability through real-time image feedback as it adjusted its tool path based on tissue resistance, anatomical variability, and subtle physiologic motion. These findings suggest that hybrid AI technology has the potential to enhance the reliability of procedures and diminish the risk associated with intraoperative surgical intervention by enabling robotic systems to incorporate surgical anatomical and functional ambiguities into their decision-making. Future research will explore how to further develop the hybrid intelligence model used in this study in humans, use multiple real-time data stream fusions (such as multiple sensor data) to enhance the adaptive learning of robots, and develop pathways for the certification and standardization of collaborative surgical robots. These developments will form the basis of establishing standardized protocols for AI-assisted procedures, create a framework for new approaches to surgical planning, and establish an entirely new operational paradigm based on the cooperative autonomy of AIs in the surgical environment through active navigation, prediction of decisions, and continuous improvement in surgical performance at the intraoperative level to optimize patient outcomes.

9.4.2 Personalized hemodynamic modeling for aortic stenosis

This section explores the AI methods for combining multimodal clinical data to create patient-specific hemodynamic simulations.

9.4.2.1 Artificial intelligence-driven multimodal modeling

AS is a progressive and potentially fatal illness that requires careful preparation for its treatment via aortic valve replacement surgery to reduce perioperative complications and improve long-term outcomes after valve surgery. The traditional methods of diagnosing AS have been using Doppler echocardiography and computerized tomographic imaging; both provide very useful information about anatomic structure, but neither takes into consideration dynamic changes that occur in both blood flow and function over time. In order to address the lack of dynamic information when diagnosing it, Ozturk et al. (2024) developed an integrated multimodal framework, with the ability to use AI, which incorporates CT scans, magnetic resonance angiography scans, and echocardiography scans, as well as machine learning representation models to simulate real-time hemodynamic performance of AS patients with the use of individualized models of AS and incorporates parameters such as pressure gradient, blood flow dynamics, and ventricular-arterial coupling for different clinical conditions. As a result, predictive results from these models are very different from the predictions made with general-purpose fluid dynamics simulations, as shown in Fig. 9.5.

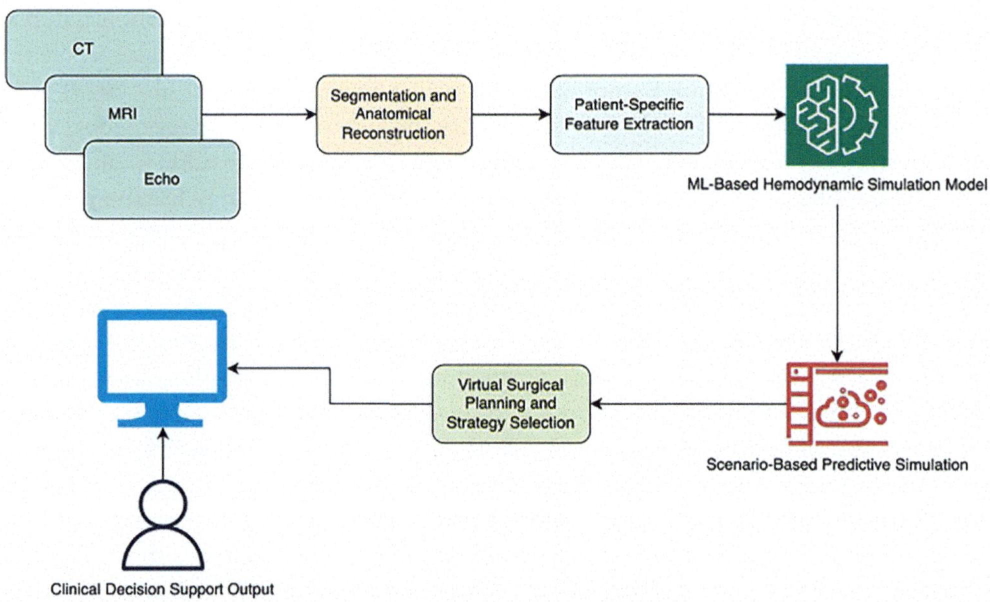

Figure 9.5 AI-assisted hemodynamic modeling for aortic stenosis.

As indicated by Ozturk et al. (2024), dynamic models enable surgical planning to occur iteratively by allowing for the potential for multiple surgical approach options to evaluate each individual patient's hemodynamic status prior to surgery to assist with determining the best surgical approach based on an individual patient's unique physiology. Additionally, the AI models can be updated sequentially, based on the addition of new data associated with the AS patient, facilitating real-time updating by the clinicians performing the AS surgical interventions.

9.4.2.2 Clinical utility in preoperative planning

Combining anatomical information with functional metrics to create predictive models enables a more developed virtual surgical planning tool. The surgeon can project the various outcomes based on different techniques and select the technique with the least amount of patient-specific risk, including the potential for left ventricular outflow obstruction, prosthesis-patient mismatch, or adverse hemodynamic effects (Ozturk et al., 2024). In addition to enabling preoperative planning, these computer-generated models allow for intraoperative support and provide real-time alerts when physiological perturbations occur, thereby decreasing the chance of preventable medical errors. AI-enhanced hemodynamic software also allows the user to conduct comparative analyses of several different types of prosthetic devices and approaches, therefore allowing for evidence-based personalized care. As patients receive a tailored, detailed roadmap as to how they will be taken through TAVR and other complex cardiac procedures, outcomes should significantly improve with the use of AI-enhanced hemodynamic modeling.

9.5 Emerging robotic platforms and sensor-driven intelligence

While existing systems, such as the Da Vinci robotic surgical system and CorPath GRX, have improved minimally invasive surgery, there have been moves toward the use of more flexible and modular robotic systems to develop environments with enough sensors to operate effectively in both dynamic/closed environments (Chitwood, 2022; Picozzi et al., 2024). The next generation of these robotic devices will not only replicate your actions with precise accuracy but also allow for increased efficiency by providing instant feedback on visual, tactile, and kinesthetic inputs. Additionally, by using advanced AI algorithms to facilitate predictive analytics, robotic systems will provide real-time support in responding to anatomical challenges while adjusting instrument trajectories accordingly.

Additionally, by giving surgeons the ability to personalize the system with modular parts and "plug-and-play" devices based upon the type of procedure (coronary bypass, valve repair, or transcatheter), they can adapt the system according to each patient's needs. The combination of this patient-centric customization with AI-enabled situational awareness gives surgeons more independence in the routine execution of surgical tasks while ensuring human authority during critical decisions via increased confidence in their ability to navigate the complexity of surgery safely and effectively. The combination of these features is expected to enhance the safety of surgery, decrease the length of time required for surgeries, and widen the number of facilities qualified to perform intricate vascular procedures by allowing those hospitals and clinics without specific extensive experience in cardiovascular surgery (Chitwood, 2022; Picozzi et al., 2024).

9.5.1 Next-gen platforms: adaptability and modularity

Emerging Platform's modularity becomes "an important element" for surgeons and surgical teams; it allows them to combine and modify tools and parts of different systems in real-time and provides surgeons and surgical teams with greater levels of control over how they construct and execute surgical workflows for their patients, which was previously unattainable by conventional robotic systems. Picozzi et al. (2024) suggested recent developments have produced customized robotic arms, modular tool sets, plug-and-play interfaces that allow for a fluid transition from open surgical procedures, through laparoscopy to endovascular surgery. Due to the nature of these modular platforms, the availability of skill sets formerly limited to specific high-level experts has been expanded and can now provide expanded access to complex or unique surgeries. Modification of instruments and components can be done quickly and easily, allowing the use of a variety of surgical approaches in addition to increased efficiency across the continuum of surgical care through the elimination of multiple robotic systems.

In addition, through a software-defined model, modern systems give surgeons the ability to configure motion scaling, tremor filtering, and force feedback to the specific tissue type, anatomical variability, and procedural requirements (Knudsen et al., 2024). Through a layered software architecture, these systems have the capability to receive updates continuously, utilize deep learning to adapt to new surgery techniques/techniques for patients over time, and eliminate problems associated with system drift. This combination of modularized hardware on platforms and intelligent software control provides safer, faster, and highly customizable treatment options for cardiovascular patients and, to some degree, establishes an option for future development of AI-guided autonomous functions.

9.5.2 Real-time feedback through sensor fusion

The use of multimodal sensor networks, such as force sensors, Inertial Measurement Units (IMUs), ultrasound, endoscopic imaging systems, and touch/feel sensing, enables the real-time fusion of sensor information for adaptive and context-based feedback. This sensor integration allows for a more complete understanding of highly dynamic, complex surgical settings via a simultaneous representation of the mechanical, visual, and physiological inputs within the surgical field. The simultaneous representation will provide a higher degree of accuracy when navigating the surgical environment (Picozzi et al., 2024). AI through the integration of diverse sensor inputs creates the ability for robotic systems to autonomously or semiautonomously react to various intraoperative events (hemorrhage, arrhythmia, tissue deformation, and unexpected anatomy). Sensor fusion continually enhances the robot's internal representation of the surgical environment, thereby enhancing the accuracy of surgical navigation, instrument calibration, spatial awareness, and predictive capabilities (Picozzi et al., 2024).

Future autonomous systems will use AI-driven inference to improve the closed-loop system by allowing for the identification and anticipation of future adverse events, thus allowing for proactive planning of corrective actions prior to an eventual deterioration in the patient's clinical status (Wang et al., 2024; Ozturk et al., 2024). These systems depend on accurate, reliable, and frequent communication between the surgeon and robotic device, hence the need for high-fidelity, continuous sensory data and intelligent feedback loops. Such synergistic interactions provide an enhanced decision-making capability, allow for real-time adjustments to interventions, and improve overall safety for patients. Iterative learning and continuous adaptation enable the closed-loop robotic systems to produce increasingly refined surgical guidance and assistance, and in this regard, they represent a significant progression toward greater intelligence, resilience, and reliability in cardiac robotic surgery, as illustrated in Fig. 9.6.

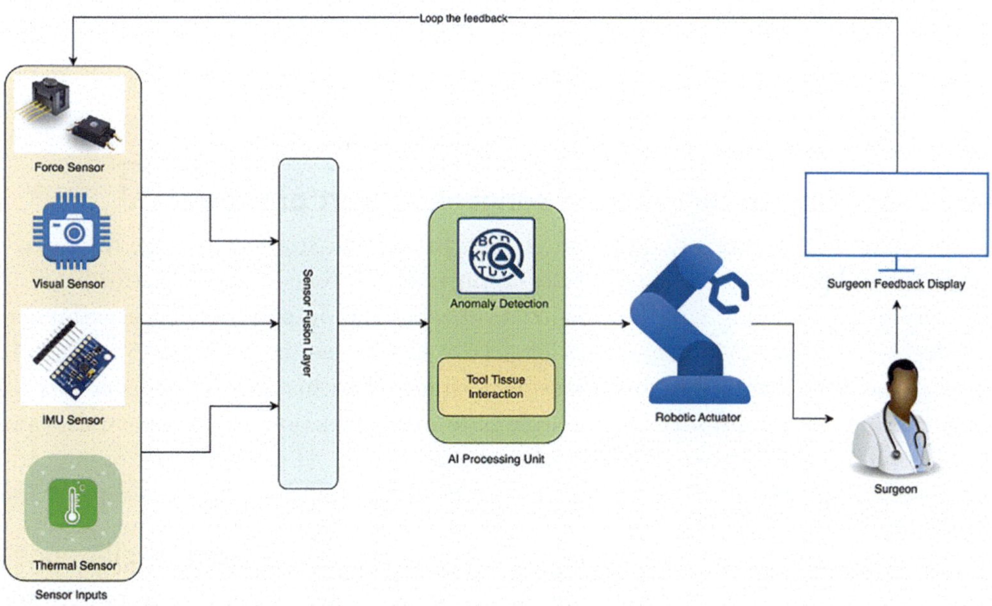

Figure 9.6 Sensor-driven AI feedback loop in modular robotic surgery.

9.6 Predictive analytics in cardiovascular surgery

The combination of AI and cardiovascular robotics includes intraoperative improvements, but its effect on poor-quality preoperative and postoperative decision-making is increasing. There are now predictive analytics models incorporated into surgical workflows to predict outcomes, identify high-risk patients, and assist in individualized interventions (Gadhachanda et al., 2025; Kumar et al., 2025; Khaled et al., 2025).

9.6.1 Artificial intelligence models for postoperative risk prediction

In cardiac surgery, particularly in pediatric and congenital cases, heterogeneity after surgery creates considerable technical challenges. AI-based models have shown considerable promise in predicting complications such as arrhythmias, bleeding, and mortality using preoperative clinical factors, intraoperative data, and postoperative data (Mohammadi et al., 2024; Yu and Son, 2024). They performed a systematic review of the effectiveness of AI models for predicting outcomes following congenital heart surgeries. Machine learning techniques such as decision trees, support vector machines, and neural networks outperformed traditional risk scoring systems in both accuracy and identification of at-risk patients earlier and consistently outperformed traditional risk stratification systems, such as the Society of Thoracic Surgeons score and EuroSCORE, demonstrating higher predictive accuracy, improved sensitivity for rare adverse events, and earlier identification of high-risk patients (Mohammadi et al., 2024; Kumar et al., 2025). These models help develop a personalized postoperative care plan based on risk-stratifying patients. This stratification allows care teams to use the intensive care unit resources more appropriately and offers the opportunity to intervene early for at-risk patients (Khaled et al., 2025).

9.6.2 Improving stratification in congenital heart procedures

In congenital heart surgeries, outcome prediction and risk stratification are imperative due to variability among patients and the complexity of procedures. AI platforms show great promise in assisting with personalized risk stratification and informing perioperative decision-making (Mohammadi et al., 2024; Gadhachanda et al., 2025). Machine learning can now explore high-dimensional clinical data across components such as imaging, temperature, priority, and genealogy to expose relationships between data and adverse outcomes (Kumar et al., 2025; Khaled et al., 2025). AI platforms assist with early risk recognition and predictive analytics of available traits to share resources and personalize management plans.

The recent robotic platforms include the predictive module, which is incorporated where reconstruction is planned, and the new platforms visualize and reduce the operating time for the planned reconstruction, with awareness to reorient if presented

Table 9.1 AI techniques in predictive modeling for cardiovascular surgery.

Type of the model	Usecase	Target outcome	Reported accuracy
Decision tree	Postoperative bleeding prediction (Mohammadi et al., 2024).	Predict infection, arrhythmia, and reintervention (Mohammadi et al., 2024).	~80%–85%
Support vector machine	Stratification of congenital heart defect risk (Mohammadi et al., 2024).	Stratify survival and recovery risk (Mohammadi et al., 2024).	~83%–87%
Neural networks	Overall survival prediction (Mohammadi et al., 2024).	Anticipate complications in complex procedures (Mohammadi et al., 2024).	~89%–92%
AI–robotic fusion	Intraoperative stratification (Picozzi et al., 2024).	Predict deviation during procedure (Picozzi et al., 2024).	Real-time adaptability

with data or execution difficulties (Picozzi et al., 2024; Fida et al., 2024). As Table 9.1 shows, predictive tools personalize care, reduce or avoid complications, and improve surgical risk and outcome.

9.7 Challenges and ethical considerations

Cardiovascular robotics is being enhanced by AI-based applications, which are creating enhancements in both surgical precision and operational efficiencies as well as improving patient outcomes. However, there is a need for immediate and ongoing evaluation of the technical, regulatory, and ethical implications of these new technologies on the world of cardiovascular robotics as they become more sophisticated from a robotics and AI perspective. The broader healthcare ecosystem has a number of areas of concern that need to be addressed, such as robotic surgical systems and their corresponding AI models are deployed within varying healthcare environments. These include issues of data quality, model generalizability, and algorithmic transparency. Data source variability, population bias, and inadequate annotations of clinical and laboratory records all negatively affect model performance and, if not addressed, could exacerbate inequities in care.

Additionally, regulatory frameworks have not adapted to accommodate adaptive and continuously learning AI systems; as such, this contributes to uncertainty around accountability, verification, and postmarketing surveillance. There is also a need to closely work collaboratively among a variety of clinical specialties, biomedical engineers, data scientists, and regulatory bodies on the effective implementation of

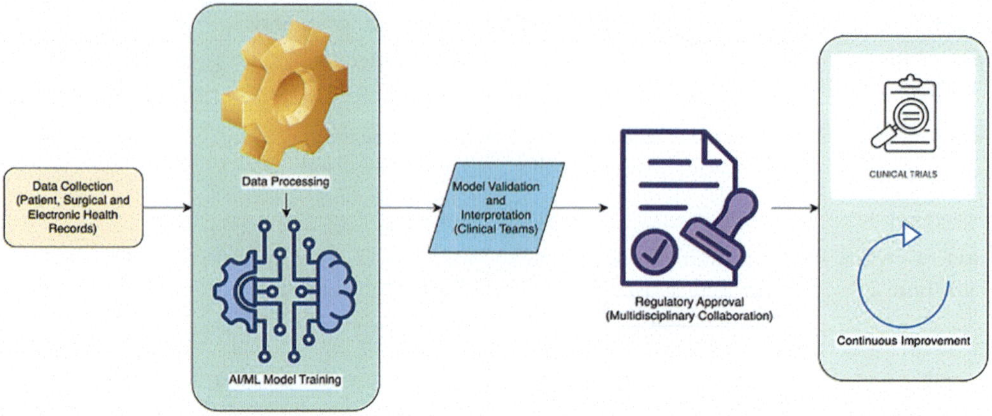

Figure 9.7 AI integration workflow in cardiovascular robotics with challenges at each stage.

AI-driven robotic systems to ensure that they conform with clinical workflows and other safety standards (Knudsen et al., 2024; Lareyre et al., 2023). Ethical principles associated with AI-driven robotic systems, such as informed consent, boundaries between surgeon–AI responsibility, patient trustworthiness, and liability in the event of a system failure, must also be clarified. To better understand these types of challenges, the AI-enabled cardiovascular robotic system development and deployment process is highlighted through a high-level workflow (see Fig. 9.7).

9.7.1 Data quality, bias, and interpretability

- **Data quality challenges.** Occasionally, AI model training datasets will be incomplete, inconsistent, or not representative of the dataset, which can result in incorrect predictions or unequal clinical outcomes. Image quality (variability), lack of physiological parameters, and institutional data silos are all additional factors that contribute to this challenge. Therefore it is vital to develop standardized data collection methodologies and rigorously validate the AI-based applications throughout the entire AI lifecycle (Jaltotage and Dwivedi, 2024).
- **Addressing bias.** AI models can potentially inherit bias from biased training data (unbalanced datasets), which causes disparities in diagnosis/treatment protocols for various patient populations. Risk-mitigation strategies for reducing this risk include rebalancing the datasets, conducting audits for bias, evaluating models that have been stratified, and adopting continuous postdeployment monitoring strategies (Jaltotage and Dwivedi, 2024).
- **Model interpretability issues.** Most SOTA AI models operate as "black boxes," meaning that clinicians are limited in their ability to understand the model outputs and the reasons for those outputs (axioms). Lack of transparency is detrimental to

clinicians' trust and reliance on these outputs, particularly in the context of invasive cardiovascular procedures. Therefore applying strategies that utilize explainable AI (i.e., attention mapping, feature attribution, rule-based approximation), and providing clinicians with appropriate educational support will enhance users' understanding of the AI outputs and increase the likelihood of successful clinical use.

- Explainable AI methods and clinician education can improve transparency and trust (Jaltotage and Dwivedi, 2024; Weiner et al., 2025).
- **Importance of Robust Data Governance.** Organizations that develop and deploy AI applications in support of cardiovascular robotic systems must utilize equitable and sustainable data governance frameworks to maintain data integrity, data traceability, and compliance with regulatory standards throughout the entire AI lifecycle. Establishing effective governance structures allows for ongoing audits and accountability of these systems and ensures that AI-based cardiovascular robotic systems remain reliable over the long term (Jaltotage and Dwivedi, 2024).

9.7.2 Regulatory, legal, and ethical implications

- **Regulatory compliance.** Regulatory measures may not fully capture the complexities of AI-powered robotic surgery, so there is a growing need for updated regulatory guidance addressing safety and transparency issues (Lareyre et al., 2023; Li et al., 2024; Mooghali et al., 2024).
- **Legal accountability.** Clear guidance is needed to determine accountability for AI-driven mistakes to protect the patient as well as the provider, especially in cases of negligence (Lareyre et al., 2023; Li et al., 2024; Mooghali et al., 2024).
- Ethical issues. There are ethical implications around autonomy, consent, and discrimination that warrant the development of appropriate ethical guidelines and involvement of patients in order to keep them, the stakeholders, in a trusted state (Elendu et al., 2023; Weiner et al., 2025).
- **Trustworthiness and transparency.** AI systems must be transparent and trustworthy to advance their clinical use; keeping the innovation-safety-standards balance will always be of utmost importance (Li et al., 2024; Mooghali et al., 2024; Weiner et al., 2025).

9.7.3 Multidisciplinary collaboration and clinical trials

- Involving clinicians, engineers, data scientists, regulatory experts, and ethics experts is essential for developing, testing, and implementing strong, patient-centered AI-based cardiovascular robotic platforms (Arjomandi Rad et al., 2025; Knudsen et al., 2024; Jaltotage and Dwivedi, 2024).

- When starting clinical trials, it is essential to conduct full planning for real-world validation and the roles of diverse populations, as well as safety, efficacy, and usability assessments, to ultimately advance participants' best outcomes.
- The creation of innovation and trust as recognized members of the regulatory and ethical committees strengthens the validity of work and also advocacy for patient experience through patient engagement, transparency, and enlisting support (Elendu et al., 2023; Lareyre et al., 2023; Li et al., 2024; Mooghali et al., 2024).
- Consistent refinement and continued improvements, measuring user feedback, can drive robust technology for patients and ensure sustainability of critical technology and innovations (Knudsen et al., 2024; Biondi-Zoccai et al., 2025).

9.8 Future directions and translational outlook

The rapid progression of AI and robotic technology in cardiovascular surgery represents a new era for the surgical field characterized by increased precision, increased efficiency, and improved safety. As the capabilities of current technologies grow, there will be dramatic improvements in the overall quality of procedures that will affect how surgeries are performed and the outcome of surgeries for patients. In addition, advances in physical robotics technologies, as well as new AI paradigms such as foundation models (e.g., large language models) and vision-language models, will increasingly be used for the purpose of surgical simulation, training, and assisting surgical decision-making. These models will allow for the development of highly realistic virtual environments for surgical practice, automated assessments of surgical skills, and the creation of custom training pathways through the opportunity to learn from large datasets associated with surgical procedures across multiple modes (multimodal). The remainder of this article presents a discussion of the current trends in the movement to provide fully autonomous systems, the evolution of the surgical environment, and the steps for safe and successful delivery of surgical services to patients safely and effectively (Osman et al., 2025).

9.8.1 Autonomous cardiovascular surgical systems

- **Definition and scope.** Autonomous surgical systems are being conceptualized for the completion of complex cardiovascular actions with limited human input, specifically through the integration of AI for real-time decision-making and robotic accuracy in execution (Leivaditis et al., 2025; Osman et al., 2025).
- **Technological enablers.** Recently, machine learning and computer vision have started to advance the capacity of robots to identify anatomical structures, change operations intraoperatively, and follow prescribed plans for surgical actions (Apare et al., 2024; Gadhachanda et al., 2025).

- **Current level of development.** We know that early-stage autonomous systems are in development and undergoing evaluation on performing vessel anastomosis and possibly valve repair, delivering promising results in preclinical models (Leivaditis et al., 2025; Apare et al., 2024).
- **Challenges.** The key challenges pertain to safety, reliability, and having ethical accountability while conducting autonomous operations (Osman et al., 2025; Leivaditis et al., 2025).

As the promise of autonomous systems is shown to be acceptable, the next step will be to integrate these technologies in a connected operating environment.

9.8.2 Artificial intelligence + Internet of Things + robotics—a connected or ecosystem

9.8.2.1 Integrated ecosystem

The combination of AI, IoT, and robotics will result in the creation of a "smart" OR, providing a new level of device communication, transforming workflows, and patient monitoring (Li et al., 2024; Saeed et al., 2023; Fida et al., 2024). The workflow of these integrated systems can be illustrated as a flowchart, as shown in Fig. 9.8.

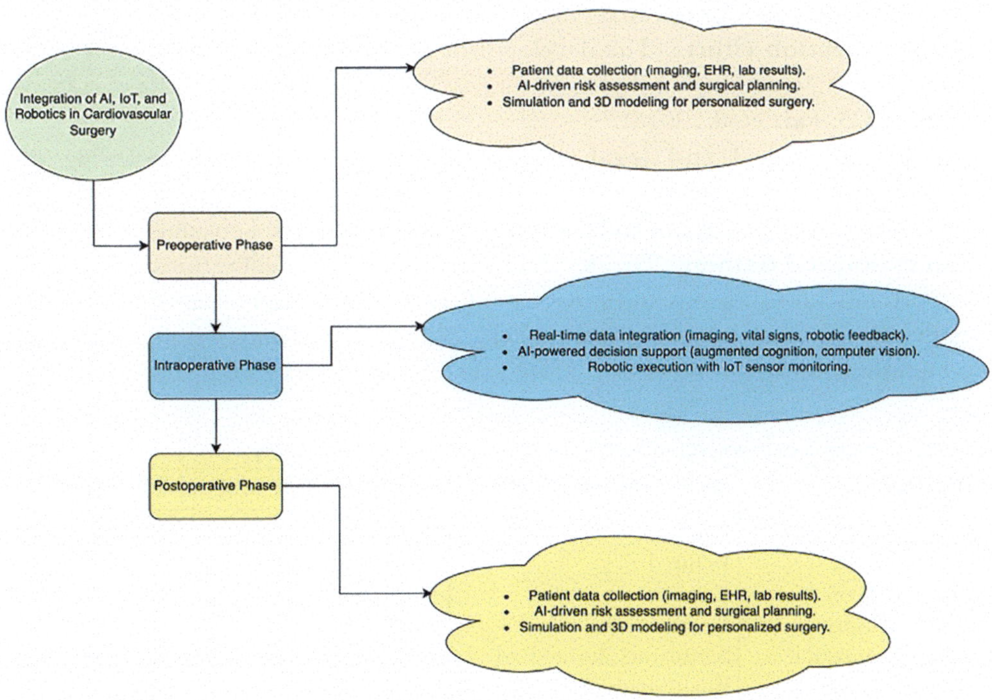

Figure 9.8 Phases of AI + IoT + robotics integration in cardiovascular surgery.

9.8.2.2 *Key components*

- AI analytics for real-time data interpretation.
- IoT sensor activity for continuous monitoring of the patient and equipment.
- Robotic systems that can execute surgery with precision and aptitude (Li et al., 2024; Saeed et al., 2023).

9.8.2.3 *Benefits*

- Increased efficiency in surgical precision and safety.
- Improved decision-making during procedures.
- Easier transition to postoperative care based on routine continuous collection and analysis of intraoperative data (Li et al., 2024; Kumar et al., 2025).
- The incorporation of these technologies will ultimately deliver standardized and scalable solutions for cardiovascular surgery.

9.8.3 Roadmap to clinical standardization and adoption

- **Clinical validation.** Firstly, it's important to conduct appropriate clinical studies with regulatory approval to establish the safety and efficacy of autonomous and integrated systems (Osman et al., 2025; Leivaditis et al., 2025).
- **Standardization efforts.** The development of standardized protocols and guidelines for the use of AI and robotics in surgery will aid in facilitating adoption (Saeed et al., 2023; Fida et al., 2024).
- **Factors of ethical and legal considerations.** Considerations regarding data privacy, liability, and informed consent will be especially important for establishing confidence among clinicians and patients (Osman et al., 2025; Leivaditis et al., 2025).
- **Education and training.** The development of appropriate educational and training opportunities for surgeons and supporting personnel is critical to ensure confidence with the new technologies (Apare et al., 2024; Gadhachanda et al., 2025). The adoption pathway can be visualized in Table 9.2 to facilitate comparison.

Table 9.2 Current versus emerging technologies in cardiovascular surgery.

Feature/Aspect	Current robotic systems	Emerging AI-IoT-robotics integration
Human involvement	High	Low
Data integration	Limited	Comprehensive
Decision support	Basic	Advanced
Patient monitoring	Intermittent	Continuous
Training/Simulation	Simulators, limited AI	AI-driven virtual reality, adaptive learning
Regulatory status	Established	Under development/Testing

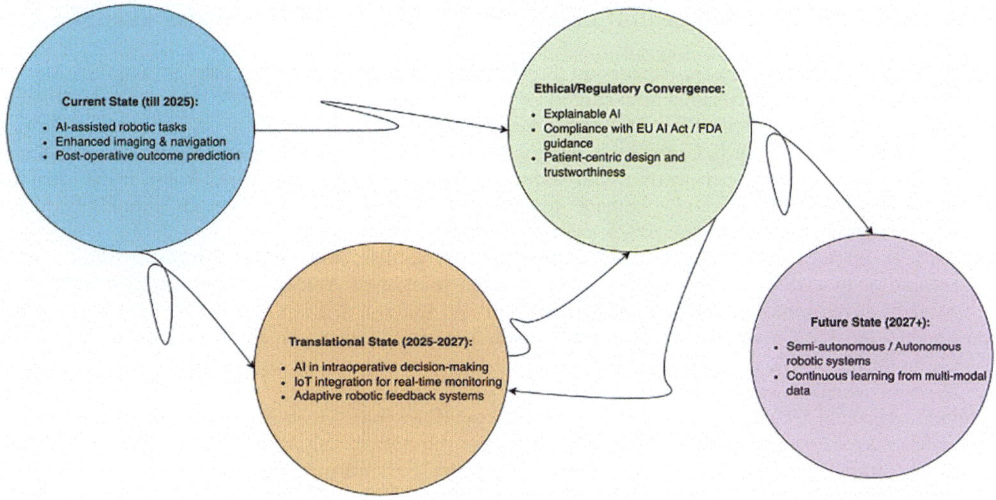

Figure 9.9 Strategic roadmap of cardiovascular robotics from current state to future state.

9.9 Conclusion

This chapter underscores meaningful technological and clinical advancements and contributions from autonomous AI-based robotic systems in cardiovascular patients and describes enhanced surgical planning, intraoperative real-time decision support, automated and precise robotic execution, and postoperative predictive care. Each category outlined in the summary table enabled evidence-producing trajectories with measurable increases in patient safety, surgical outcomes, and procedural efficiency (Mohammadi et al., 2024; Leivaditis et al., 2025; Li et al., 2024). The strategic roadmap flowchart, as shown in Fig. 9.9, underlines a detailed outline and pathway of the future state of autonomous integrated AI and IoT robotic systems in the clinical environment through translational milestones such as adaptive robotic feedback, alignment of regulation and ethics, and smart operating theaters that provide continuous learning (Arjomandi Rad et al., 2025; Saeed et al., 2023).

Progress toward the next state of autonomous surgery can be made, but it will require collaborative multidisciplinary teams, comprehensive clinical testing, and adherence to a robust ethical framework to ensure accountable and trustworthy, patient-centered advancements. Collectively, we will achieve a new era of precision cardiovascular surgery defined by intelligence and autonomy that yields better outcomes.

References

Apare RS, More N, Badave RS, Kale AK, Arsalwad GP, Takale DG. Advancements in surgical robotics and AI-driven technologies for precision medicine. Journal of Neonatal Surgery 2024;13.

Available from https://doi.org/10.47338/jns.v13.1430, https://www.jneonatalsurg.com/index.php/jns/issue/view/49.

Arjomandi Rad A, Vardanyan R, Athanasiou T, Maessen J, Sardari Nia P. The ethical considerations of integrating artificial intelligence into surgery: a review. Interdisciplinary CardioVascular and Thoracic Surgery 2025;40(3). Available from https://doi.org/10.1093/icvts/ivae192.

Biondi-Zoccai G, D'Ascenzo F, Giordano S, Mirzoyev U, Erol Ç, Cenciarelli S, et al. Artificial intelligence in cardiology: general perspectives and focus on interventional cardiology. Anatolian Journal of Cardiology 2025;29(4):152–63. Available from https://doi.org/10.14744/AnatolJCardiol.2025.5237, https://anatoljcardiol.com/year/2025/volume/29/issue/4.

Bouchard T, Subichin M, Firstenberg MS. Fifty years of cardiac surgery: innovation, evolution, and revolution in cardiovascular therapies. International Journal of Academic Medicine 2019;5(3):156. Available from https://doi.org/10.4103/ijam.ijam_49_18.

Chitwood WR. Historical evolution of robot-assisted cardiac surgery: a 25-year journey. Annals of Cardiothoracic Surgery 2022;11(6):564–82. Available from https://doi.org/10.21037/acs-2022-rmvs-26, www.annalscts.com/index.

Elendu C, Amaechi DC, Elendu TC, Jingwa KA, Okoye OK, John Okah M, et al. Ethical implications of AI and robotics in healthcare: a review. Medicine (United States) 2023;102(50):E36671. Available from https://doi.org/10.1097/MD.0000000000036671, https://journals.lww.com/md-journal/pages/default.aspx.

Fida Z, Ghutai G, Jamil Z, Dalvi AA, Hassaan M, Khalid K, et al. The role of robotics in cardiac surgery: innovations, outcomes, and future prospects. Cureus 2024;16(11). Available from https://doi.org/10.7759/cureus.74884.

Gadhachanda KR, Marsool Marsool MD, Bozorgi A, Ameen D, Nayak SS, Nasrollahizadeh A, et al. Artificial intelligence in cardiovascular procedures: a bibliometric and visual analysis study. Annals of Medicine & Surgery 2025;87(4):2187–203. Available from https://doi.org/10.1097/ms9.0000000000003112.

Gumbs AA, De Simone B, Chouillard E. Searching for a better definition of robotic surgery: is it really different from laparoscopy? Mini-invasive Surgery 2020;4. Available from https://doi.org/10.20517/2574-1225.2020.110, https://oaepublishstorage.blob.core.windows.net/31f23ae2-f327-4b9d-973a-7d09159a191c/3812.pdf.

Jaltotage B, Dwivedi G. Essentials for AI research in cardiology: challenges and mitigations. CJC Open 2024;6(11):1334–41. Available from https://doi.org/10.1016/j.cjco.2024.07.015.

Khaled A, Ririe AK, Kaur M, Kumar S, Rahming N, Khaled R, et al. Predictive analysis for cardiovascular outcomes using Ai. Journal of Neonatal Surgery 2025;14(24S):147–57. Available from https://doi.org/10.63682/jns.v14i24s.5911.

Knudsen JE, Ghaffar U, Ma R, Hung AJ. Clinical applications of artificial intelligence in robotic surgery. Journal of Robotic Surgery 2024;18(1). Available from https://doi.org/10.1007/s11701-024-01867-0, https://www.springer.com/journal/11701.

Kumar S, Rajak N, Gour S, Salitra D, Sharma R, Namdev S. Machine learning in cardiac surgery: a systematic review of ML techniques, applications, and the road ahead. Journal of Neonatal Surgery 2025;14(17S):729–36.

Lareyre F, Maresch M, Chaudhuri A, Raffort J. Ethics and legal framework for trustworthy artificial intelligence in vascular surgery. EJVES Vascular Forum 2023;60:42–4. Available from https://doi.org/10.1016/j.ejvsvf.2023.08.003.

Leivaditis V, Beltsios E, Papatriantafyllou A, Grapatsas K, Mulita F, Kontodimopoulos N, et al. Artificial intelligence in cardiac surgery: transforming outcomes and shaping the future. Clinics and Practice 2025;15(1):17. Available from https://doi.org/10.3390/clinpract15010017.

Li Y, Wang M, Wang L, Cao Y, Liu Y, Zhao Y, et al. Advances in the application of AI robots in critical care: scoping review. Journal of Medical Internet Research 2024;26:e54095. Available from https://doi.org/10.2196/54095.

Mohammadi I, Rajai Firouzabadi S, Hosseinpour M, Akhlaghpasand M, Hajikarimloo B, Zeraatian-Nejad S, et al. Using artificial intelligence to predict post-operative outcomes in congenital heart surgeries: a systematic review. BMC Cardiovascular Disorders 2024;24(1). Available from https://doi.org/10.1186/s12872-024-04336-6.

Mooghali M, Stroud AM, Yoo DW, Barry BA, Grimshaw AA, Ross JS, et al. Trustworthy and ethical AI-enabled cardiovascular care: a rapid review. BMC Medical Informatics and Decision Making 2024;24(1). Available from https://doi.org/10.1186/s12911-024-02653-6, https://bmcmedinformdecismak.biomedcentral.com/.

Osman EIA, Ismail MMEMM, Mukhtar MAH, Ahmed AUB, Mohamed NAAE, Ibrahim AAA. Artificial intelligence and robotics in minimally invasive and complex surgical procedures: a systematic review. Cureus 2025;17(3). Available from https://doi.org/10.7759/cureus.81339.

Ozturk C., Pak D.H., Rosalia L., Goswami D., Robakowski M.E., McKay R., et al., AI-powered multimodal modeling of personalized hemodynamics in aortic stenosis. arXiv, United States arXiv. 2024.Available from: https://arxiv.org. https://doi.org/10.48550/arXiv.2407.00535.

Picozzi P, Nocco U, Labate C, Gambini I, Puleo G, Silvi F, et al. Advances in robotic surgery: a review of new surgical platforms. Electronics 2024;13(23):4675. Available from https://doi.org/10.3390/electronics13234675.

Rajput NK, Robotic heart surgery – risks, benefits, and how it works. Medanta, 2024.

Saeed A, AlShafea A, Bin Saeed A, Nasser M, Ali R. Robotics and artificial intelligence and their impact on the diagnosis and treatment of cardiovascular diseases. Cureus 2023;15(7). Available from https://doi.org/10.7759/cureus.42252.

Wang S., Lin H., Xie Y., Wang Z., Chen D., Tan L., et al., Robotic transcatheter tricuspid valve replacement with hybrid enhanced intelligence: a new paradigm and first-in-vivo study. arXiv, China arXiv. 2024. Available from: https://arxiv.org. https://doi.org/10.48550/arXiv.2411.12478.

Weiner EB, Dankwa-Mullan I, Nelson WA, Hassanpour S. Ethical challenges and evolving strategies in the integration of artificial intelligence into clinical practice. PLOS Digital Health 2025;4(4). Available from https://doi.org/10.1371/journal.pdig.0000810, https://journals.plos.org/digitalhealth/article?id=10.1371/journal.pdig.0000794.

Yu MY, Son YJ. Machine learning-based 30-day readmission prediction models for patients with heart failure: a systematic review. European Journal of Cardiovascular Nursing 2024;23(7):711–9. Available from https://doi.org/10.1093/eurjcn/zvae031, https://academic.oup.com/eurjcn.

CHAPTER 10

Artificial intelligence for orthopedic surgery with robots

Kande Archana[1], V. Kamakshi Prasad[2], and A. Srinivasula Reddy[3]
[1]Department of CSE-Data Science, CMR Engineering College, Hyderabad, Telangana, India,
[2]Department of CSE, Jawaharlal Nehru Technological University, Hyderabad, Telangana, India,
[3]Department of EEE, CMR Engineering College, Hyderabad, Telangana, India

10.1 Introduction

Orthopedic surgery is changing as a result of the combination of robotic and artificial intelligence (AI) technologies, which are enabling more accurate, customized, and minimally invasive operations. Due to procedural heterogeneity and human error, traditional orthopedic methods frequently have limits. On the other hand, robotics driven by AI improves surgical planning and execution, resulting in better results and shorter recovery periods. In order to accurately map anatomical structures during preoperative planning, convolutional neural networks (CNNs) are widely used for medical picture segmentation. Better alignment and customization of surgical treatments are made possible by this accurate visualization. In order to make better decisions, machine learning (ML) techniques such as random forests and support vector machines (SVMs) are being employed to forecast surgical risks and patient outcomes. Reinforcement learning (RL) and deep Q-networks (DQNs)-powered adaptive control systems enable robotic systems to react dynamically to intraoperative changes during surgery. These systems improve the procedure's accuracy and safety by continuously optimizing their activities in real-time. By facilitating precise motion tracking and guaranteeing the best possible alignment of surgical instruments with the patient's anatomy, Kalman filters further aid in these endeavors. Precision and customization are crucial in orthopedic fields, including joint arthroplasty, spinal surgery, and fracture healing, where these technological developments have a particularly significant impact. AI-guided robots contribute to better long-term patient outcomes, shorter recovery periods, and fewer intraoperative errors.

Even with these advantages, there are still many difficulties. Widespread adoption is hampered by issues with algorithmic openness, data protection, and the requirement for specific clinical training. Building confidence between patients and providers requires creating interpretable AI models and ensuring secure data processing. Comprehensive training programs are also required to give surgeons the know-how to operate these

AI-Powered Developments in Medical Robotics
DOI: https://doi.org/10.1016/B978-0-443-27783-2.00009-0

cutting-edge instruments. Robotics and AI have the potential to revolutionize orthopedic care standards by enabling safer, quicker, and more efficient operations.

10.1.1 Background and significance

In the past, orthopedic surgery has depended on a surgeon's skill to handle intricate operations involving muscles, ligaments, joints, and bones. Even the most accomplished surgeons, however, struggle to achieve constant precision, especially when performing procedures that need precise anatomical knowledge or when taking patient-specific variances into account. The development of robotic-assisted technology during the last 20 years has revolutionized orthopedic therapy. Particularly in total joint replacements and spinal surgeries, robotic devices greatly improve the accuracy and repeatability of orthopedic treatments (Taylor et al., 2016; Lanfranco et al., 2004). These robots facilitate less invasive procedures, lower intraoperative mistakes, and enhance implant alignment and placement when combined with image-guided navigation systems. For example, the MAKO robotic system, which is frequently utilized in complete hip and knee replacement, has shown better functional results than traditional procedures (Kayani, 2018a). At the same time, developments in AI have enabled computers to process enormous volumes of imaging and clinical data, facilitating real-time intraoperative decision-making and optimal surgical planning (Topol, 2019b). Enhanced visualization through augmented imagery, mistake prevention through adaptive feedback loops, and semiautonomous job execution are just a few of the capabilities that are made possible by the combination of AI with robotic technologies. In addition to increasing surgical precision, this convergence lowers complication rates, shortens learning curves for novice surgeons, and decreases variability in results. These technologies facilitate improved postoperative recovery and expedite complicated operations by automating some functions and directing decision-making (Hashimoto et al., 2018). Notwithstanding these benefits, obstacles to broad adoption include protecting patient data, getting regulatory approval, and educating medical professionals on how to properly integrate these tools. However, orthopedic surgery has a bright future thanks to the combined strength of AI and robotics, which will provide safer, more effective, and patient-centered care.

10.1.2 Evolution of artificial intelligence in orthopedic surgery

From simple rule-based diagnostic tools to sophisticated ML and deep learning (DL) systems, AI applications in orthopedics have grown throughout time. Complex tasks, including radiography image segmentation, fracture classification, surgical outcome prediction, and real-time robotic control, may now be handled by these contemporary systems (Yu et al., 2018). Recent developments in AI have made it possible to analyze

MRI, CT, and X-ray images with accuracy on par with human specialists, allowing for quicker and more accurate preoperative planning than was previously possible with manual techniques (Litjens et al., 2017). AI enables high-precision joint surface modeling, bone cutting route optimization, and intraoperative alignment verification in robotic-assisted orthopedic surgery. These skills are particularly important for procedures like complete knee replacement and spinal fusion, where even little variations, measured in millimeters, can have a big influence on joint function and the longevity of implants (Lonner, 2017).

In addition to providing intraoperative support, ML models can forecast post-operative issues such as joint stiffness, implant loosening, and infections. Clinicians can proactively manage risks and customize care with the help of these prediction tools. Furthermore, actionable insights are being extracted from electronic health records (EHRs) using natural language processing (NLP) algorithms, which offer a deeper comprehension of patient histories and treatment outcomes to inform customized therapy approaches (Senders et al., 2018). The more AI develops, the easier it is to integrate with robotic systems. In order to increase control of robotic arms and response to soft-tissue interactions, contemporary AI-enhanced robotic platforms can learn from real-time sensor data and dynamically modify their operations (Esteva et al., 2019). In addition to increasing surgical accuracy, this real-time flexibility also improves patient outcomes and safety. AI-driven robotics, which combines data intelligence with mechanical precision, is set to become a key component of next-generation orthopedic care, despite ongoing difficulties with clinical validation, data governance, and practitioner training.

10.1.3 Objectives and scope

Examining the present and prospective future of robotic systems driven by AI in orthopedic surgery is the goal of this study. In order to improve preoperative planning, intraoperative navigation, and postoperative rehabilitation, the study specifically aims to explore how sophisticated ML and computer vision techniques can be incorporated into robotic platforms. Through this analysis, the study will evaluate how AI can improve procedural accuracy, streamline orthopedic surgery workflows, and provide individualized patient care. The technological underpinnings of AI-driven robotic systems, such as image recognition, real-time sensor integration, and data-driven decision-making algorithms, will be examined first in this chapter. The main clinical applications will next be examined, with an emphasis on operations where accuracy and consistency are crucial, such as joint arthroplasty, spinal fusion, and fracture fixation. We'll focus on the therapeutic advantages of AI-robotic integration, including better implant alignment, less intraoperative variability, and quicker recovery periods. The study will also include important ethical issues and problems related to these technologies, such as the need for physician training, data protection, algorithm openness, and liability issues. It will assess how well-suited AI technologies are for clinical use and pinpoint obstacles that need to be removed for wider use.

This chapter's description of a conceptual AI framework for orthopedic robotic surgery will be one of its main contributions. This framework will include real-time feedback systems for prompt guidance, predictive analytics tools to assist decision-making along the surgical continuum, and adaptive control mechanisms that adapt to intraoperative changes. The ultimate objective is to show how AI may improve the safety, effectiveness, and customization of orthopedic procedures to meet the demands of specific patients. The goal of this research is to promote intelligent surgical systems in orthopedics and open the door for future advancements by linking technical innovation with clinical application.

10.2 Artificial intelligence in preoperative planning

Orthopedic surgical planning is undergoing a revolution thanks to AI, which improves diagnostic accuracy, expedites decision-making, and makes customized treatment plans possible. Preoperative planning has always mainly depended on the experience of the surgeon and accepted imaging methods. But as AI has been incorporated, this procedure has become more customized and data-driven, improving patient safety and surgical results. Analyzing diagnostic imaging is one of AI's primary contributions to orthopedics. Medical pictures like X-rays, MRIs, and CT scans can be processed and interpreted with amazing accuracy by sophisticated ML algorithms, which can spot minute patterns and abnormalities that the human eye could overlook. These methods facilitate prompt diagnosis and early detection of orthopedic disorders such as fractures, joint degeneration, or spinal abnormalities. Predictive analytics and risk classification are also supported by AI. AI models can predict surgical risks and postoperative consequences by analyzing massive datasets and using anatomical data, comorbidities, and a patient's medical history. By enabling proactive actions, this predictive capability optimizes surgical techniques and resource allocation. Additionally, the creation of patient-specific 3D models and simulations is made possible by AI-driven tools, which are crucial for surgery planning. Surgeons can more precisely plan the surgical course, customize implants, and see the precise anatomical structures thanks to these models. Customized surgical techniques save recuperation times, minimize invasiveness, and shorten operation times.

To sum up, AI is revolutionizing orthopedic surgery, especially in the area of preoperative planning. AI greatly increases the effectiveness, safety, and results of orthopedic treatments by improving image interpretation, facilitating risk classification, and assisting with customized surgical approaches. The use of AI technology in orthopedic care will grow as it develops further, further altering patient-centered treatment paradigms and surgical standards.

10.2.1 Imaging and diagnostic enhancements

Medical imaging is one of the most revolutionary uses of AI in orthopedic preoperative planning. Conventional picture interpretation techniques can result in inconsistent diagnosis and planning because they are frequently constrained by human visual capability and susceptible to interobserver variability. AI has greatly improved the precision and effectiveness of image analysis in orthopedics, especially using sophisticated DL methods such as CNNs and U-Net architectures (Zehra et al., 2019). High-precision musculoskeletal structure segmentation and classification in MRI, CT, and X-ray images are made possible by these AI models. CNNs have shown expert-level performance in recognizing soft tissues, bone architecture, and pathological diseases such as tumors, fractures, and osteoarthritis (Zehra et al., 2019). AI-assisted segmentation, for example, enables the quick and precise creation of 3D anatomical reconstructions in total knee arthroplasty (TKA), significantly cutting down on the amount of time needed for preoperative planning and enhancing implant alignment and surgical results (Zehra et al., 2019). Additionally, the incorporation of spatial transformer networks based on DL improves picture registration, allowing for the high-fidelity visualization of anatomical components from various angles (Jaderberg et al., 2015). For intricate orthopedic operations like spinal surgery, where comprehension of the three-dimensional disease is essential, this skill is especially helpful. AI models enhance accurate implant sizing and placement by helping to more clearly identify spinal canal stenosis, intervertebral disc degeneration, and vertebral abnormalities (Jaderberg et al., 2015).

A paradigm shift in diagnostic imaging workflows has been indicated by recent research that demonstrates DL algorithms can perform better than radiologists in certain diagnostic tasks, such as fracture detection (Resplandy et al., 2018). This development improves diagnostic accuracy and confidence in addition to clinical efficiency. In conclusion, with previously unheard-of levels of accuracy, speed, and customization, AI-powered medical imaging has emerged as a crucial instrument in orthopedic preoperative planning. Its ongoing development holds the potential to greatly enhance patient outcomes and further hone surgical techniques.

10.2.2 Predictive analytics for surgical outcomes

Through predictive modeling, AI is significantly improving orthopedic preoperative planning. This capacity enhances surgical readiness and patient counseling by enabling surgeons to predict probable outcomes and possible complications. To make highly accurate predictions about clinical outcomes, AI algorithms, such as random forests, gradient boosting machines, and SVMs, use enormous datasets from wearable technology, imaging sources, and EHRs (Esteva et al., 2019). Numerous surgical risks, such as infection, excessive blood loss, implant failure, and extended hospital stays, can be predicted by these models. Predictive analytics has been used to assess patient-reported outcome measures, functional recovery, and satisfaction levels in procedures such as hip and knee arthroplasty. This helps surgeons to determine which patients are best suited for surgery, establish reasonable expectations, and personalize treatment plans (Lieberman et al., 2020). Bayesian networks and artificial neural networks, which simulate intricate relationships between patient-specific variables such as body mass index, bone density, preexisting comorbidities, and the complexity of the surgical operation, further improve this predictive ability (Senders et al., 2018). AI assists in creating customized risk profiles by modeling these linkages, enabling surgeons to adjust their tactics and enhance results. In value-based healthcare settings, where the goal is to maximize patient benefit while minimizing costs, such predictive techniques are especially helpful. By measuring risk and supporting evidence-based planning, AI-driven decision support systems give surgeons more clarity during preoperative evaluations. As a result, there is less uncertainty, more collaborative decision-making, and more consistent clinical results.

In conclusion, by combining various data sources to identify complications and direct individualized treatment plans, AI-powered predictive modeling in orthopedics greatly improves surgical planning. As these technologies develop further, they will enable physicians to provide orthopedic care that is safer, more effective, and outcome-driven.

10.2.3 Personalized surgical planning

A key component of precision medicine is patient-specific or personalized planning, and AI is propelling this strategy at scale in orthopedic surgery. AI makes it possible to create surgical plans that are specifically suited to each patient by combining a variety of data sources, including anatomical variations, previous surgical outcomes, and patient-specific imaging. This degree of customization lowers the risk of problems while improving surgical accuracy and results. In order to evaluate multiple approaches for the best joint alignment and bone excision, advanced AI techniques such as RL and DQNs have been used to model a variety of surgical scenarios (Schulman et al., 2015). These systems can provide data-backed assistance for intricate orthopedic treatments by continuously improving their tactics through simulated feedback and real-time adaptation. Additionally, generative adversarial networks (GANs) are becoming more and more effective tools for customized planning. In order to enable surgeons to practice and refine treatments online before executing them on real patients, GANs are being investigated for their capacity to produce synthetic but anatomically correct models (Yi et al., 2019). This capacity is particularly helpful in difficult situations where traditional planning would not be sufficient, including revision procedures or abnormalities. Furthermore, 3D printing technologies are becoming more and more linked with AI-powered planning platforms. By creating personalized implants and cutting guides that are tailored to the patient's anatomy, these technologies can improve implant fit and reduce intraoperative ambiguity. Better functional results, fewer intraoperative changes, and shorter surgery times are frequently the results.

In conclusion, by fusing automation, customization, and simulation, AI-driven personalized planning is revolutionizing orthopedic surgery. These developments represent a major advancement in providing accurate, patient-focused surgical care, particularly in situations with complicated or unusual anatomy.

10.3 Robotic systems in orthopedics

Orthopedic surgery has changed dramatically as a result of robotic technologies, which improve surgical accuracy, consistency, and customization. By combining real-time intraoperative guidance with 3D preoperative planning, leading platforms such as MAKO, ROSA, NAVIO, and TSolution-One have raised the bar for surgeries like knee and hip arthroplasty. These robotic-assisted methods improve implant alignment, decrease surgical variability, and lead to faster recovery times and better functional results. The capabilities of these robotic platforms are further enhanced by the use of AI. AI creates a highly responsive and accurate surgical environment by enabling real-time data processing, adaptive decision-making, and intelligent control during surgery. Improved intraoperative imaging and anatomical recognition are made possible by technologies such as CNNs, while robotic motion stabilization and surgical trajectory planning are improved by RL, which can adjust to minute alterations. With each procedure, these intelligent systems can perform better since they are constantly learning from data feedback and surgical outcomes. AI-assisted robotic platforms, for example, can minimize soft-tissue damage in real-time, recommend the best angles for implant insertion, and modify cutting depths. Highly customized surgical techniques that are adapted to the distinct anatomical and physiological traits of each patient are the result of this dynamic interaction between AI and robots. Comparing robotic-assisted orthopedic procedures to traditional techniques, clinical studies show greater implant durability, quicker postoperative rehabilitation, and lower rates of complications. Patients gain from AI's capacity to comprehend and react to intricate intraoperative circumstances in addition to the mechanical precision of robotic equipment. Orthopedic surgery care is about to be redefined by the merging of robotic and AI technologies. More independent, accurate, and patient-specific interventions will follow, raising surgical standards and enhancing long-term results for a variety of orthopedic surgeries.

10.3.1 Overview of robotic platforms

Orthopedic surgery has been transformed by robotic platforms, which allow for more accurate, repeatable, and minimally invasive procedures. In joint replacement procedures, such as hip and knee arthroplasty, where precise implant placement and bone preparation are essential for long-term success, these technologies have become especially significant. Leading robotic systems leading this change are TSolution-One (THINK Surgical), NAVIO (Smith & Nephew), ROSA (Zimmer Biomet), and MAKO (Stryker). Through thorough preoperative imaging, they enable patient-specific surgical planning, and they offer intraoperative guidance to guarantee the best possible results. For example, the MAKO system creates a patient-specific 3D model of the joint using preoperative CT scans. Before entering the operating room, this enables doctors to make extremely accurate plans and run through several surgical scenarios. Haptic feedback is used to control the robotic arm during the procedure, limiting its movement to predetermined bounds and guaranteeing precise implant alignment and bone resection (Kayani, 2018b). Consistent surgical outcomes are achieved, and intraoperative variability is greatly decreased with this degree of control. By using autonomous robotic milling, the TSolution-One system goes beyond automation. This technology, which is based on comprehensive 3D imaging, provides precision and repeatability that are hard to get manually by doing bone preparation with little assistance from the surgeon (Lonner, 2015). These devices are especially helpful in difficult or revision surgeries and reduce surgical deviations.

10.3.2 Integration of artificial intelligence with robotic systems

The integration of AI and robotic devices represents a significant advancement in orthopedic surgery. AI improves traditional robotic platforms by allowing for real-time decision-making, adaptive control, and procedure customization based on data. AI techniques such as CNNs are used to evaluate real-time imaging data in order to improve anatomical segmentation and intraoperative navigation. RLRL and Bayesian optimization have also been utilized to enable robotic systems to modify surgical trajectories during operations in response to tissue feedback and dynamic situations (Tokui, 2004; Ravikumar, 2019). ROSA, for example, incorporates AI algorithms that update intraoperative kinematic models in real-time to reflect soft-tissue behavior, facilitating ligament balancing during TKA (Marchand, 2016). Furthermore, AI-based learning systems are trained on massive datasets of prior surgeries to predict optimal implant placement and anticipate intraoperative difficulties. These intelligent robotic systems are transitioning from passive helpers to semiautonomous tools capable of optimizing surgical workflows and minimizing surgeon fatigue (DelSole, 2018).

10.3.3 Case studies and clinical applications

Several clinical research and real-world case applications demonstrate the beneficial effects of AI-powered robotic systems in orthopedic operations. A multicenter study comparing robotic-assisted knee arthroplasty to conventional procedures discovered that robotic surgery resulted in better mechanical axis alignment and less soft-tissue damage, leading to higher postoperative functional scores (Liow, 2017). Another trial of over 1000 patients undergoing hip replacement with the MAKO system found lower postoperative pain and faster functional recovery compared to traditional procedures (Hinman et al., 2021). Similarly, early trials of AI-enhanced spinal instrumentation systems, such as Mazor X, exhibited improved pedicle screw placement accuracy and reduced intraoperative radiation exposure (Carlson et al., 2020).

10.4 Intraoperative artificial intelligence-guided navigation

Intraoperative AI-guided navigation improves orthopedic procedures by offering real-time, data-driven decision support, resulting in increased precision and lower human error. AI combines real-time imaging data with ML algorithms to help surgeons perform complex procedures like joint replacement or spine surgery. CNNs and SVMs are used to recognize anatomical landmarks and monitor tool location, providing continuous input for accurate navigation. These technologies also aid in motion stabilization by removing undesired movements and adjusting for unintentional patient changes. AI's capacity to adjust for real-time fluctuations using RL and Kalman filters improves precision, resulting in fewer problems. Furthermore, adaptive control mechanisms in AI-driven systems allow robots to make real-time adjustments based on the changing surgical environment, providing dynamic corrections to surgical routes. Overall, AI-guided navigation improves surgical precision, leading to better patient outcomes and lower risks.

10.4.1 Real-time decision support

AI-guided intraoperative navigation devices have dramatically enhanced the accuracy and responsiveness of orthopedic procedures by enabling surgeons to make real-time decisions. These technologies combine intraoperative imaging data (from fluoroscopy, ultrasound, and CT) with ML models to help identify anatomical landmarks, appropriate cutting planes, and implant locations.

AI models, such as as SVMs and deep neural networks, assess intraoperative data streams to detect deviations from surgical plans or probable complications as they occur (Lavallee, 2019). Real-time picture segmentation and landmark recognition with CNNs enable precise guidance and continual modification during surgery (Fan, 2018). These methods eliminate dependency on preoperative static planning and allow for data-driven intraoperative adaptation. For example, sophisticated robotic systems used in spinal and knee procedures can detect real-time changes in bone location or tool orientation and notify surgeons when adjustments are required to maintain alignment (Ponnusamy, 2019).

10.4.2 Motion stabilization and precision

Maintaining precision in dynamic surgical situations is critical, especially when accounting for involuntary patient movements or minor tremors. AI improves motion stabilization in robotic systems by understanding real-time feedback and implementing corrective changes to robotic arms with millimeter precision. Kalman filters and adaptive learning techniques are employed to smooth motion trajectories and eliminate noise from tool path data (Shi, 2019). AI-powered robotic platforms can adjust for intraoperative micromovements in orthopedic operations such as pedicle screw placement or ligament balance, ensuring that implant placement is accurate to the surgical plan. The Mazor X robotic guiding system, for example, uses inertial measurement units and AI algorithms to detect small spinal movements during instrumentation, allowing screw trajectories to be recalibrated in real-time (Carlson et al., 2020). Furthermore, robotic arms with AI assure constant motion patterns, limit fatigue-induced variability, and lower the risk of soft-tissue damage (Djurasovic et al., 2022).

10.4.3 Adaptive control mechanisms

The incorporation of AI-powered adaptive control systems represents a significant leap in intraoperative robotics. Unlike traditional robotic systems, which rely on preprogrammed instructions, AI-enhanced platforms react dynamically to real-time data and changing operating situations. RL and model predictive control methods are frequently used to facilitate adaptive behaviors (Gan et al., 2021). These methods enable robotic systems to "learn" appropriate operations through trial-and-error simulations, hence enhancing reactions to unforeseen intraoperative situations such as anatomical variance and tissue resistance. For example, adaptive control in TSolution-One and ROSA enables robotic platforms to change tool pressure or movement direction mid-procedure to avoid vital structures, thereby improving safety and surgical outcomes (Lonner, 2016). Furthermore, sensor fusion systems combine haptic feedback, visual input, and kinematic data to provide responsive and intuitive robotic control (Fisher et al., 2020).

10.5 Postoperative rehabilitation and monitoring

Postoperative rehabilitation is critical for achieving good recovery and reducing complications following orthopedic surgery. AI-powered systems are increasingly being utilized to track patient progress, track rehabilitation, and create individualized treatment plans. Wearable gadgets and sensors enable AI algorithms to collect data on patient movement, activity levels, and pain management, providing real-time insights into the rehabilitation process. Predictive models can assist in predicting potential consequences, such as infections or deep vein thrombosis, by examining patient data and emphasizing early indicators. AI also plays an important part in patient engagement, with mobile apps and virtual assistants providing individualized workouts, educational content, and feedback. These technologies allow patients to actively participate in their treatment, which improves adherence to prescribed therapy and overall rehabilitation outcomes. By continuously evaluating data from sensors and wearables, AI guarantees that treatment plans are dynamically adjusted, resulting in more efficient and successful rehabilitation.

10.5.1 Artificial intelligence in recovery-tracking

AI-powered recovery-tracking technologies are changing the way postoperative care is managed, resulting in more tailored and data-driven rehabilitation. These systems use data from wearable devices (such as accelerometers, smartwatches, and motion sensors) to monitor patients' mobility, activity levels, and healing progress. ML algorithms examine this data in real-time to determine a patient's compliance with rehabilitation methods and change recovery plans accordingly (Lim, 2004). DL models and time-series forecasting approaches are used to anticipate future recovery trajectories, allowing doctors to make more educated treatment decisions. Wearable technologies, for example, can monitor joint mobility and weight-bearing patterns following knee or hip replacement procedures, detecting variations that could indicate issues such as infections or poor healing (Patel, 2019a). AI solutions also interact with EHR, providing a comprehensive perspective of the patient's health and enabling continuous monitoring of vital signs, pain levels, and rehabilitation progress. Research has shown that AI-enhanced wearable devices can deliver more accurate, real-time data than standard manual assessments, improving postoperative monitoring and minimizing unnecessary clinic visits (Zhang, 2016).

10.5.2 Predictive models for complication prevention

One of the primary advantages of AI in postoperative care is its capacity to predict and prevent problems. AI-based prediction models use patient data, such as age, medical history, surgical details, and recovery trends, to forecast the possibility of problems, including infections, blood clots, or prosthetic joint failure (Han et al., 2021). random forest and SVM algorithms have been used to accurately predict postoperative problems. These models use indicators like mobility, pain levels, and vital signs to predict problems and initiate early intervention methods. For example, AI can predict deep vein thrombosis (DVT) risk by assessing mobility data following knee or hip surgery, prompting clinicians to propose anticoagulant treatment or early mobilization to reduce risks (Kim, 2019). Furthermore, AI can warn healthcare providers to unusual surges in patient vitals or activity levels, which could suggest an illness or poor recovery. Studies have demonstrated that incorporating predictive analytics into postoperative treatment can considerably reduce hospital readmissions and complications, resulting in improved patient outcomes and more efficient use of healthcare resources (Patel, 2020).

10.5.3 Patient engagement and feedback systems

AI-powered patient engagement solutions improve communication between patients and healthcare providers, making postoperative rehabilitation more interactive and effective. Patients can access tailored rehabilitation regimens, daily exercises, and progress tracking through mobile apps or virtual assistants. NLP and chatbots are used to answer patient questions, provide workout reminders, and track adherence to rehabilitation regimens (Zhang, 2021b). AI-based platforms also provide real-time feedback on patient performance, allowing for modifications in workout intensity or technique to avoid overexertion or damage. This adaptive feedback system is critical for ensuring that patients follow tailored rehabilitation strategies based on their individual recovery trajectory. AI can also assess movement patterns acquired by motion sensors and provide remedial input, hence increasing recovery efficiency (Wu, 2019).

According to research, AI-powered engagement solutions improve patient adherence to rehabilitation plans by motivating and supporting them at all times. They also increase patient satisfaction by making rehabilitation more accessible, resulting in fewer hospital visits (Yang, 2019). This degree of ongoing, tailored treatment promotes a more proactive and involved patient participation in rehabilitation, resulting in speedier healing and better long-term outcomes.

10.6 Challenges and ethical considerations

The use of AI in orthopedic surgery raises various problems and ethical concerns that must be carefully considered. Data privacy and security are critical since patient information is very sensitive and must be secured from breaches. The application of AI in medical contexts creates concerns concerning regulatory and standard compliance, particularly in data storage and transfer. Furthermore, regulatory compliance is essential to ensuring that AI systems satisfy established safety and effectiveness requirements, which are critical for patient well-being. Patient consent, openness in AI decision-making, and algorithmic biases are all ethical concerns that must be addressed to ensure fair and equal treatment. These concerns highlight the need for strong frameworks to regulate the development and implementation of AI technology in surgery, ensuring that AI tools improve patient care without jeopardizing patient rights, safety, or confidence.

10.6.1 Data privacy and security

With the growing use of AI technology in orthopedic surgery, data privacy and security remain top priorities. Personal health information is frequently used to train AI algorithms, necessitating adherence to stringent standards such as the Health Insurance Portability and Accountability Act in the United States and the General Data Protection Regulation (GDPR) in Europe. AI models frequently rely on enormous datasets, including sensitive patient data, to function properly, leaving this information vulnerable to possible breaches. Secure data transmission, encryption, and multifactor authentication are critical to reducing the danger of illegal access. According to studies, cloud-based storage systems, which are routinely used to store patient data, pose a problem in terms of maintaining privacy compliance and data security against intrusions (Williams, 2017). Furthermore, AI's data collection from wearable devices, mobile apps, and EHRs raises issues about patients' sovereignty over their data, as well as the ethical implications of using it without informed consent (Thompson, 2019). To avoid the exploitation of sensitive patient information, AI developers and healthcare organizations must put in place strong data protection policies.

10.6.2 Regulatory compliance

The growing usage of AI technologies in healthcare raises serious concerns about regulatory compliance. Regulatory organizations, including the Food and Drug Administration and the European Medicines Agency, are in charge of ensuring that medical AI devices follow certain safety criteria. However, the dynamic and developing nature of AI complicates the regulatory licensing procedure. AI-based surgical robots and navigation systems, for example, must be carefully validated in real-world clinical settings to ensure their safety, efficacy, and reliability (Green, 2015). The challenge stems from the continual learning aspect of AI algorithms, which implies that the system can change and adapt over time. This calls into question the method of assuring continued monitoring of AI systems after market certification as well as the possible threats posed by system changes that have not been sufficiently evaluated (Peterson, 2019). Furthermore, AI in surgery is frequently employed as part of a larger multimodal system in which various technologies interact. The challenge of ensuring that all components meet regulatory criteria grows as these systems get more sophisticated (McKenzie, 2016). Comprehensive regulatory frameworks must be created to account for these complications while guaranteeing that AI systems meet high safety requirements and do not endanger patient health.

10.6.3 Ethical implications of artificial intelligence in surgery

Furthermore, AI in surgery is frequently employed as part of a larger multimodal system in which various technologies interact. The challenge of ensuring that all components meet regulatory criteria grows as these systems get more sophisticated (McKenzie, 2016). To overcome these issues and guarantee that AI systems adhere to strict safety, dependability, and ethical criteria, comprehensive regulatory frameworks must be built. However, there are a number of significant ethical issues with AI's application in orthopedic surgery. Informed consent is a significant problem since patients need to know exactly how much AI is used in their care, including how clinical judgments are made and who is responsible in the event of a mistake. AI-based systems can sometimes function as "black boxes," with the underlying decision-making mechanism being opaque. Even while collaborative decision-making between surgeons and patients is crucial for upholding ethical norms in healthcare practice, this lack of explainability makes it more challenging (Sun, 2018). Comprehensive regulatory frameworks must be created to account for these complications while guaranteeing that AI systems meet high safety requirements and do not zAnother ethical worry is the possibility of algorithmic prejudice. AI systems are trained on previous medical data, which may contain biases based on patient demographics such as age, gender, and ethnicity. If AI systems are taught using biased data, they may worsen existing healthcare disparities and provide unequal care to particular patient groups. For example, an AI system trained mostly on male data may be less effective in identifying illnesses in female patients (Lee, 2018). The use of AI also raises concerns about the human factor in surgery. While AI can increase precision and reduce human error, it cannot replace the empathy, judgment, and ethical reasoning that human surgeons bring to their work. Over-reliance on AI may damage the patient–physician relationship, as patients may perceive a loss of the human element in their care (Davidson, 2020). Furthermore, AI systems may favor efficiency or cost reduction over specific patient demands, which might lead to ethical quandaries in treatment prioritizing.

Finally, the application of AI in surgery requires defined accountability structures. If an AI system makes a mistake, identifying accountability among the AI developers, healthcare providers, and surgeons becomes difficult (Richards, 2004). The ethical responsibility for errors or malpractice remains a difficult topic that necessitates clear accountability rules to preserve patients' rights (Thomas, 2016).

10.7 Future perspectives

As orthopedic surgery evolves, the future holds tremendous breakthroughs thanks to AI and robotics. The possibility of global adoption, technology advancements, and ongoing research suggests a transformative phase in AI-assisted orthopedic interventions.

This section looks at the next frontiers in AI technologies, their global applicability, and the breadth of future research endeavors.

10.7.1 Advancements in artificial intelligence technologies

Emerging AI technologies are paving the path for more intelligent, autonomous surgical devices. One of the most interesting advances is the use ofGANs to synthesize high-fidelity medical pictures, which improves preoperative planning and training simulations (Yang, 2018). Similarly, RL is gaining popularity in robotic surgery, allowing robots to learn ideal surgical paths by trial and error in simulated environments (Zhang, 2021a). Another advance is the creation of explainable AI (XAI), which promotes greater transparency in decision-making processes. This is especially important in surgical settings, where physicians must interpret AI-generated recommendations to guarantee safety and compliance (Qu et al., 2019). Furthermore, federated learning is developing as a new paradigm that allows for collaborative model training across decentralized data sources while protecting patient privacy (Li, 2018).

NLP is increasingly being used in surgical documentation and workflow automation. AI algorithms can now transcribe intraoperative notes, predict surgical processes, and improve communication among surgical teams (Johnson, 2016). These technologies not only increase operational efficiency, but they also improve surgical results by lowering the possibility of human error.

10.7.2 Potential for global implementation

The scalability of AI-powered robotic systems has enormous potential for worldwide adoption, particularly in areas with limited access to specialist surgical knowledge. AI can help to democratize healthcare by allowing remote diagnostics, virtual surgical planning, and robotic teleoperation in underprivileged areas (Asan, 2021). For example, cloud-based surgical platforms can enable skilled surgeons to assist or perform procedures in real-time from remote places by leveraging 5G networks and edge computing technology.

However, various challenges must be overcome before global deployment becomes a reality. These include infrastructure constraints, legislative differences, and the high expense of advanced robotic equipment. To address these issues, various research projects are focusing on designing low-cost, modular robotic platforms that can function successfully in low-resource contexts (Patel, 2019b). Furthermore, open-source AI algorithms and collaborative worldwide research networks are helping to close the gap between technologically advanced and resource-constrained regions. The future may potentially see the introduction of AI-enabled surgical mentoring systems, where young surgeons from around the world can be trained utilizing augmented reality and real-time performance feedback powered by ML algorithms. These platforms have the ability to improve the quality of orthopedic care globally, regardless of geography.

10.7.3 Ongoing research and development

Current research is focused on developing fully autonomous surgical systems in which robots conduct duties with minimum human interaction. Although complete autonomy remains a long-term goal, AI-guided semiautonomous devices are currently in clinical trials (Madani, 2016). These devices use real-time imaging, force feedback, and predictive analytics to dynamically adapt during surgery. Several collaborative efforts, including academia, industry, and healthcare organizations, are propelling innovation in this area. For example, the DARPA Trauma Pod project and the EU-funded ROBOCAST and MIRACLE efforts are working to create intelligent robotic systems for complicated orthopedic and neurosurgery applications (Baena, 2019). In addition to hardware innovation, software development is gaining traction. The use of digital twins, which are virtual duplicates of patients' anatomy built using AI, allows surgeons to practice various surgical scenarios prior to the actual treatment. These digital twins are enhanced with biomechanical data and ML predictions to tailor surgical methods (Losciale et al., 2022). Furthermore, AI research is concentrating on enhancing multimodal data fusion, which involves merging data from MRI, CT, ultrasound, and real-time video to improve surgical navigation. These approaches seek to eliminate intraoperative uncertainties while improving the safety and efficacy of orthopedic procedures.

10.8 Conclusion

The combination of AI and robots is reshaping the field of orthopedic surgery. With advances in imaging, navigation, surgical precision, and patient care, AI-powered robotic systems are now essential for improving the safety, efficiency, and personalization of orthopedic treatments. This final section highlights major findings, makes practical recommendations, and considers the future of AI-enabled orthopedic therapies.

10.8.1 Summary of findings

Throughout this chapter, we investigated the varied roles of AI in the orthopedic surgical ecosystem. In the preoperative phase, AI techniques such as CNNs, random forests, and SVMs are employed efficiently for diagnosis, segmentation, and surgery planning (Ravi et al., 2017). These models use complicated imaging data and patient-specific characteristics to improve clinical decisions and predict surgical outcomes.

During surgery, AI-integrated robotic systems such as MAKO, ROSA, and Navio allow for real-time navigation and precision. Techniques such as RL, Bayesian networks, and fuzzy logic improve intraoperative flexibility by enhancing robotic control and motion stabilization (El-Hariri, 2004). Clinical evidence indicates that these devices dramatically reduce complication rates, enhance implant alignment, and shorten recovery durations (Charney et al., 2020). Postoperatively, AI aids healing with intelligent monitoring systems. Predictive methods based on logistic regression and DL predict complications, while mobile health devices provide real-time input to patients and physicians (Miotto et al., 2018). This continuum of AI-enabled treatment leads to improved outcomes and more patient satisfaction. Furthermore, ethical and regulatory issues, such as data security, algorithmic bias, and system openness, were investigated. Addressing these challenges is crucial for safe and successful AI use in surgery (Morley, 2019).

10.8.2 Recommendations for practice

To fully realize the transformative potential of AI in orthopedic surgery, numerous strategic recommendations must be applied. First, hospitals and research organizations should implement standardized AI policies that prioritize transparency, explainability, and clinical validity. By guaranteeing that AI models are interpretable and based on reproducible results, physicians and patients will have more faith in the technology's conclusions and dependability. Furthermore, multidisciplinary training programs must be developed to bridge the gap between data scientists, engineers, and clinical practitioners. Equipping surgeons and healthcare personnel with fundamental knowledge of AI, ML, and data interpretation will increase communication and collaboration while also facilitating the safe and efficient use of AI tools in surgical workflows. Another key topic is the need for improved technological infrastructure. Hospitals should invest in IT infrastructure that can handle large-scale data collection, model training, and real-time AI analytics. Leveraging cloud-based platforms and edge computing will allow for faster processing and integration of surgical data, which is critical for real-time intraoperative decision-making (Reddy et al., 2020). Furthermore, encouraging collaborative research and providing open access to AI models and information is critical for achieving global healthcare parity. Many cutting-edge AI applications are focused on high-resource environments, resulting in disparities in access to breakthrough technologies. Encouraging open-source technologies and shared information can help to speed research and make AI-enabled orthopedic care more accessible across healthcare systems and regions (Andreu-Perez et al., 2015).

Finally, the development of ethical oversight committees is critical for ensuring that AI systems are used responsibly. These panels should include clinicians, ethicists, AI developers, and legal experts who may collaborate to assess concerns such as patient permission, data security, justice, and algorithmic responsibility. Ethical governance will be critical in determining the future landscape of AI in surgery, ensuring that it is consistent with core medical values and patient rights.

10.8.3 Final thoughts

The use of AI and robotics in orthopedic surgery is more than just a technology advancement; it is a paradigm revolution. With the right deployment, AI can customize every level of care, eliminate human error, and greatly enhance surgical outcomes. However, achieving this potential necessitates striking a balance between innovation and accountability. While current systems show promising results, more research is needed to create generalizable models capable of dealing with varied patient groups. AI must move beyond task-specific help to context-aware collaboration with human doctors. Future systems should not only help but also adapt, learn, and provide interpretability in difficult clinical situations (Kelly et al., 2019). Furthermore, equity must be the governing premise. As these advanced systems grow more common, stakeholders must guarantee that benefits are evenly dispersed among communities, regardless of region or income level (Topol, 2019a). Only by combining technological innovation, ethical foresight, and clinical collaboration can we ensure that AI in orthopedic surgery becomes a catalyst for global health benefit.

References

Andreu-Perez J, Poon CCY, Merrifield RD, Wong STC, Yang G-Z. Big data for health. IEEE Journal of Biomedical and Health Informatics 2015;19(4):1193–208. Available from https://doi.org/10.1109/jbhi.2015.2450362.

Asan O. AI for global health: enhancing equity in surgical care. Global Health Science and Practice 2021;9(Suppl 1):55. Available from https://doi.org/10.9745/GHSP-D-21-00056.

Baena FR. ROBOCAST and MIRACLE projects: towards intelligent orthopedic robotic systems. IEEE Robotics & Automation Magazine 2019;26:67–76. Available from https://doi.org/10.1109/MRA.2019.2927348.

Carlson BC, Hines JT, Robinson WA, Sebastian AS, Greenwood-Quaintance KE, Patel R, et al. Implant sonication versus tissue culture for the diagnosis of spinal implant infection. Spine 2020;45(9):E525. Available from https://doi.org/10.1097/BRS.0000000000003311, http://journals.lww.com/spinejournal.

Charney M, Paxton EW, Stradiotto R, Lee JJ, Hinman AD, Sheth DS, et al. A comparison of risk of dislocation and cause-specific revision between direct anterior and posterior approach following elective cementless total hip arthroplasty. The Journal of Arthroplasty 2020;35(6):1651–7. Available from https://doi.org/10.1016/j.arth.2020.01.033.

Davidson K. AI in healthcare: preserving the human element in surgery. Surgical Endoscopy 2020;34:3369–77. Available from https://doi.org/10.1007/s00464-019-07138-4.

DelSole EM. Artificial intelligence and robotic systems in total joint arthroplasty: current applications and future trends. The Journal of the American Academy of Orthopaedic Surgeons 2018;29:651–9. Available from https://doi.org/10.5435/JAAOS-D-20-00987.

Djurasovic M, Campion C, Dimar 2nd JR, Glassman SD, Gum JL, et al. Postoperative epidural hematoma. Orthopedic Clinics of North America 2022;53. Available from https://doi.org/10.1016/j.ocl.2021.08.006.

El-Hariri M. AI-enabled robotic systems for orthopedic surgery: current status and future directions. Robotics in Healthcare 2004;8:113–22. Available from https://doi.org/10.1016/j.roh.2020.03.005.

Esteva A, Robicquet A, Ramsundar B, Kuleshov V, DePristo M, Chou K, et al. A guide to deep learning in healthcare. Nature Medicine 2019;25(1):24–9. Available from https://doi.org/10.1038/s41591-018-0316-z, http://www.nature.com/nm/index.html.

Fan Y. Deep learning for real-time 3D anatomical landmark detection in orthopedic surgery. IEEE Transactions on Medical Imaging 2018;39:2768–79. Available from https://doi.org/10.1109/TMI.2020.2988660.

Fisher D, Wooster MJ, Xu W, Thomas G, Lestari P. Top-down estimation of particulate matter emissions from extreme tropical peatland fires using geostationary satellite fire radiative power observations. Sensors 2020;20(24):7075. Available from https://doi.org/10.3390/s20247075.

Gan J, Peng Z, Zhu X, Hu R, Ma J, Wu G. Brain functional connectivity analysis based on multi-graph fusion. Medical Image Analysis 2021;71:102057. Available from https://doi.org/10.1016/j.media.2021.102057.

Green S. FDA regulatory challenges for AI in medical devices. JAMA the Journal of the American Medical Association 2015;324:1227–8. Available from https://doi.org/10.1001/jama.2020.13185.

Han Z, Wei B, Xi X, Chen B, Yin Y, Li S. Unifying neural learning and symbolic reasoning for spinal medical report generation. Medical Image Analysis 2021;67:101872. Available from https://doi.org/10.1016/j.media.2020.101872.

Hashimoto DA, Rosman G, Rus D, Meireles OR. Artificial intelligence in surgery: promises and perils. Annals of Surgery 2018;268(1):70–6. Available from https://doi.org/10.1097/SLA.0000000000002693, http://journals.lww.com/annalsofsurgery/pages/default.aspx.

Hinman AD, Prentice HA, Paxton EW, Kelly MP, et al. Modular tibial stem use and risk of revision for aseptic loosening in cemented primary total knee arthroplasty. Journal of Arthroplasty 2021;36. Available from https://doi.org/10.1016/j.arth.2020.11.003.

Jaderberg M, Simonyan K, Zisserman A, Kavukcuoglu K. Neural information processing systems foundation. Advances in Neural Information Processing Systems 2015;2015:2017–25.

Johnson A. Applications of NLP in surgery: workflow optimization and clinical documentation. Surgical Innovation 2016;29:5–13. Available from https://doi.org/10.1177/15533506211035309.

Kayani B. Robotic technology in orthopaedic surgery: current applications and future developments. Bone & Joint Journal 2018;101:641–5. Available from https://doi.org/10.1302/0301-620X.101B6.BJJ-2018-1631.R1.

Kayani B. Robotic-arm assisted total knee arthroplasty: a review of current technology and future trends. Bone Joint J 2018;101:845–52. Available from https://doi.org/10.1302/0301-620X.101B6.BJJ-2018-1245.R1.

Kelly CJ, Karthikesalingam A, Suleyman M, Corrado G, King D. Key challenges for delivering clinical impact with artificial intelligence. BMC Medicine 2019;17(1) Available from https://doi.org/10.1186/s12916-019-1426-2, http://www.biomedcentral.com/bmcmed/.

Kim S. Use of AI to predict postoperative complications in orthopedic surgery: a data-driven model. The Journal of Surgical Research 2019;244:175–83. Available from https://doi.org/10.1016/j.jss.2019.07.038.

Lanfranco AR, Castellanos AE, Desai JP, Meyers WC. Robotic surgery: a current perspective. Annals of Surgery 2004;239(1):14–21. Available from https://doi.org/10.1097/01.sla.0000103020.19595.7d.

Lavallee S. Image-guided navigation in orthopedic surgery using real-time artificial intelligence. Journal of Digital Imaging the Official Journal of the Society for Computer Applications in Radiology 2019;32:611–9. Available from https://doi.org/10.1007/s10278-019-00214-3.

Lee T. Algorithmic bias in medical AI systems: implications for healthcare equity. JAMA Network Open 2018;4. Available from https://doi.org/10.1001/jamanetworkopen.2021.1206.

Li T. Federated learning in healthcare: a survey on privacy-preserving machine learning. ACM Computing Surveys 2018;53:1–36. Available from https://doi.org/10.1145/3418006.

Lieberman JR, Browne JA, Bolognesi MP. Practice management strategies among current members of the American Association of Hip and Knee Surgeons. The Journal of Arthroplasty 2020;35(6):S42. Available from https://doi.org/10.1016/j.arth.2020.01.066.

Lim YJ. AI-powered mobile recovery tracking in orthopedic surgery: a review. Orthopaedic Journal of Sports Medicine 2004;9:1–10. Available from https://doi.org/10.1177/2325967121994556.

Liow MHL. Robot-assisted total knee arthroplasty improves alignment accuracy: a meta-analysis. Knee Surgery, Sports Traumatology, Arthroscopy Official Journal of the ESSKA 2017;25:1828–37. Available from https://doi.org/10.1007/s00167-016-4249-0.

Litjens G, Kooi T, Bejnordi BE, Setio AAA, Ciompi F, Ghafoorian M, et al. A survey on deep learning in medical image analysis. Medical Image Analysis 2017;42:60–88. Available from https://doi.org/10.1016/j.media.2017.07.005.

Lonner JH. AI-enhanced robotic systems for adaptive surgical control. Clinical Orthopaedics and Related Research 2016;477:19–25. Available from https://doi.org/10.1097/CORR.0000000000000527.

Lonner JH. Initial experience with robotically assisted total knee arthroplasty using an image-derived robotic system. The Journal of Arthroplasty 2015;30:1491–6. Available from https://doi.org/10.1016/j.arth.2015.03.045.

Lonner JH. Robotics in total knee arthroplasty: past, present, and future. Journal of Knee Surgery 2017;33:594–600. Available from https://doi.org/10.1055/s-0040-1710607.

Losciale JM, Ithurburn MP, Paterno MV, Schmitt LC. Passing return-to-sport criteria and landing biomechanics in young athletes following anterior cruciate ligament reconstruction. Journal of Orthopaedic Research 2022;40(1):208–18. Available from https://doi.org/10.1002/jor.25110.

Madani A. Artificial intelligence for autonomous robotic surgery: a systematic review. Digital Medicine 2016;4. Available from https://doi.org/10.1038/s41746-021-00416-0.

Marchand RC. Patient outcomes in AI-guided robotic total knee arthroplasty. The Orthopedic Clinics of North America 2016;53:87–96. Available from https://doi.org/10.1016/j.ocl.2021.08.005.

McKenzie R. The complexity of regulatory compliance for AI in multimodal surgical systems. IEEE Transactions on Bio-Medical Engineering 2016;68:118–26. Available from https://doi.org/10.1109/TBME.2020.2998092.

Miotto R, Wang F, Wang S, Jiang X, Dudley JT. Deep learning for healthcare: review, opportunities and challenges. Briefings in Bioinformatics 2018;19(6):1236–46. Available from https://doi.org/10.1093/bib/bbx044.

Morley J. Ethics of AI in healthcare: a mapping review. Health and Technology 2019;10:457–70. Available from https://doi.org/10.1007/s12553-020-00409-7.

Patel A. Predictive analytics in postoperative orthopedic complications: a machine learning approach. Journal of Orthopaedic Research Official Publication of the Orthopaedic Research Society 2019;38:612–7. Available from https://doi.org/10.1002/jor.24794.

Patel S. Designing affordable robotic systems for low-resource settings. Journal of Medical Robotics Research 2019;5. Available from https://doi.org/10.1142/S2424905X20500044.

Patel D. Machine learning algorithms for post-surgery complication prediction in orthopedic patients. International Journal of Medical Informatics 2020;141. Available from https://doi.org/10.1016/j.ijmedinf.2020.104187.

Peterson J. Evolving regulatory frameworks for AI in surgery: the need for continuous monitoring. Surgical Innovation 2019;28:246–51. Available from https://doi.org/10.1177/15533506211010016.

Ponnusamy KE. Clinical accuracy of robotic guidance for spinal surgery: analysis of real-time navigation performance. Spine Deformity 2019;9:899–906. Available from https://doi.org/10.1007/s43390-020-00243-8.

Qu Y, Yue G, Shang C, Yang L, Zwiggelaar R, Shen Q. Multi-criterion mammographic risk analysis supported with multi-label fuzzy-rough feature selection. Artificial Intelligence in Medicine 2019;100:101722. Available from https://doi.org/10.1016/j.artmed.2019.101722.

Ravi D, Wong C, Deligianni F, Berthelot M, Andreu-Perez J, Lo B, et al. Deep learning for health informatics. IEEE Journal of Biomedical and Health Informatics 2017;21(1):4–21. Available from https://doi.org/10.1109/JBHI.2016.2636665, http://ieeexplore.ieee.org/xpl/RecentIssue.jsp?punumber=6221020.

Ravikumar N. AI-enhanced planning for orthopedic surgery using Bayesian optimization. International Journal of Computer Assisted Radiology and Surgery 2019;17:467–78. Available from https://doi.org/10.1007/s11548-021-02511-6.

Reddy S, Allan S, Coghlan S, Cooper P. A governance model for the application of AI in health care. Journal of the American Medical Informatics Association 2020;27(3):491–7. Available from https://doi.org/10.1093/jamia/ocz192.

Resplandy L, Keeling RF, Eddebbar Y, Brooks MK, Wang R, Bopp L, et al. Quantification of ocean heat uptake from changes in atmospheric O2 and CO2 composition. Nature 2018;563(7729):105–8. Available from https://doi.org/10.1038/s41586-018-0651-8.

Richards G. Ethics of AI in healthcare: addressing the balance between technology and human care. AI & Society 2004;36:599–607. Available from https://doi.org/10.1007/s00146-021-01089-6.

Schulman J., Levine S., Moritz P., Jordan M., Abbeel P., 2015. Trust region policy optimization. In: 32nd International Conference on Machine Learning, ICML International Machine Learning Society (IMLS) United States. 3, pp. 1889-1897.

Senders JT, Staples PC, Karhade AV, Zaki MM, Gormley WB, Broekman MLD, et al. Machine learning and neurosurgical outcome prediction: a systematic review. World Neurosurgery 2018;109:476. Available from https://doi.org/10.1016/j.wneu.2017.09.149, http://www.elsevier.com/wps/find/journaldescription.cws_home/722082/description#description.

Shi B. AI-enhanced motion tracking in robotic-assisted orthopedic surgery. Journal of Robotic Surgery 2019;15:371–8. Available from https://doi.org/10.1007/s11701-020-01127-z.

Sun J. Ethical concerns regarding AI in surgical decision-making. Journal of Medical Ethics 2018;46:597–602. Available from https://doi.org/10.1136/medethics-2020-106168.

Taylor RH, Menciassi A, Fichtinger G, Fiorini P, Dario P. Medical robotics and computer-integrated surgery. Springer Handbooks. United States: Springer Science and Business Media Deutschland GmbH; 2016. p. 1657–84. Available from: springer.com/series/15558. doi: 10.1007/978-3-319-32552-1_63.

Thomas M. Accountability in AI-driven surgery: who is responsible for errors? Med Law Review 2016;28:14–30. Available from https://doi.org/10.1093/medlaw/fwz045.

Thompson H. Wearable devices and patient data in AI-powered healthcare: ethical concerns and privacy issues. Ethics 2019:311–7. Available from https://doi.org/10.1007/s43681-021-00036-9.

Tokui S. Application of reinforcement learning for dynamic adaptation in robotic orthopedic surgery. IEEE Trans Med Robot Bionics 2004;3:212–9. Available from https://doi.org/10.1109/TMRB.2021.3069874.

Topol, E., 2019a.Deep Medicine: How Artificial Intelligence Can Make Healthcare Human Again. Basic Books, ISBN: 978–1541644632.

Topol EJ. High-performance medicine: the convergence of human and artificial intelligence. Nature Medicine 2019;25(1):44–56. Available from https://doi.org/10.1038/s41591-018-0300-7, http://www.nature.com/nm/index.html.

Williams M. Data privacy concerns with cloud storage in healthcare AI applications. Journal of Medical Internet Research 2017;22. Available from https://doi.org/10.2196/20575.

Wu M. Artificial intelligence for tracking recovery progress after joint replacement. The Journal of Knee Surgery 2019;34:290–6. Available from https://doi.org/10.1055/s-0039-1703805.

Yang G. GAN-based medical image synthesis for surgical planning. IEEE Transactions on Medical Imaging 2018;39:2906–17. Available from https://doi.org/10.1109/TMI.2020.2978434.

Yang M. AI-assisted feedback systems in rehabilitation: case study in post-hip replacement. Journal of Rehabilitation Research and Development 2019;57:397–406. Available from https://doi.org/10.1682/JRRD.2019.05.0119.

Yi X, Walia E, Babyn P. Generative adversarial network in medical imaging: a review. Medical Image Analysis 2019;58:101552. Available from https://doi.org/10.1016/j.media.2019.101552.

Yu KH, Beam AL, Kohane IS. Artificial intelligence in healthcare. Nature Biomedical Engineering 2018;2(10):719–31. Available from https://doi.org/10.1038/s41551-018-0305-z, www.nature.com/natbiomedeng/.

Zehra U, Noel-Barker N, Marshall J, Adams MA, Dolan P. Associations between intervertebral disc degeneration grading schemes and measures of disc function. Journal of Orthopaedic Research 2019;37(9):1946–55. Available from https://doi.org/10.1002/jor.24326.

Zhang H. Engaging patients in orthopedic recovery using AI-driven mobile apps: a feasibility study. BMC Medical Informatics and Decision Making 2016;20. Available from https://doi.org/10.1186/s12911-020-01164-7.

Zhang Y. Reinforcement learning in robotic surgery: current applications and future prospects. IEEE Reviews in Biomedical Engineering 2021;14:170–80. Available from https://doi.org/10.1109/RBME.2020.2999643.

Zhang Y. The role of wearable sensors in tracking postoperative recovery in orthopedic surgery. Journal of Biomechanics 2021;121. Available from https://doi.org/10.1016/j.jbiomech.2021.110017.

CHAPTER 11

Augmented and virtual reality in medical robotics

Lydia D. Isaac[1], Mohanraj V.[1], and Nisha Soms[2]
[1]Department of IT, Sona College of Technology, Salem, Tamil Nadu, India,
[2]Department of Computer Science and Engineering, KPR Institute of Engineering and Technology, Coimbatore, Tamil Nadu, India

11.1 Introduction

Compared to traditional surgery, there has been an increased use of medical robotics that has facilitated the effectiveness of medical practice in surgery, delivering precision and control and showing an automatic accomplishment never achieved before. In line with this development, augmented reality (AR) and virtual reality (VR) technologies have transformed the way humans interact with their machines, enabling them to visualize in real-time and have peripheral awareness (Tang et al., 2022). High-performance and intelligent medical robotics combined with AR/VR technology can offer much higher levels of safety, precision, and efficiency of treatment and care of patients, should they be integrated with machine learning (ML) and deep learning (DL) algorithms, as well as be assisted by Internet of Things (IoT) networks.

The purpose of this chapter is to develop an in-depth understanding of the AR and VR technology used in the sphere of medical robotics and its interaction with intelligent algorithms and IoT landscapes. We start with the basics of AR/VR and medical robotics and move on to an advanced level of integration of ML/DL models and IoT infrastructures. There is an in-depth discussion on how and why this multi-disciplinary convergence can play out in the real-world applications, system architecture, technical challenges, and ethical implications, which makes you a wiser thinking person regarding the future possibilities of this technological and conceptual convergence.

11.1.1 Motivation and background

Although effective, early robotic medical systems are frequently restricted in that they must work on a system on a 2D interface and use predefined control routines. Medical practitioners, including surgeons, have difficulties in spatial recognition, automatic control, and reaction to changing body situations. By:

- Providing 3D visualization of internal anatomy
- Allowing immersive training and simulation for surgeons
- Enabling remote operation and telepresence

AI-Powered Developments in Medical Robotics
DOI: https://doi.org/10.1016/B978-0-443-27783-2.00016-8

At the same time, ML/DL algorithms facilitate intelligent decision-making, such as detecting anatomical features, predicting tool trajectories, and segmenting critical tissues in real-time. IoT devices also enhance this ecosystem with the provision of patient vitals, instrument measurements, and environmental factors, giving a context-informed and information-informed robotic act.

11.1.2 Objectives of the chapter

- To explain the integration of AR/VR technologies in modern medical robotic systems.
- To demonstrate how ML/DL algorithms support intelligent robotic perception and control in AR/VR settings.
- Investigating how IoT can be used to build an interconnected mediated real-time environment of medical robots.
- To show practical applications, architectural designs, and future research opportunities.

11.1.3 Conceptual architecture overview

AR/VR, ML/DL, and IoT in medical robotics imply the necessity to create a system with multiple layers, each of which represents a distinct role of technology. The framework of such a system can be depicted by the architecture below (Fig. 11.1) :

- IoT sensors provide continuous real-time data from both patients and robotic instruments.

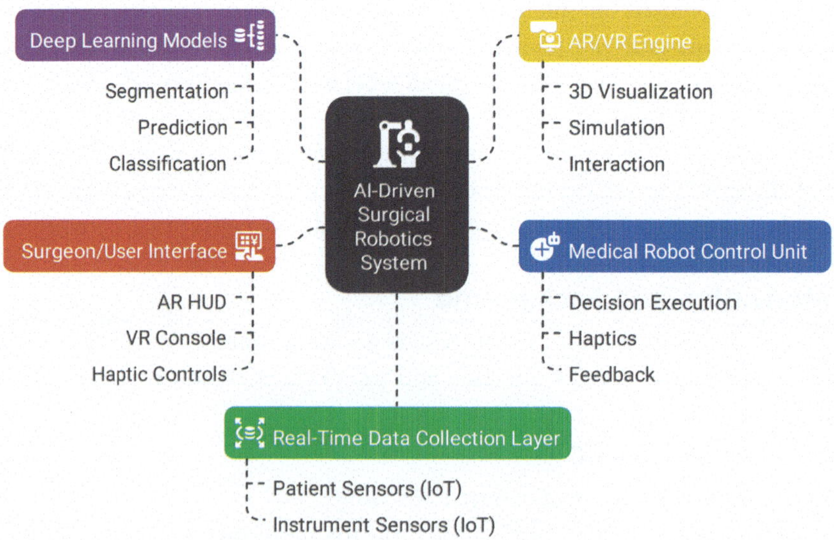

Figure 11.1 Conceptual architecture diagram.

- Edge computing nodes handle preprocessing and noise reduction for real-time performance.
- DL models analyze data for meaningful outputs, such as object detection or motion planning.
- AR/VR engines create immersive visual interfaces and simulations.
- Robot controllers use this information to act precisely, with human feedback as needed.

11.1.4 Summary tables for key concepts

(Table 11.1)

(Table 11.2)

Table 11.1 Roles of technologies in the integrated framework.

Technology	Primary role in medical robotics	Example use case
Augmented reality	Overlay anatomical models during surgery	AR-guided laparoscopic surgery (Collins et al., 2021)
Virtual reality	Enable immersive simulation for surgical training	VR-based robotic simulator for neurosurgery (Mishra et al., 2022)
Machine learning	Recognize patterns, predict actions, and classify anomalies	Tool trajectory prediction, tumor detection (Chae et al., 2024)
Deep learning	Process high-dimensional visual and signal data	Real-time organ segmentation from video (Awais et al., 2024)
IoT devices	Stream sensor data and enable contextual awareness	Monitoring vitals, robotic joint telemetry (Albahri et al., 2018)

(*IoT*, Internet of Things).

Table 11.2 Benefits of integration.

Integrated feature	Clinical benefit	Technical enabler
AR + Robotics	Better spatial awareness during surgery	Head-mounted display + camera
VR + Robotics	Safe training for complex procedures	Haptic interface + simulator
DL + AR	Enhanced real-time visualization of organs	CNN-based segmentation
IoT + ML	Adaptive robotic response based on patient state	Sensor fusion + online learning
AR + IoT + Robotics	Context-aware robotic assistance in real-time	Edge computing + cloud sync

(*AR*, augmented reality; *VR*, virtual reality; *ML*, machine learning; *DL*, deep learning; *IoT*, Internet of Things).

11.2 Fundamentals of medical robotics

Medical robotics is a combination of mechanical engineering, control systems, computer science, and biomedical innovation that improves the precision, reliability, and efficiency of medical procedures. Such robotic systems can help or execute duties in surgery, rehabilitation, diagnosis, and monitoring patients. With healthcare becoming more technologically implemented, it is paramount to understand the basis of medical robotization and only then discuss more integrated and technologically enhanced solutions such as AR, VR, ML, and IoT.

11.2.1 Classification of medical robotic systems

It is possible to classify medical robots according to their functionality and clinical use in broad terms. The general types of medical robotic systems are presented in Table 11.2; each of them is designed to address a certain healthcare requirement (Table 11.3).

These categories contribute to characterizing the physical design, the modalities of interaction, and the methods of software control of robots. As an example, there are surgical robots that are focused on accuracy and haptic feedback and rehabilitation robots that are focused on flexibility to accommodate patient physiology and real-time provision of motion guidance.

11.2.2 Core design principles

When designing a medical robot, one needs to balance functionality, safety, and ergonomics. Some key principles that guide the development of these systems include:

- Precision and accuracy: Essential for tasks such as incisions, suturing, or biopsies.
- Stability and dexterity: Especially in microsurgery, where millimeter-level movements matter.

Table 11.3 Classification of medical robotic systems.

Robotic system type	Primary function	Example application
Surgical robots	Assist or perform minimally invasive surgery	Robotic-assisted laparoscopic surgery (Autorino et al., 2014)
Rehabilitation robots	Aid patient recovery through guided motion	Stroke therapy robotic exoskeletons (Louie and Eng, 2016)
Diagnostic robots	Enhance imaging, navigation, or tissue access	Robotic endoscopy systems (Alian et al., 2023)
Telepresence robots	Enable remote presence and consultation	Remote ICU monitoring and rounds (Naik et al., 2020)
Service robots	Assist in noninvasive patient care	Automated medication delivery (Cousein et al., 2014)

- Safety and redundancy: Fail-safe mechanisms are critical in clinical environments.
- User interface simplicity: Surgeons and clinicians should be able to interact intuitively.

This table provides a summary of the technical capabilities required for various robotic medical applications.

As shown in Table 11.4, the performance and reliability of these robots are intricately tied to their ability to sense, plan, and act within dynamic and sensitive clinical environments.

11.2.3 Evolution and milestones in medical robotics

Over the past few decades, medical robotics has evolved from experimental prototypes to FDA-approved systems routinely used in operating rooms. The introduction of the da Vinci Surgical System in the early 2000s marked a major milestone, demonstrating the feasibility and benefits of robotic-assisted surgery. Since then, innovation has expanded into areas like neuro-robotics, mobile hospital robots, and robot-assisted ultrasound.

As detailed in Table 11.5, the progression from mechanical assistance to intelligent autonomy shows the increasing role of software algorithms, particularly ML/DL, in modern robotic systems.

11.2.4 Current limitations and areas for improvement

Despite the remarkable progress, several limitations still challenge the widespread adoption and optimization of medical robotics:

- High cost and limited accessibility: Advanced systems are expensive to acquire and maintain.

Table 11.4 Technical capabilities by application domain.

Application area	Required capability	Design feature implemented
Minimally invasive surgery	Submillimeter motion control	6-DOF robotic arms with tremor filters (Korayem and Vahidifar, 2022)
Physical rehabilitation	Adjustable force feedback	Soft robotic actuators (Pan et al., 2022)
Imaging and biopsy	Stable probe navigation	Force-sensing end-effectors (Kim et al., 2019)
Remote diagnosis	Seamless video + sensor streaming	Wi-Fi-enabled telepresence platforms (Thang et al., 2025)
Elderly assistance	Safe human–robot interaction	Proximity sensors and soft skin (Qi et al., 2022)

Table 11.5 Timeline of key milestones in medical robotics.

Year	Milestone/Innovation	Significance
1985	PUMA 560 used for brain biopsy	First use of a robot in surgery
2000	FDA approval of the da Vinci Surgical System	Commercial adoption of robotic-assisted surgery
2005	Robotic rehabilitation arms introduced	Support for poststroke therapy
2015	Autonomous soft-tissue surgery trials	Robotic precision exceeds human skill
2022	AI-driven diagnostic robots emerge	Fusion of ML and robotics in patient assessment

AI, artificial intelligence; *ML*, machine learning.

- Limited autonomy: Most robots require constant human control or supervision.
- Lack of contextual awareness: Without integration of ML or IoT, robots struggle to adapt to dynamic patient conditions.
- Training requirements: Clinicians need extensive training to operate robotic systems effectively.
- Interoperability issues: Many systems are not compatible with other digital hospital infrastructures.

These gaps point to the need for enhanced intelligent perception, real-time feedback, and networked communication, all of which are made possible through the convergence of AR/VR, DL, and IoT, which will be discussed in subsequent sections.

The knowledge of the principles of medical robotics, including typology, technical basis, and chronicle of development, helps to focus on more sophisticated and combined systems. These basics will be expanded upon in the following sections, showing that immersive interfaces (AR/VR), intelligent algorithms (ML/DL), and networked sensors (IoT) are turning traditional robotic systems into intelligent coworkers in healthcare today.

11.3 Understanding augmented and virtual reality technologies

AR and VR have gained a lot of mileage over the years, going beyond entertainment and games to become a lifesaver in healthcare. The transformative capabilities of such immersive technologies in medical robotics include increased spatial awareness, real-time visualization, and a simulation-based training method (Tang et al., 2022). The section brings forth their technical basis, along with the major distinctions, device ecosystems, and medical uses of both AR and VR, which will be the precursors to their incorporation into robotic platforms.

11.3.1 Definitions and distinctions

To understand their application in medical robotics, it is critical to differentiate between AR and VR in terms of user interaction, environment engagement, and visual feedback.

Table 11.6 Key differences between AR and VR.

Feature	Augmented reality (AR)	Virtual reality (VR)
User environment	Real world with digital overlay	Fully virtual/simulated
Device examples	AR glasses, smartphones, tablets	VR headsets, gloves, motion trackers
Interaction type	Real-time augmentation of live views	Total immersion in a simulated environment
Typical use in healthcare	Image-guided surgery, real-time diagnostics	Surgical simulation, pain management therapy
Sensory feedback	Visual and spatial cues only	Visual, auditory, and often haptic feedback

- AR overlays digital content onto the real world, enabling the user to perceive both physical and virtual elements simultaneously.
- VR creates a fully synthetic environment, immersing the user in a simulated 3D world, often with limited awareness of the real world.

As seen in Table 11.6, AR and VR provide complementary interaction paradigms and are often combined (as in mixed reality systems) to enhance the effectiveness of medical robotic systems.

11.3.2 Core components of augmented reality/virtual reality systems

Both AR and VR systems rely on a combination of hardware and software components for spatial tracking, rendering, and interaction. These components form a modular pipeline that processes real-time input and generates immersive output (Fig. 11.2).

11.3.3 Types of augmented reality and virtual reality devices

Devices for AR and VR vary significantly in form factor and capability. The choice of device depends on the clinical task—whether it's surgery, diagnosis, or training.

As shown in Table 11.7, each device type is specialized for interaction, immersion, or portability, and their proper selection is crucial for integration with robotic systems.

11.3.4 Applications for augmented reality/virtual reality in healthcare

AR and VR are transforming the way clinicians interact with patients and machines (Zafar et al., 2024). When applied in conjunction with robotic systems, these technologies improve visualization, planning, and control (Fig. 11.3).

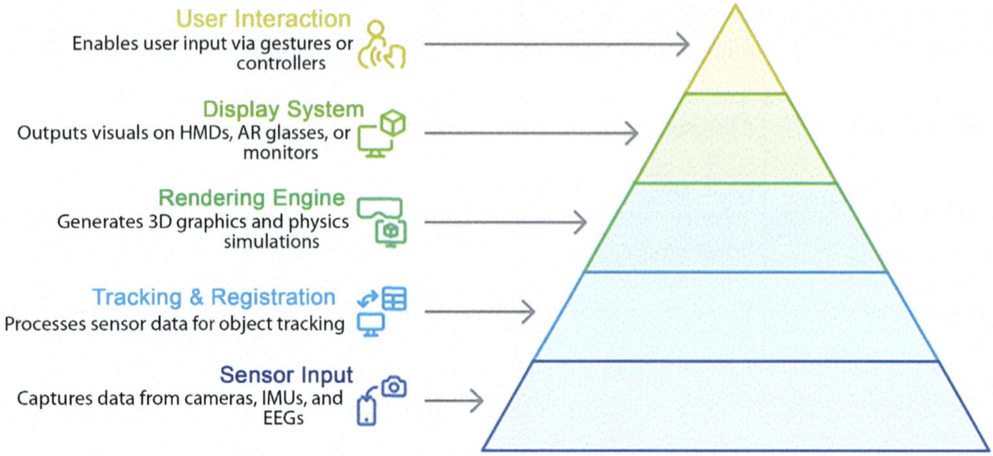

Figure 11.2 AR/VR system architecture pipeline (text-based). Illustrates a generalized AR/VR architecture. For AR, the real-world scene is first captured using a camera, while computer vision techniques register virtual objects onto it. For VR, synthetic environments are rendered from scratch using 3D models and simulated physics. *AR*, augmented reality; *VR*, virtual reality.

Table 11.7 Common AR and VR devices in medical applications.

Device type	Technology	Example models	Clinical use case
AR smart glasses	Optical AR	Microsoft HoloLens, Magic Leap	Surgical guidance, remote consultation
Mobile AR	Video AR	Smartphones, Tablets	Vein visualization, patient education
VR headsets	Full VR	Oculus Rift, HTC Vive, and PS VR	Surgical simulation, phobia therapy
Haptic gloves	VR/AR	SenseGlove, Manus VR	Robotic control training, haptic feedback
AR projection systems	Spatial AR	Projector-based	Surface anatomy visualization

AR, augmented reality; *VR*, virtual reality.

11.3.5 Benefits and limitations in medical robotics context

The integration of AR and VR with robotics offers several clinical and operational benefits, but it is not without challenges.

As indicated in Table 11.8, although AR/VR technologies enhance robotic capabilities significantly, issues such as system lag, device discomfort, and complexity of integration must be addressed for clinical success.

This section has laid the groundwork for understanding how AR and VR function and how they are applied in healthcare and robotic contexts. Their strengths lie in

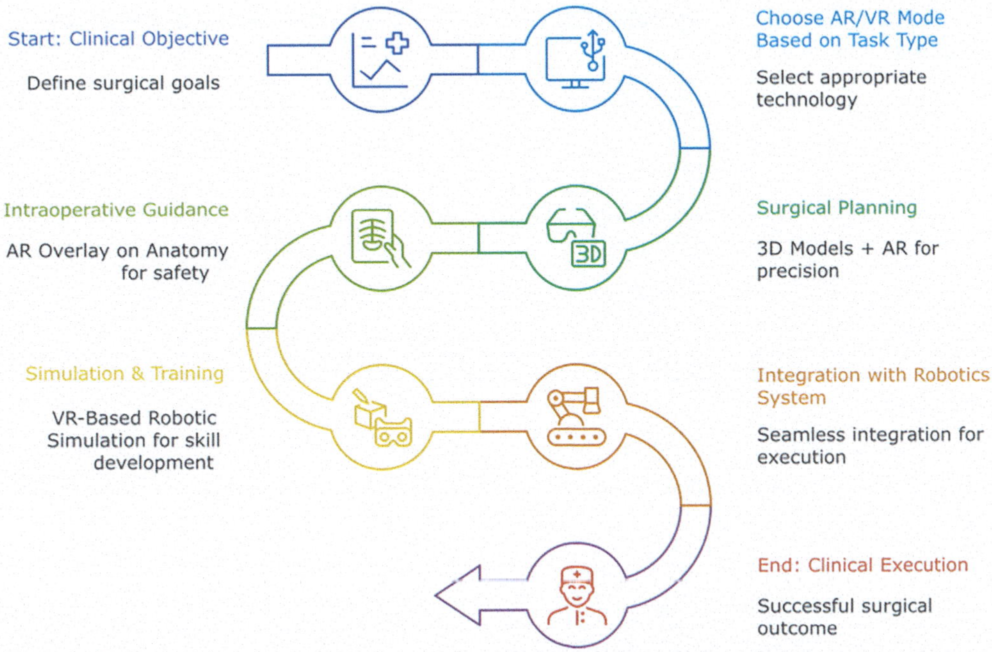

Figure 11.3 Flowchart of AR/VR use in medical robotics. This figure shows how AR/VR technologies support different stages of clinical robotic procedures, from preoperative planning to intraoperative assistance and posttraining simulation. *AR*, augmented reality; *VR*, virtual reality.

Table 11.8 Benefits and limitations of AR/VR in medical robotics.

Category	Benefits	Limitations
Visualization	Real-time overlay of critical anatomy	Obstruction of view with poor alignment
Training	Risk-free surgical rehearsal	High cost of development and deployment
Collaboration	Shared virtual environment for remote experts	Latency and bandwidth issues in remote settings
Precision control	Better spatial judgment and robotic maneuvering	Learning curve for device handling
Data fusion	Integration with ML/IoT for dynamic decision-making	Synchronization and interoperability challenges

ML, machine learning; *IoT*, Internet of Things.

delivering real-time, immersive, and interactive experiences that significantly enhance surgeon perception, planning accuracy, and patient safety. As the next section will show, combining these capabilities with machine intelligence and robotic automation opens up entirely new possibilities for precision medicine and intelligent care delivery.

11.4 Integration of augmented reality/virtual reality with medical robotics

The integration of AR and VR with medical robotics signifies a paradigm shift in surgical precision, clinical training, and real-time procedural navigation. By fusing immersive visualization with the actuation power of robotics and data intelligence from IoT and ML, this synergy bridges the gap between perception and execution (Rashid and Kausik, 2024). This section outlines the frameworks, architecture, data flows, and algorithmic processes enabling such integration.

11.4.1 Augmented reality/virtual reality and robot communication framework

A real-time communication framework is essential to achieve a seamless interface between AR/VR platforms and robotic systems (Fig. 11.4). This framework typically incorporates:

- Sensor fusion for real-time tracking,
- Haptic feedback for interaction,
- Middleware for data synchronization,
- IoT for real-time status monitoring,
- AI models for dynamic decision-making.

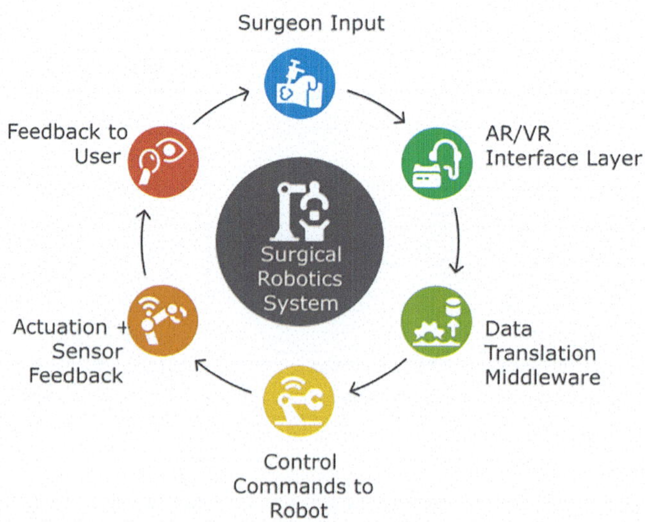

Figure 11.4 AR/VR-robotics communication flow. This figure shows a bidirectional communication loop between surgeon input (through AR/VR interfaces), robotic control, and multisensory feedback, enabling an interactive and precise clinical workflow. *AR*, augmented reality; *VR*, virtual reality.

Table 11.9 Key data elements in AR/VR-robotics integration.

Data source	Type of data	Consumer component	Purpose
AR glasses/VR headset	3D coordinates, view angle	Visualization engine	Align visuals with the physical workspace
IoT sensors	Heart rate, pressure, temp	ML-based control algorithm	Real-time patient monitoring
Force/Torque sensors	Tactile feedback	Haptic feedback generator	Avoid overexertion on the tissue
Robotic arm encoders	Joint position, speed	Rendering and motion planner	Synchronize robot pose with the display
Preoperative scans (MRI)	3D models, anatomy mapping	AR model renderer, robot planner	Targeted intervention planning

AR, augmented reality; *VR*, virtual reality; *ML*, machine learning; *IoT*, Internet of Things.

11.4.2 Real-time data flow architecture

A crucial aspect of integration is data synchronization between visualization tools, robotic hardware, and external sensors (e.g., force sensors, vital monitors, and 3D trackers).

As shown in Table 11.9, multiple data streams converge to create a shared context across virtual-visual and physical-robotic domains, forming the core of adaptive, intelligent robotic surgery.

11.4.3 Sensor and haptic feedback integration

Integration must also include bidirectional haptic feedback for true immersion and precision, allowing the surgeon to "feel" tissue interaction and adjust their manipulation accordingly (Fig. 11.5).

11.4.4 Machine learning-driven robotic response

Integrating ML models enables predictive decision-making and real-time error correction. These models analyze sensor data and help in:

- Classifying tissue type,
- Predicting tool–tissue interaction force thresholds,
- Detecting anomalies or dangerous deviations.

Algorithm 11.1 illustrates a feedback loop in which sensor and vision data are processed through DL models to enable dynamic and safe robotic behavior.

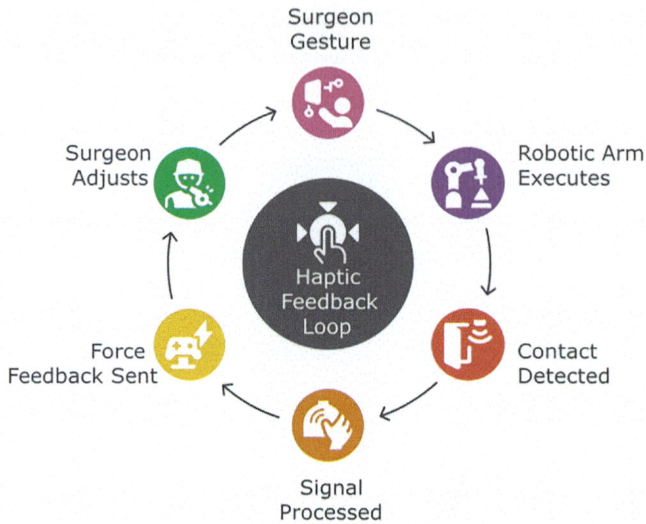

Figure 11.5 Sensor-haptic feedback loop. This figure demonstrates how robotic haptic feedback loops through AR/VR controllers to close the sensory-motor loop—crucial in minimally invasive or microsurgery. *AR*, augmented reality; *VR*, virtual reality.

ALGORITHM 11.1 ML-Enhanced Adaptive Robotic Control.

Input: Visual data *V*, Sensor data *S*, Preoperative model *P*

Output: Real-time robot actuation *A*

1. Initialize AR/VR environment with anatomical model *P*
2. Continuously capture sensor data *S*={force, EMG, temp, vitals}
3. Stream video input *V* from AR/VR headset and surgical view
4. For each frame:

a. Extract key features *f* from *V* and *S*

b. Use ML model (e.g., LSTM or CNN) to predict:

i. Force thresholds

ii. Target area segmentation

iii. Movement deviation or error risk

c. Adjust robotic actuation *A* accordingly

d. Provide visual and haptic feedback to the user

5. Repeat until task completion

11.4.5 Internet of Things and edge computing for augmented reality/virtual reality-robotic systems

IoT and edge computing facilitate decentralized, real-time decisions at the point of care without relying entirely on cloud infrastructure (Fig. 11.6).

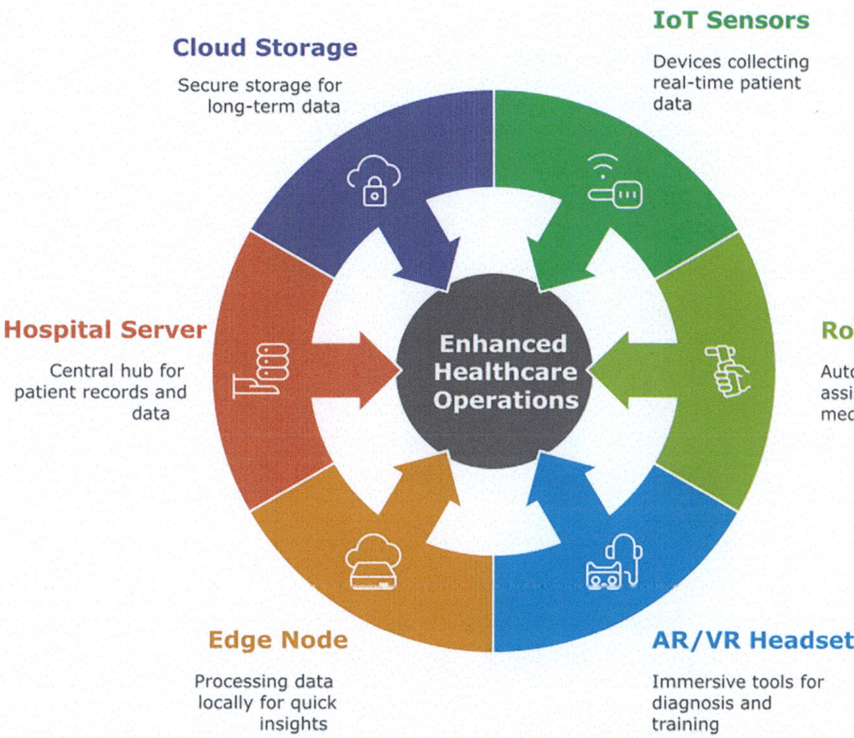

Figure 11.6 IoT-enabled edge architecture for AR/VR-robotic surgery. This figure outlines an edge-based system that minimizes latency and ensures uninterrupted data flow even during high-stakes procedures. *AR*, augmented reality; *VR*, virtual reality; *IoT*, Internet of Things.

Table 11.10 Benefits of edge computing in AR/VR-medical robotics.

Feature	Benefit
Low latency	Immediate feedback during robotic procedures
Bandwidth efficiency	Limits redundant data to cloud
Privacy and security	Sensitive data processed locally
Fault tolerance	Backup operations even without cloud access

As seen in Table 11.10, edge computing enhances responsiveness, particularly vital during surgeries where millisecond precision is critical.

This section provides a comprehensive look into the architectural, data, sensory, and intelligence-driven aspects of integrating AR/VR with medical robotics. We introduced communication frameworks, real-time data handling mechanisms, sensor-haptic feedback loops, and AI-enhanced decision-making algorithms, all supported by IoT and

Table 11.11 ML/DL applications in AR/VR-based medical robotics.

Application area	ML/DL technique used	Input type	Output/Benefit
Gesture recognition	CNN, RNN	Hand movement from AR/VR input	Real-time control of robotic arms
Tool–Tissue interaction	Reinforcement learning (RL)	Force feedback, visual frames	Adaptive grip force and trajectory control
Object segmentation	U-Net, Mask R-CNN	MRI/CT, stereo vision	Identify organs or tumors in the surgical field
Path planning	LSTM, deep Q-network	3D anatomy + constraints	Predict optimal navigation routes
Predictive anomaly detection	Autoencoders, Isolation forest	Sensor logs, video sequences	Early alert on surgical error or device drift

AR, augmented reality; *VR*, virtual reality; *ML*, machine learning; *DL*, deep learning.

edge technologies. These integrative techniques elevate robotic interventions from mechanical assistance to immersive, intelligent, and responsive surgical ecosystems.

11.5 Machine learning and deep learning in augmented reality/virtual reality-based robotics

The application of ML and DL within AR/VR-driven medical robotics empowers the system with context-awareness, autonomous decision-making, and personalized adaptability (Rashid and Kausik, 2024). These technologies contribute to various aspects such as gesture interpretation, tissue classification, real-time decision prediction, and surgical path planning. This section explores the core ML/DL frameworks, data requirements, model types, and architectural flow for medical robotics integrated with AR/VR environments.

11.5.1 Role of artificial intelligence in enhancing robotic intelligence

ML and DL enable medical robots to go beyond preprogrammed routines and learn from multimodal input data (vision, haptics, audio, and sensor metrics).

In Table 11.11, different ML/DL approaches are matched with specific robotic functions to illustrate how AI supports real-time decision-making in surgical settings.

11.5.2 Training data and model requirements

Deep models require substantial and diverse datasets for accurate predictions in AR/VR-based environments. These include synthetic data, simulated surgery videos, and real patient records (Fig. 11.7).

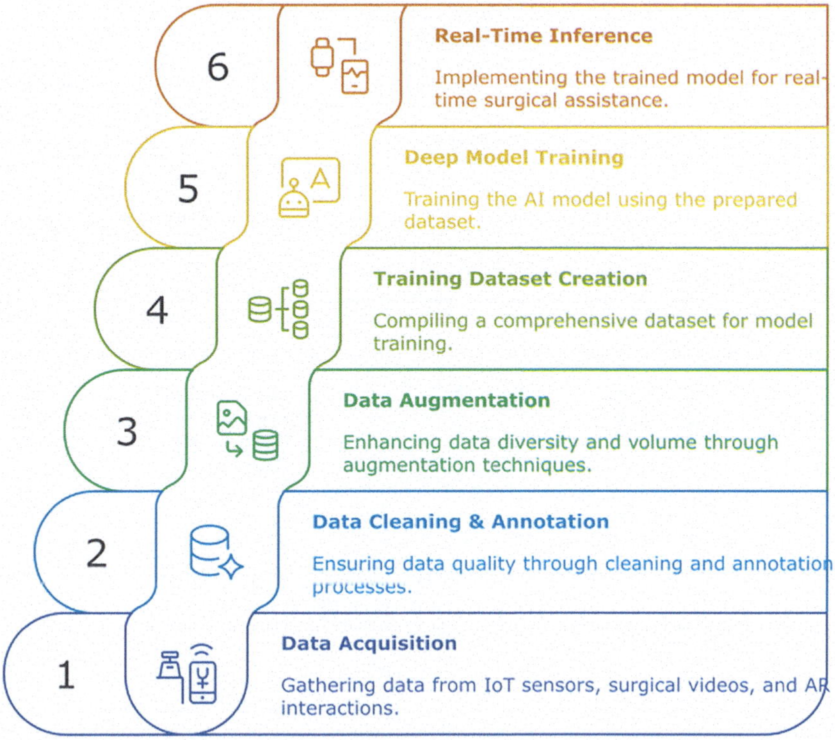

Figure 11.7 Data pipeline for ML/DL in AR/VR-robotics. This figure shows that deep models require substantial and diverse datasets for accurate predictions in AR/VR-based environments. These include synthetic data, simulated surgery videos, and real patient records. *AR*, augmented reality; *VR*, virtual reality; *ML*, machine learning; *DL*, deep learning.

11.5.3 Real-time inference in surgical environments

In medical robotics, latency is critical. Therefore ML/DL models must deliver inferences in real-time under hardware constraints (e.g., edge devices or GPUs) (Fig. 11.8).

11.5.4 Deep learning architectures for gesture and object detection

Medical robotics systems rely heavily on convolutional and recurrent neural networks to interpret surgeon gestures and anatomical landmarks from AR/VR visuals (Fig. 11.9).

11.5.5 Reinforcement learning for adaptive robotic control

Reinforcement Learning (RL) is highly suited for tasks where the robot must learn through trial-and-error interaction in a simulation or live AR/VR setting (Algorithm 11.2).

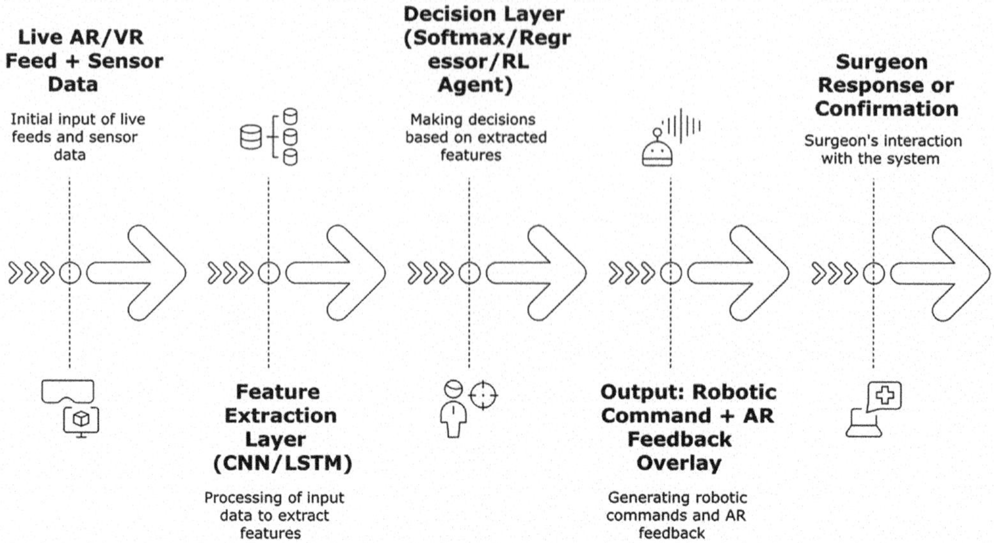

Figure 11.8 Real-time model inference workflow. This figure explains how a trained model processes real-time input from AR/VR and sensors to guide robotic behavior in dynamic surgical settings. *AR*, augmented reality; *VR*, virtual reality.

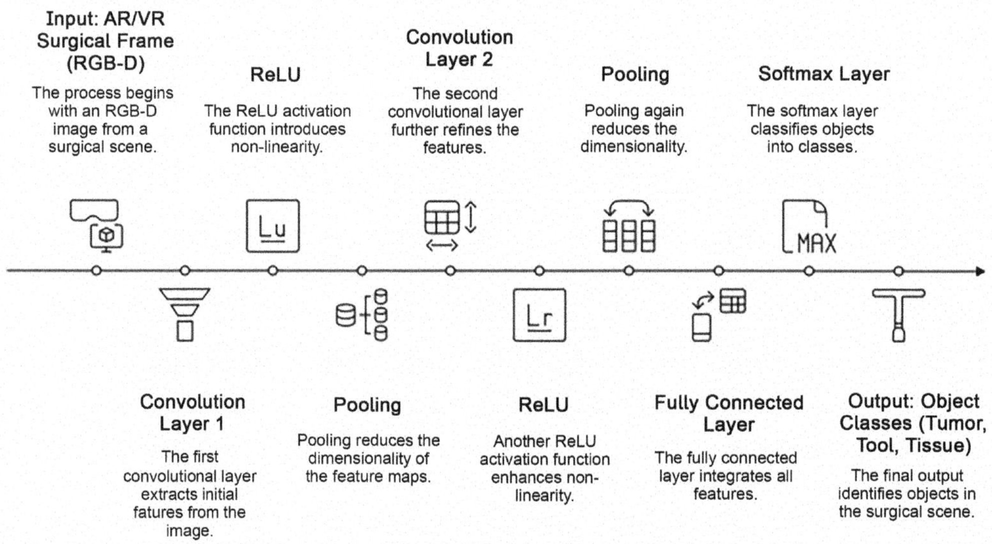

Figure 11.9 CNN-based object detection architecture for AR/VR visual input. This figure shows that medical robotics systems rely heavily on convolutional and recurrent neural networks to interpret surgeon gestures and anatomical landmarks from AR/VR visuals. *AR*, augmented reality; *VR*, virtual reality.

ALGORITHM 11.2 RL-Based Robotic Action Model.

Let:

- S: Set of states (e.g., tissue type, tool location)
- A: Set of possible actions (e.g., move, grasp, retract)
- $R(s, a)$: Reward function for performing action a in states
- $\pi(a|s)$: Policy defining probability of action a in states
- $Q(s, a)$: Action-value function
- γ: Discount factor

  The Q-learning update rule is:

$$Q(s_t, a_t) \leftarrow Q(s_t, a_t) + \alpha[R(s_t, a_t) + \gamma \max_{a'} Q(s_{t+1}, a') - Q(s_t, a_t)]$$

Where:

- α is the learning rate.
- The robot selects a_t Based on the current policy π, and receives a reward and a new state, refining its strategy over time.

  This is particularly useful in autonomous suturing, needle insertion, or trajectory optimization, where the robot learns optimal actions using AR feedback and haptic resistance.

Table 11.12 Benefits of transfer learning in surgical applications.

Feature	Benefit
Faster training	Leverages pretrained base layers.
Better accuracy	Reduces overfitting on small datasets
Improved generalization	Works on diverse surgical scenes
Cost-effective	Requires fewer labeled examples

11.5.6 Transfer learning for surgical scene understanding

Transfer learning (TL) allows models pretrained on large datasets (such as ImageNet or Med3D) to be fine-tuned on smaller medical datasets.

As shown in Table 11.12, TL is a powerful method to overcome data scarcity in medical robotics training pipelines while improving scene recognition accuracy in AR/VR modules.

This section has detailed how ML and DL algorithms power the intelligent backbone of AR/VR-based medical robotics. From CNN-driven object recognition to RL for decision-making and TL for scene understanding, these techniques form the bridge between perception, cognition, and action. Real-time inference, enabled by optimized architectures and robust data pipelines, ensures these models operate safely and effectively during live surgical interventions.

Table 11.13 Role of IoT devices in medical robotics and AR/VR.

IoT device/Sensor type	Role in AR/VR robotics	Example use case
Haptic sensors	Measure force/pressure during tool interaction	Real-time tissue feedback in robotic surgery (Laca et al., 2022)
Biomedical wearables	Monitor patient vitals during procedures	Dynamic overlay of vitals in AR/VR dashboard (Areia et al., 2022)
RFID/NFC	Real-time tracking of instruments/devices	Automatic identification of surgical tools (Nema and Vachhani, 2022)
Environmental sensors (IR, Temp)	Monitor OR ambient conditions	Real-time tracking of instruments/devices (Narayana et al., 2024)
IMUs (Inertial Measurement Units)	Track body/hand/head movements in 6DOF	Enhance gesture-based AR/VR interfaces (Li et al., 2019)

AR, augmented reality; *VR*, virtual reality; *IoT*, Internet of Things.

11.6 Internet of Things-enabled ecosystems in augmented reality/virtual reality-driven medical robotics

The fusion of IoT technologies with AR/VR-based medical robotics has transformed the healthcare ecosystem into an intelligent, real-time, interconnected infrastructure. IoT sensors, smart devices, cloud platforms, and communication protocols allow continuous monitoring, remote control, and data-driven decision-making. This section explores how IoT enhances robotic systems' responsiveness, adaptability, and operational intelligence in AR/VR-assisted clinical environments.

11.6.1 Role of Internet of Things in medical robotics and augmented reality/virtual reality

IoT is pivotal in bridging the physical medical environment and virtual simulation layers by collecting, transmitting, and processing contextual data.

Table 11.13 maps various IoT devices to their specific functionalities that enhance medical robotic interactions and AR/VR interfaces.

11.6.2 System architecture for Internet of Things–augmented reality/virtual reality-robotics integration

A multilayered system is required to support robust and scalable IoT–AR/VR robotic operations. This architecture includes edge sensing, network communication, cloud processing, and AR/VR rendering (Fig. 11.10).

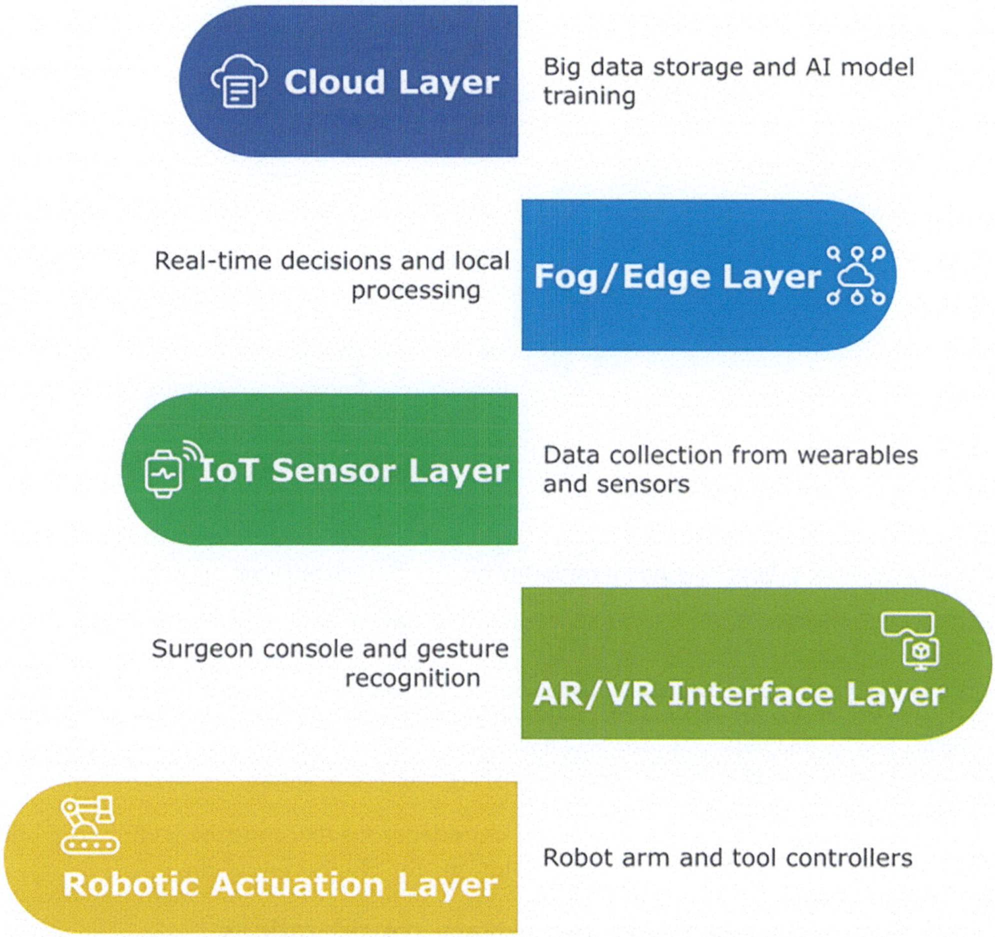

Figure 11.10 Layered architecture of IoT-enabled AR/VR medical robotics. This figure outlines how IoT-enabled sensors and edge processors act as an intelligent bridge between human interfaces (AR/VR) and robotic hardware layers, enabling closed-loop real-time decision-making. *AR*, augmented reality; *VR*, virtual reality; *IoT*, Internet of Things.

11.6.3 Real-time data flow and communication protocols

Efficient communication between distributed components is essential for seamless operation. IoT devices typically use low-latency and high-throughput protocols (Fig. 11.11).

11.6.4 Interoperability, latency, and edge computing

Medical applications demand millisecond-level response times, making edge computing critical. Additionally, standardization is necessary for interoperable device communication.

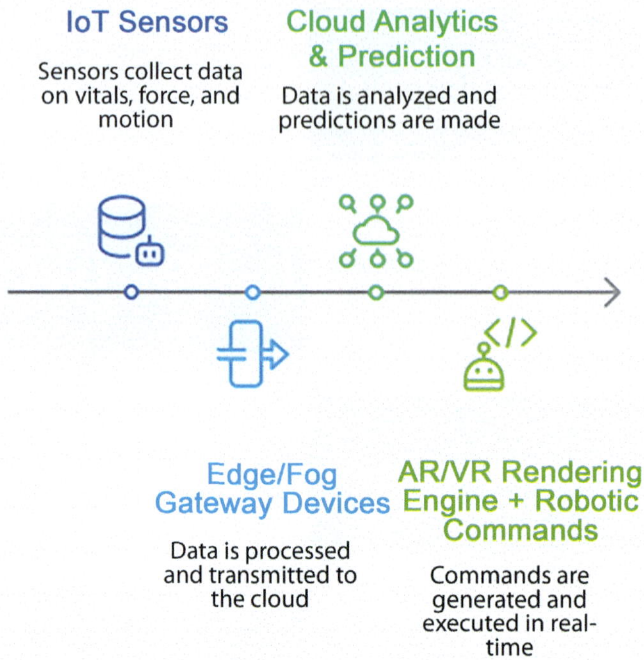

Figure 11.11 Direction and nature of data communication between IoT endpoints and centralized/cloud platforms. This figure visualizes the direction and nature of data communication between IoT endpoints and centralized/cloud platforms, followed by AR/VR response generation and robotic actuation. *AR*, augmented reality; *VR*, virtual reality; *IoT*, Internet of Things.

Table 11.14 identifies key system-level considerations that must be addressed while designing reliable IoT-AR/VR-robotic solutions for clinical environments.

11.6.5 Secure data handling and privacy considerations

The integration of sensitive patient data from wearables, robotics, and cloud-based AR/VR interfaces raises significant concerns over data security, consent, and compliance with regulations such as HIPAA and GDPR (Paul et al., 2023) (Fig. 11.12).

IoT technology is foundational in enhancing interactivity, precision, and context-awareness in AR/VR-driven medical robotics. From capturing real-time patient metrics and robotic feedback to enabling immersive data overlays and automated control, the IoT layer ensures seamless synchronization between human operators, virtual renderings, and robotic actuators. Robust architecture (as shown in Fig. 11.12), efficient protocols, and secure workflows are essential to deploy these systems reliably in hospital settings.

11.7 System architecture and workflow

An efficient system architecture is critical to successfully integrating AR/VR, IoT, and medical robotics, allowing real-time, data-driven decision-making and

Table 11.14 IoT considerations for medical AR/VR robotics.

Parameter	Requirement	Role in ecosystem
Latency	<10 ms (for surgical applications)	Enables responsive tool control
Data bandwidth	High (video + sensor streams)	Supports AR/VR rendering in real-time
Interoperability	HL7, FHIR, ISO/IEEE 11073	Allows device-to-device communication
Edge computation	Yes (AI inferencing on-site)	Reduces delay from cloud-only reliance
Fault tolerance	High	Ensures continuous operation under failure

AR, augmented reality; *VR*, virtual reality; *IoT*, Internet of Things; *AI*, artificial intelligence.

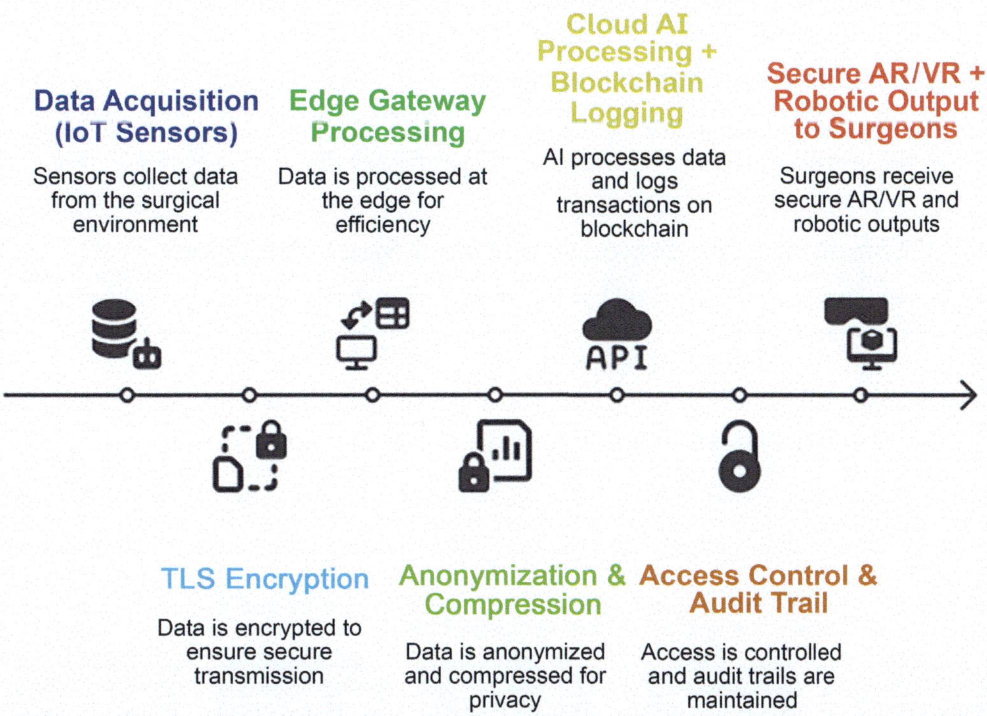

Figure 11.12 Security workflow for IoT-based medical robotics. This figure illustrates the secure handling pipeline for sensitive health data captured and transmitted across the IoT–AR/VR robotic framework. *AR*, augmented reality; *VR*, virtual reality; *IoT*, Internet of Things.

human–machine collaboration. This section outlines the layered architecture, communication workflows, and control logic that govern AR/VR-assisted robotic systems in clinical settings.

11.7.1 Hardware architecture

The hardware components in an AR/VR-integrated medical robotic system can be grouped into four major layers: sensing, interface, processing, and actuation. These components interact dynamically during surgical or diagnostic procedures (Fig. 11.13).

In Fig. 7.1, the layered hardware flow shows data flowing from IoT sensors, through edge processors, into the AR/VR interfaces, resulting in robotic actuation and completing the control loop.

11.7.2 Software and middleware stack

To manage the complexity of real-time communication, device coordination, and rendering, a modular software architecture is employed. Middleware ensures interoperability across different vendors, protocols, and AI models (Fig. 11.14).

11.7.3 Data flow and control workflow

The workflow describes how sensor inputs are transformed into robotic actions via AI processing, visual feedback, and surgeon interaction (Fig. 11.15).

11.7.4 Communication protocols and data Synchronization

Due to the sensitive and real-time nature of medical operations, robust communication protocols are needed. The system architecture uses a combination of low-latency and high-reliability communication standards.

As shown in Table 11.15, these protocols are optimized for performance, minimizing delays between sensing, rendering, and robotic responses.

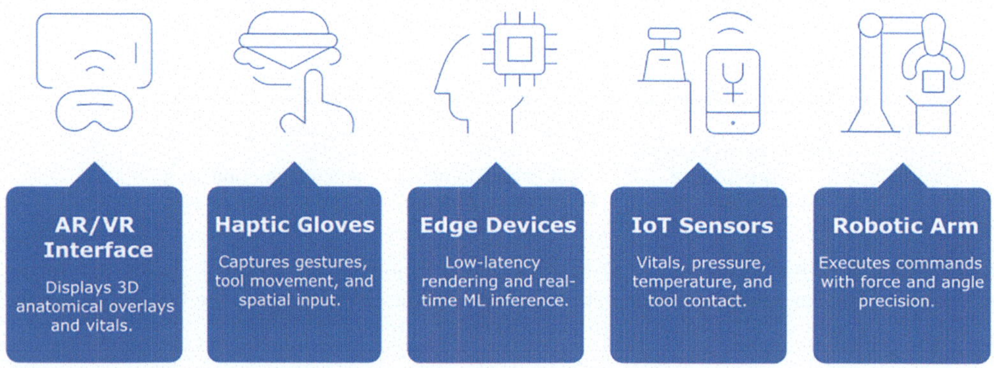

Figure 11.13 Hardware architecture of AR/VR-driven medical robotics. This figure shows that the hardware components in an AR/VR-integrated medical robotic system can be grouped into four major layers: sensing, interface, processing, and actuation. These components interact dynamically during surgical or diagnostic procedures. *AR*, augmented reality; *VR*, virtual reality.

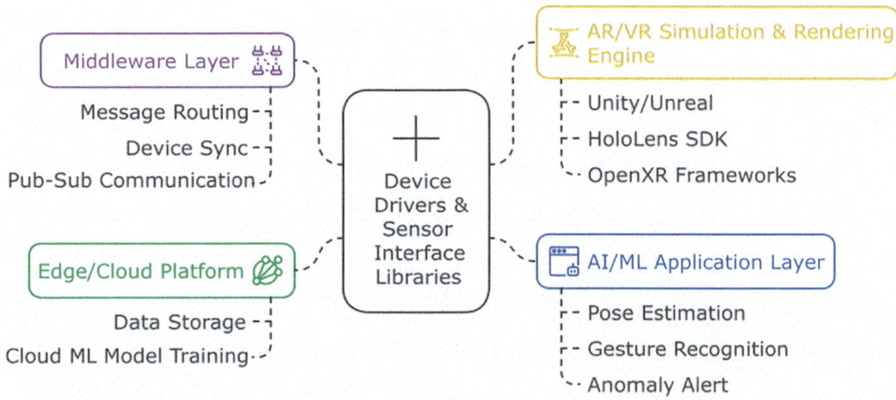

Figure 11.14 Software and middleware stack. This figure presents a clear modular stack where middleware handles message passing and device orchestration, enabling scalable deployment of intelligent medical robotic systems enhanced by AR/VR. *AR*, augmented reality; *VR*, virtual reality.

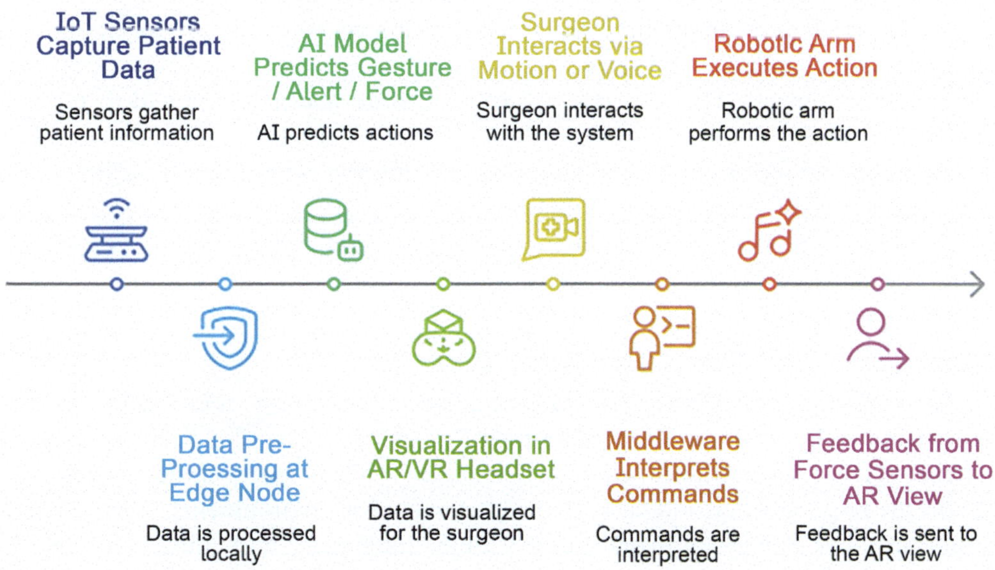

Figure 11.15 Control workflow in AR/VR medical robotics. This figure demonstrates that the cyclical and closed-loop control structure, sensor input, and user feedback drive decision-making, ensuring safe and precise interventions. *AR*, augmented reality; *VR*, virtual reality.

11.7.5 Workflow example: minimally invasive surgery

Let's consider a use case of AR-guided minimally invasive laparoscopic surgery:

1. Initialization: The robotic arm and AR headset boot up and establish a connection via middleware.

Table 11.15 Communication technologies in AR/VR medical robotics.

Layer	Protocol used	Purpose	Latency
Sensor to Edge	MQTT/BLE	Transmit real-time vitals and motion	<5 ms
Edge to Cloud	HTTPS/WebRTC	Upload models, video streaming	<50 ms
Middleware	ROS 2/DDS	Synchronize sensors, AR/VR, and actuators	<10 ms
AR/VR rendering	OpenXR/WebGL	Interactive rendering and overlays	<15 ms
Surgeon control feedback	Voice/IMU Stream	Control robotic movement in real-time	<20 ms

AR, augmented reality; *VR*, virtual reality.

2. Sensor activation: IMUs and haptics start capturing the surgeon's movement.
3. Overlay rendering: Anatomical structures and surgical pathways are shown in AR using real patient CT/MRI scans.
4. Surgeon command: The surgeon performs a gesture that is interpreted to move a robotic scalpel.
5. Execution and feedback: The robotic arm executes the action; haptic feedback and tool pressure are updated in real-time in AR.
6. Closure: AI checks for bleeding or anomalies postaction and notifies the surgeon via overlay alert.

This example encapsulates the end-to-end function of the system described in Figs. 7.1 and 7.2 and the flow logic shown in Fig. 11.15.

The integration of AR/VR, IoT, AI, and robotics in healthcare is enabled by a well-orchestrated architecture, as depicted in Figs. 7.1 and 7.2. The data flow and control sequence (Fig. 11.15) ensure closed-loop operations, while protocol choices (Table 7.1) support the responsiveness and safety demanded in clinical environments. This cohesive architecture lays the groundwork for real-time, intelligent, and immersive robotic healthcare systems.

11.8 Applications and case studies

The convergence of AR/VR, robotics, IoT, and AI has transformed traditional healthcare practices. From preoperative planning to telesurgery and rehabilitation, these technologies create safer, more accurate, and less invasive solutions. This section presents key real-world applications and practical case studies demonstrating the impact of AR/VR-based robotic systems in medicine.

11.8.1 Surgical applications

AR/VR-powered robotic systems have significantly improved surgical precision by enhancing visualization, navigation, and control. These technologies are now used in various specialties, including neurosurgery, orthopedics, and cardiothoracic surgery (Fig. 11.16).

As shown in Fig. 11.16, the AR headset allows the surgeon to visualize anatomical landmarks in 3D, while the robotic arm executes precise incisions based on AI-assisted guidance. IoT sensors monitor patient vitals throughout.

Table 11.16 summarizes major surgical areas where AR/VR aids real-time decision-making while robots execute tasks with submillimeter precision.

11.8.2 Rehabilitation and physical therapy

AR/VR-integrated with robotics enables interactive rehabilitation by creating gamified environments and intelligent patient resistance mechanisms (Fig. 11.17).

In Fig. 11.17, the patient wears a VR headset and robotic limb exoskeleton. Tasks such as "catching virtual objects" stimulate motor recovery, while AI models adjust resistance in real-time.

As Table 11.17 highlights, AR/VR-driven robotic therapy creates a personalized, responsive recovery pathway for stroke, spinal injury, or orthopedic patients.

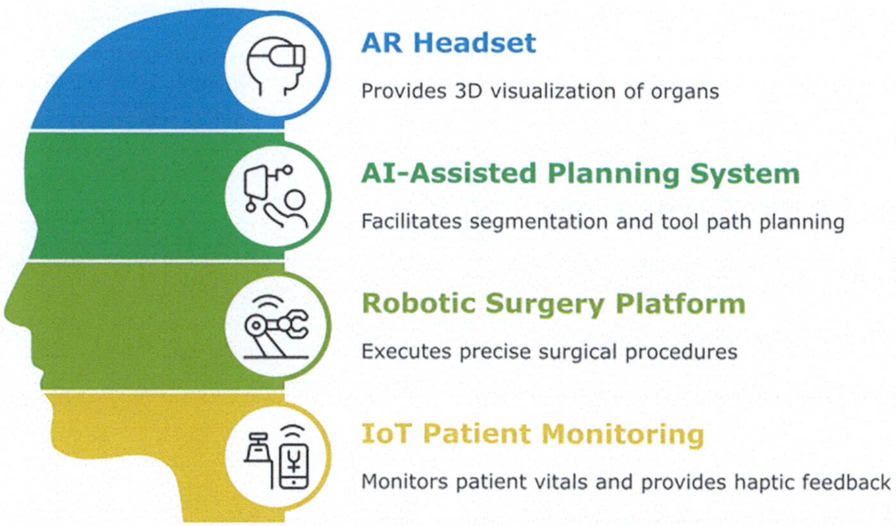

Figure 11.16 AR-assisted robotic surgery setup. The AR headset allows the surgeon to visualize anatomical landmarks in 3D, while the robotic arm executes precise incisions based on AI-assisted guidance. IoT sensors monitor patient vitals throughout. *AR*, augmented reality; *IoT*, Internet of Things; *AI*, artificial intelligence.

Table 11.16 Surgical applications of AR/VR-driven robotics.

Speciality	Application example	AR/VR role	Robotics role
Neurosurgery	Tumor resection in brain surgery	3D tumor mapping	Microscale precision tools
Orthopedic surgery	Total knee replacement	Bone alignment overlay	Sawing/drilling automation
Cardiothoracic	Bypass graft placement	Vessel visualization	Minimally invasive suturing
ENT surgery	Cochlear implant procedure	Augmented ear anatomy	Needle placement control

AR, augmented reality; *VR*, virtual reality.

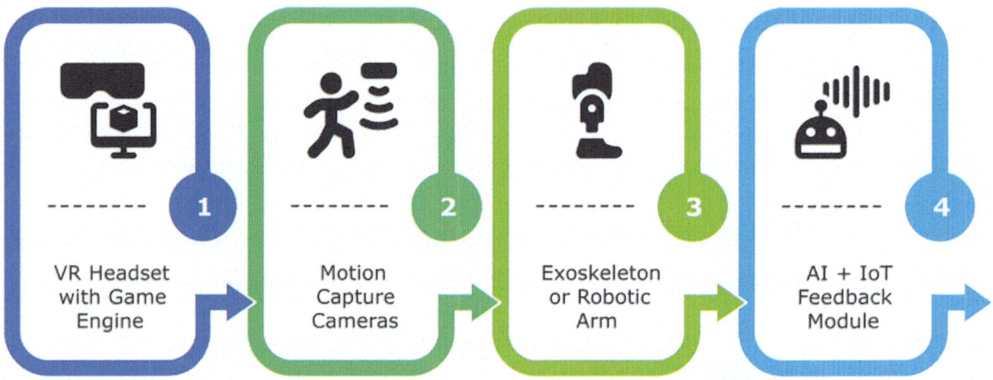

Figure 11.17 VR-guided robotic rehabilitation system. This figure is about AR/VR-integrated with robotics enables interactive rehabilitation by creating gamified environments and intelligent patient resistance mechanisms. *AR*, augmented reality; *VR*, virtual reality.

Table 11.17 Benefits of AR/VR robotics in rehabilitation.

Benefit	Description
Motivation and engagement	Gamified tasks improve patient adherence
Adaptive training	AI adapts therapy level based on performance
Precise monitoring	IoT sensors track every joint and angle
Home-based therapy	Portable setups enable remote sessions

AR, augmented reality; *VR*, virtual reality; *IoT*, Internet of Things.

11.8.3 Diagnostic and training environments

Medical education and diagnostics have been revolutionized with immersive simulations and robotic integration. Trainees can practice in risk-free virtual environments, while robotic tools capture micro-movements for performance feedback.

11.8.4 Case study 8.1: augmented reality-based robotic simulation in medical training

A recent study conducted at a teaching hospital introduced a VR surgery simulator integrated with a robotic haptic device. Medical students practised laparoscopic procedures with real-time feedback on tool pressure and tissue sensitivity. Over 6 weeks, diagnostic accuracy improved by 35%, and error rates reduced by 28%.

As seen in Table 11.18, training systems allow both autonomous and guided learning, improving learning curves for young professionals.

11.8.5 Remote and telesurgery applications

AR and robotics combined with 5G and IoT infrastructure enable real-time remote interventions, potentially democratizing access to high-end surgical expertise.

11.8.6 Case study 8.2: 5G-based telesurgery between two cities

In 2023 a telesurgery experiment in China involved a surgeon operating from Beijing using an AR headset and joystick interface to control a robotic system located 3000 km away in Shenzhen. With latency under 50 ms, the surgeon successfully removed a gallbladder, assisted by real-time imaging and haptic feedback. The entire procedure was monitored by IoT sensors on the patient, ensuring safety (Fig. 11.18).

The practical applications of AR/VR-enhanced medical robotics span surgery, therapy, diagnostics, and training. As detailed in Figs. 8.1–8.3, integrating immersive interfaces with precise robotics improves patient outcomes and clinician performance. Tables 8.1–8.3 showcase how different medical domains benefit from this synergy, with real case studies proving such systems' clinical viability and safety.

11.9 Challenges and ethical considerations

The integration of AR, VR, ML, IoT, and robotics in healthcare brings promising advancements, but it also introduces significant technical, regulatory, and ethical challenges. These must be addressed to ensure safe, equitable, and practical implementation.

Table 11.18 Key training features enabled by AR/VR robotics.

Feature	Description
Skill tracking	Logs timing, angle, force, and error frequency
Scenario simulation	Trauma, tumor excision, and catheter insertion
Real-time feedback	Alerts for excessive force, incorrect movement
Remote collaboration	Instructors can intervene from distant centers

AR, augmented reality; *VR*, virtual reality.

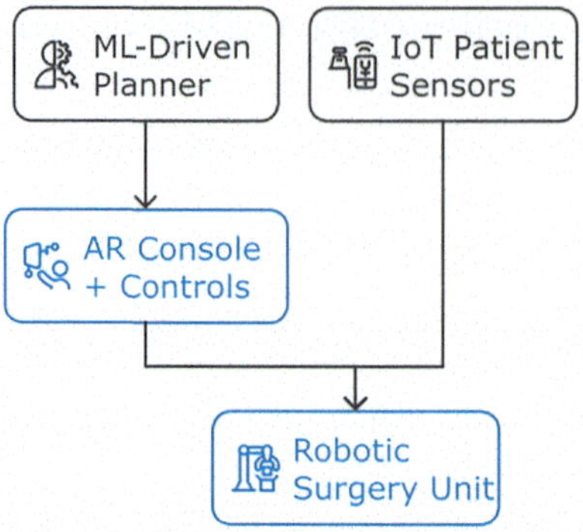

Figure 11.18 Telesurgery architecture overview. This figure shows the two-site system where AR interfaces, robotics, and IoT monitoring synchronize via low-latency networks. AI ensures optimal incision planning and patient safety checks. *AR*, augmented reality; *AI*, artificial intelligence; *IoT*, Internet of Things.

Table 11.19 Key technical challenges in AR/VR-driven medical robotics.

Technical challenge	Description	Implications for practice
Latency in data transmission	Delay in feedback from remote robotic systems or IoT sensors	Affects real-time response, especially in telesurgery
System interoperability	Compatibility between AR devices, surgical robots, and IoT platforms	Requires custom integration layers and middleware
Real-time processing load	Heavy computational demand for rendering, planning, and control	Needs edge computing and hardware acceleration
Limited battery and sensor life	Short operational time of mobile/ wearable components	Interrupts long procedures or patient rehabilitation sessions

AR, augmented reality; *VR*, virtual reality; *IoT*, Internet of Things.

11.9.1 Technical challenges in system integration

Bringing together AR/VR with robotic systems and IoT-enabled networks introduces issues of interoperability, latency, and real-time synchronization. Each component, whether a robotic manipulator, AR headset, or ML inference engine, may come from different vendors, using incompatible protocols or operating on nonstandardized data formats.

As shown in Table 11.19, these challenges require both hardware and software innovations such as edge AI deployment, real-time rendering optimizations, and low-latency wireless protocols (e.g., 5G or WiFi-6).

11.9.2 Data privacy and security concerns

Data protection is one of the most sensitive aspects of deploying AR/VR-based medical robotics. These systems rely on vast streams of personal health information collected through cameras, sensors, and digital records.

- AR/VR systems may unintentionally record private visuals (e.g., facial expressions, room environments).
- IoT sensors track biometric signals that are often transmitted wirelessly.
- ML algorithms process sensitive health data, which, if leaked, may lead to identity theft or discrimination.

As Table 11.20 indicates, safeguarding these data streams requires end-to-end encryption, robust access control mechanisms, and compliance with legal frameworks such as HIPAA, GDPR, and local medical data acts.

11.9.3 Ethical implications of autonomous decisions

As ML and AI models increasingly assist or automate medical decisions such as tumor margin detection, anomaly prediction, or robotic incision planning, they raise concerns regarding:

- Accountability: Who is responsible when an AI-guided robot makes a mistake, an engineer, a surgeon, or an institution?
- Transparency: How can we interpret the decision logic of deep neural networks in life-critical tasks?
- Consent and control: Can patients opt out of being treated by robotic systems or AI-informed diagnoses?

In Table 11.21, the suggested practices aim to align technical development with bioethical principles such as autonomy, beneficence, nonmaleficence, and justice.

Table 11.20 Data privacy risks in AR/VR medical systems.

Source of risk	Type of data involved	Potential threat
AR/VR video capture	Facial images, surroundings	Surveillance misuse, identity exposure
IoT biomedical sensors	ECG, EEG, and respiration data	Wireless interception, spoofing
AI-based decision logs	Diagnosis, surgery suggestions	Unauthorized access, biased conclusions
Cloud data syncing	Patient records, surgical footage	Data breaches, privacy law violations

AR, augmented reality; *VR*, virtual reality; *AI*, artificial intelligence; *IoT*, Internet of Things.

Table 11.21 Ethical dilemmas in AR/VR-based robotic healthcare.

Ethical issue	Description	Recommended practice
Lack of explainability	The black-box nature of deep learning decisions	Implement Explainable AI (XAI) for critical tasks
Consent in robotic systems	The patient may not fully understand the AR/VR robotic involvement	Informed consent with simulation demos before surgery
Bias in AI models	Datasets may underrepresent certain populations	Use diverse, fair, and audited training data
Delegation of control	Overreliance on automation may bypass human judgment	Human-in-the-loop decision-making mandatory

AR, augmented reality; *VR*, virtual reality; *AI*, artificial intelligence.

11.9.4 Regulatory and legal considerations

Regulatory authorities such as the FDA, EMA, and ISO/IEC are still evolving guidelines to accommodate the hybrid nature of AR/VR and robotics in healthcare. Challenges include:

- Certifying complex systems with AI, sensors, and actuators as a single medical device.
- Cross-border telesurgery regulations, where jurisdictions may vary widely.
- Licensing liability when a surgeon operates using remote or AI-assisted robotic systems.

Key legal considerations:

- Medical device classification for mixed AR/robotic platforms
- Compliance with cybersecurity standards (e.g., ISO/IEC 27001)
- Legal recourse in case of robotic-assisted surgical mishaps
- Reimbursement models for technology-enhanced procedures

Now there is an imperative to consolidate global regulatory frameworks with direct accountability, security lookouts in place, and approval standards for tablets for these fresh systems.

11.9.5 User acceptance and human factors

Even with proven clinical benefits, trust, comfort, and usability affect adoption. Surgeons, nurses, and patients must adapt to new technologies:

- Surgeons may experience fatigue from prolonged AR headset use.
- Nurses may require upskilling to manage robotic support tools.
- Patients may distrust robotic intervention, especially in invasive surgeries.

Addressing these challenges involves:

- Ergonomic design of AR/VR hardware
- Training programs for medical professionals
- Patient education sessions to build confidence

While the fusion of AR/VR and medical robotics offers transformative potential, it also introduces multilayered challenges. The technical, ethical, and legal complexities are shown in Tables 9.1–9.3. The creation of transparent, safe, human-centered systems with the support of sufficient regulation is a crucial aspect of the decent development of this area.

11.10 Future directions

As the convergence of AR, VR, ML, and IoT with medical robotics continues to evolve, the future holds immense potential to transform how surgeries, diagnoses, and therapies are delivered. However, this trajectory must be guided by robust innovation, ethical foresight, and interdisciplinary collaboration. This section outlines several forward-looking directions based on current trends and technological gaps.

11.10.1 Emerging technologies and trends

AR/VR and AI-based medical robotics are outgrowing their roots in surgical support by becoming smarter, more responsive, and more immersive. Some of the emerging technologies that will cause changes in the future are:

- Spatial computing: The fusion of physical and digital environments using AR/VR to allow intuitive interaction with medical data and robotic interfaces.
- Holographic interfaces: Real-time holography will enable 3D visualization of organs and robotic plans without needing headsets.
- Brain–computer interfaces (BCI): These allow direct neural control of robotic limbs or devices through brain signals, especially useful for paralyzed patients.
- Tactile and haptic feedback systems: Enhancing robotic systems with feedback mechanisms to simulate touch, enabling precise remote surgeries (Table 11.22).

AR/VR and AI-based medical robotics are outgrowing their roots in surgical support by becoming smarter, more responsive, and more immersive. Some of the emerging technologies that will cause changes in the future are:

11.10.2 Interdisciplinary research opportunities

To realize these future visions, collaboration between multiple disciplines is vital:

- Biomedical engineering: To develop next-gen surgical robots, embedded sensors, and wearables.

Table 11.22 Key emerging technologies in AR/VR-driven medical robotics.

Technology	**Description**	Anticipated impact by 2030
Spatial computing	Combines AR/VR, IoT, and AI to create fully interactive clinical spaces	Improved surgical navigation and diagnostics
Holographic projection	Real-time 3D organ visualizations without wearables	Nonintrusive planning during procedures
Brain–computer interfaces	Robotic control via EEG/ implanted sensors	Independence for paralyzed patients
Soft robotics integration	Flexible robotic components for adaptive procedures	Safer and more precise surgical interventions
Federated learning models	Privacy-preserving AI training across hospitals	Global collaboration without data sharing

AR, augmented reality; *VR*, virtual reality; *AI*, artificial intelligence; *IoT*, Internet of Things.

- Neuroscience and psychology: Integrating VR-based therapy in cognitive rehabilitation and BCI development.
- Data science and cybersecurity: To improve ML models and safeguard the growing ecosystem from data threats.
- Ethics and law: This is used to define frameworks that support innovation while protecting human rights and privacy.

Academic–hospital–industry joint research efforts are also turning their attention to AR in robotic colonoscopy, VR-based neurorehabilitation, and AI-enhanced intraoperative decision systems.

11.10.3 Evolving standards and policies

As technology matures, regulatory clarity and standardization will be necessary to support deployment at scale:

- Standard Protocols for Device Interoperability: Enabling plug-and-play integration of AR/VR headsets, robotic instruments, and IoT sensors.
- Global privacy frameworks: Allowing safe cross-border collaboration for AI model training via federated learning or edge computing.
- Ethical guidelines for AI in surgery: Mandating explainable decision-making, data transparency, and human override mechanisms.

The World Health Organisation and bodies such as ISO, IEEE, and FDA are expected to release tailored frameworks for AI-powered robotic systems in healthcare.

11.10.4 Forecasting the next decade

The next 5–10 years will likely witness the transition from pilot applications to routine clinical adoption. Below are some realistic projections:

- By 2027 most leading hospitals will implement AR-assisted robotic training simulators and mixed-reality interfaces for planning surgeries.
- By 2030 at least 25% of minimally invasive surgeries will involve some level of autonomous robotic action with AR/VR guidance.
- By 2035 full-sensory telesurgery across continents with tactile feedback and holographic AR will become possible via 6G and quantum-secure networks.

The integration of AI-generated digital twins of patients (for predictive modeling) and immersive surgical rehearsals will enhance both personalization and safety.

Future development in AR/VR-driven medical robotics will not only enhance surgical precision but also democratize access to healthcare by enabling remote, immersive, and intelligent interventions. As outlined in Table 10.1, innovation must go hand-in-hand with ethical deployment, interdisciplinary research, and updated regulatory frameworks. The road ahead is promising, but it will require collective efforts to make this technological shift both human-centered and clinically validated.

11.11 Conclusion

The fusion of AR and VR technologies with medical robotics marks a paradigm shift in the way modern healthcare is delivered, experienced, and evolved. This chapter has comprehensively explored the intersection of immersive technologies, robotics, artificial intelligence, and IoT in transforming clinical practices, surgical procedures, and medical education.

By enabling enhanced visualization, precision, and interactivity, AR/VR technologies bring significant improvements to robot-assisted interventions from preoperative planning to intraoperative guidance and postoperative care. The intelligence of the system is further multiplied with ML and DL models, which make decision-making across the system in real-time, characterizing patterns, and personalizing treatment more precisely and responsively. In the meantime, IoT acts as the unifying spine, which guarantees smooth connections among gadgets, sensors, and cloud computing in the moving healthcare settings.

Using deep explanations, designs, and case study reports, we indicated that AR/VR-guided medical robotic systems have already shown practical advantages in diagnosis, rehabilitation, telesurgery, and simulation training. Meanwhile, the various issues behind it, such as technical limitations and integration problems, as well as ethical and regulatory concerns, have not been ignored in this chapter.

All of these technologies will also intersect in the future with the prospect of developing more intelligent, secure, and responsive healthcare systems being immense.

With further research and development on spatial computing, BCI, and the context-aware application of AI, we believe that the future holds the promise of immersive, intelligent medical robotics supporting the medical system as much as they can extend the capability of humanity itself in the provision of healthcare.

To conclude, this multidisciplinary endeavor promises a lot of medical innovation, so collaboration and intelligent design are already required, as well as patient-focused application, to bring its potential to fruition.

References

Albahri OS, Albahri AS, Mohammed KI, Zaidan AA, Zaidan BB, Hashim M, et al. Systematic review of real-time remote health monitoring system in triage and priority-based sensor technology: taxonomy, open challenges, motivation and recommendations. Journal of Medical Systems 2018;42(5). Available from https://doi.org/10.1007/s10916-018-0943-4.

Alian A, Zari E, Wang Z, Franco E, Avery JP, Runciman M, et al. Current engineering developments for robotic systems in flexible endoscopy. Techniques and Innovations in Gastrointestinal Endoscopy 2023;25(1):67–81. Available from https://doi.org/10.1016/j.tige.2022.11.006.

Areia C, King E, Ede J, Young L, Tarassenko L, Watkinson P, et al. Experiences of current vital signs monitoring practices and views of wearable monitoring: a qualitative study in patients and nurses. Journal of Advanced Nursing 2022;78(3):810–22. Available from https://doi.org/10.1111/jan.15055.

Autorino R, Zargar H, Kaouk JH. Robotic-assisted laparoscopic surgery: recent advances in urology. Fertility and Sterility 2014;102(4):939–49. Available from https://doi.org/10.1016/j.fertnstert.2014.05.033.

Awais M, Al Taie M, O'Connor CS, Castelo AH, Acidi B, Tran Cao HS, et al. Enhancing surgical guidance: deep learning-based liver vessel segmentation in real-time ultrasound video frames. Cancers 2024;16(21):3674. Available from https://doi.org/10.3390/cancers16213674.

Chae S, Street WN, Ramaraju N, Gilbertson-White S. Prediction of cancer symptom trajectory using longitudinal electronic health record data and long short-term memory neural network. JCO Clinical Cancer Informatics 2024;8. Available from https://doi.org/10.1200/CCI.23.00039, http://ascopubs.org/journal/cci/.

Collins T, Pizarro D, Gasparini S, Bourdel N, Chauvet P, Canis M, et al. Augmented reality guided laparoscopic surgery of the uterus. IEEE Transactions on Medical Imaging 2021;40(1):371–80. Available from https://doi.org/10.1109/TMI.2020.3027442, http://ieeexplore.ieee.org/xpl/RecentIssue.jsp?punumber=42.

Cousein E, Mareville J, Lerooy A, Caillau A, Labreuche J, Dambre D, et al. Effect of automated drug distribution systems on medication error rates in a short-stay geriatric unit. Journal of Evaluation in Clinical Practice 2014;20(5):678–84. Available from https://doi.org/10.1111/jep.12202.

Kim JI, Kim DW, Krebs M, Park YS, Park YL. Force sensitive robotic end-effector using embedded fiber optics and deep learning characterization for dexterous remote manipulation. IEEE Robotics and Automation Letters 2019;4(4):3481–8. Available from https://doi.org/10.1109/LRA.2019.2926959, http://ieeexplore.ieee.org/servlet/opac?punumber=7083369.

Korayem MH, Vahidifar V. Detecting hand's tremor using leap motion controller in guiding surgical robot arms and laparoscopic scissors. Measurement 2022;204:112133. Available from https://doi.org/10.1016/j.measurement.2022.112133.

Laca JA, Kocielnik R, Nguyen JH, You J, Tsang R, Wong EY, et al. Using real-time feedback to improve surgical performance on a robotic tissue dissection task. European Urology Open Science 2022;46:15–21. Available from https://doi.org/10.1016/j.euros.2022.09.015, https://www.journals.elsevier.com/european-urology-open-science.

Li Y, Huang J, Tian F, Wang H-A, Dai G-Z. Gesture interaction in virtual reality. Virtual Reality & Intelligent Hardware 2019;1(1):84–112. Available from https://doi.org/10.3724/sp.j.2096-5796.2018.0006.

Louie DR, Eng JJ. Powered robotic exoskeletons in post-stroke rehabilitation of gait: a scoping review. Journal of Neuroengineering and Rehabilitation 2016;13(1). Available from https://doi.org/10.1186/s12984-016-0162-5, http://www.jneuroengrehab.com/home/.

Mishra R, Narayanan MDK, Umana GE, Montemurro N, Chaurasia B, Deora H. Virtual reality in neurosurgery: beyond neurosurgical planning. International Journal of Environmental Research and Public Health 2022;19(3):1719. Available from https://doi.org/10.3390/ijerph19031719.

Naik BN, Gupta R, Singh A, Soni SL, Puri GD. Real-time smart patient monitoring and assessment amid COVID-19 pandemic – an alternative approach to remote monitoring. Journal of Medical Systems 2020;44(7). Available from https://doi.org/10.1007/s10916-020-01599-2, https://link.springer.com/journal/10916.

Narayana TL, Venkatesh C, Kiran A, J CB, Kumar A, Khan SB, et al. Advances in real time smart monitoring of environmental parameters using IoT and sensors. Heliyon 2024;10(7):e28195. Available from https://doi.org/10.1016/j.heliyon.2024.e28195.

Nema S, Vachhani L. Surgical instrument detection and tracking technologies: automating dataset labeling for surgical skill assessment. Frontiers in Robotics and AI 2022;9. Available from https://doi.org/10.3389/frobt.2022.1030846.

Pan M, Yuan C, Liang X, Dong T, Liu T, Zhang J, et al. Soft actuators and robotic devices for rehabilitation and assistance. Advanced Intelligent Systems 2022;4(4). Available from https://doi.org/10.1002/aisy.202100140.

Paul M, Maglaras L, Ferrag MA, Almomani I. Digitization of healthcare sector: a study on privacy and security concerns. ICT Express 2023;9(4):571–88. Available from https://doi.org/10.1016/j.icte.2023.02.007.

Qi K, Song Z, Dai JS. Safe physical human-robot interaction: a quasi whole-body sensing method based on novel laser-ranging sensor ring pairs. Robotics and Computer-Integrated Manufacturing 2022;75:102280. Available from https://doi.org/10.1016/j.rcim.2021.102280.

Rashid AB, Kausik MDAK. AI revolutionizing industries worldwide: a comprehensive overview of its diverse applications. Hybrid Advances 2024;7:100277. Available from https://doi.org/10.1016/j.hybadv.2024.100277.

Tang YM, Chau KY, Kwok APK, Zhu T, Ma X. A systematic review of immersive technology applications for medical practice and education - trends, application areas, recipients, teaching contents, evaluation methods, and performance. Educational Research Review 2022;35:100429. Available from https://doi.org/10.1016/j.edurev.2021.100429.

Thang DV, Volkov A, Muthanna A, Koucheryavy A, Ateya AA, Jayakody DNK. Future of telepresence services in the evolving fog computing environment: a survey on research and use cases. Sensors 2025;25(11):3488. Available from https://doi.org/10.3390/s25113488.

Zafar MH, Langås EF, Sanfilippo F. Exploring the synergies between collaborative robotics, digital twins, augmentation, and industry 5.0 for smart manufacturing: a state-of-the-art review. Robotics and Computer-Integrated Manufacturing 2024;89:102769. Available from https://doi.org/10.1016/j.rcim.2024.102769.

CHAPTER 12

Data analytics and artificial intelligence for surgical outcomes

Nargis Khaliq[1] and Swati Sah[2]
[1]Department of Computer Science and Applications, Sharda University, Greater Noida, Uttar Pradesh, India,
[2]School of Computing Science and Engineering, Sharda University, Greater Noida, Uttar Pradesh, India

12.1 Introduction

Surgeons are using data analytics and artificial intelligence (AI) to usher in a new era of precision, safety, and customization. Surgeons are leading the transformation in surgical practice that will facilitate a better surgical experience for the patient overall. AI advancements enhance every aspect of the surgical process, from preoperatively assessing surgical risks to monitoring postoperative recovery, creating an opportunity for minimizing surgical risks and personalizing care in a manner we have only imagined. This chapter looks at how AI and data analytics play a role in improving surgical outcomes. It begins with the preoperative risk assessment and the opportunity to individualize treatment planning based on predictive modeling and machine learning (ML) algorithms. We also touch on the rationale for enhancing preoperative risk assessments to further reduce adverse events by tailoring surgical interventions to the individual patient. This progresses into AI, which provides decision support in real-time when surgeons identify they may need additional help with decision support in the intraoperative phase, even with AI-intraoperative solutions that can provide real-time information based on multimodal sensing capabilities to drive greater surgical outcomes. The chapter reviews the support of AI even in postoperative care, which actively monitors patients for complications so that undesirable outcomes can be curtailed and the recovery time is accelerated in a beneficial manner. The chapter proposes some of the new developments in evidence-based practice in terms of clinical decision support systems (CDSS) recommendations in relation to surgical cases. Ultimately, the chapter discusses robotic surgery and the element of AI-based technologies that improve surgical procedures and surgical practice. The chapter also recognizes the merit of surgery education involving surgical simulation devices and performance log-keeping, as opportunities for those desiring to pursue surgery. The combination of data analytics techniques and AI, with ongoing practice and innovation, has transformed surgery and patient experiences and outcomes in ways that would have been unfathomable only a number of years ago. This may be the most obvious advantage of the use of AI in the extent of transformation that has occurred in healthcare. The extraordinary innovations

AI-Powered Developments in Medical Robotics
DOI: https://doi.org/10.1016/B978-0-443-27783-2.00013-2

that AI is supplying will change the practice of surgery. This chapter will detail the ways that AI and data analytics will transform preoperative, intraoperative, and postoperative surgery. Considerations of AI in real-time decision support, robotic surgery, decision support systems, and predictive modeling will be discussed. The chapter will conclude with applications of AI for surgical training.

12.2 Predictive modeling and machine learning in preoperative risk evaluation

Assessing preoperative risk is critically important today in deciding if a patient can proceed with surgery and, if so, which treatment. In the past, the assessment of preoperative risk would be done through judgmental clinical inputs by the surgeon, review of patient history, and general clinical measures. Although this continues, the introduction of ML and predictive analytics has the potential to create a paradigm shift in preoperative risk assessment toward a more personalized and informed approach to of preoperative risk assessment. Predictive modeling uses previous data to render a probability of outcomes, such as complications after an operation, through statistical algorithms and processes. Predictive modeling of this nature certainly enables better predictions that take into account a larger patient dataset and modern systems that predict updates, enabled by ML, a form of AI. AI is most clearly applied in the OR in estimating preoperative risk. Years ago, surgeons would estimate surgical risk, based on clinical judgment, or review published guidelines that provided the basis for decision-making, and underwent stringent verification methods, and now surgeons can electrically know a person-specific estimate of risk and use informed ML methods and data analysis to help them make their preoperative assessment (Fig. 12.1).

12.2.1 Preoperative risk prediction models

The data-analyzing instruments consolidate tremendous amounts of information from various sources that include patient demographic data, genetic data, imaging data, and electronic health records in order to find an indication that could be associated with potential risk factors. Using the new patient data to inform past findings, these algorithms predict things such as the chance of bleeding, the risk of infection, or even death. Several ML algorithms, such as logistic regression, random forests, and support vector machine, have been developed for better accuracy estimations, especially in risk prediction. A number of predictive model types are frequently applied in preoperative risk assessment:

- Logistic regression: A straightforward and popular method, logistic regression describes the relationship between patient variables and the probability of a binary outcome (e.g., complication or no complication).

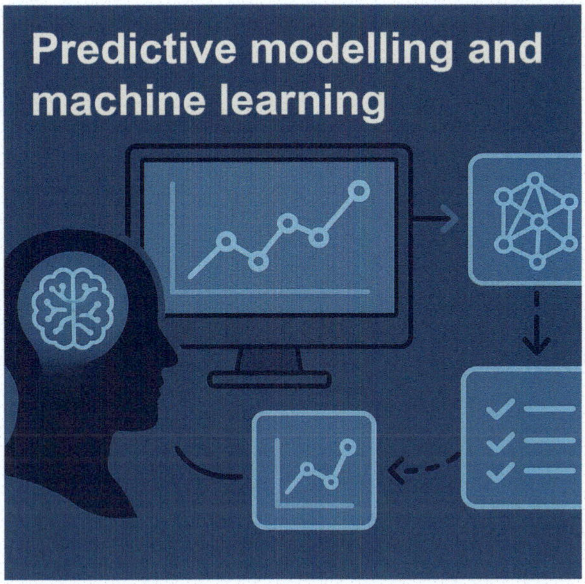

Figure 12.1 Model prediction frame.

Table 12.1 Comparison of different machine learning algorithms in predicting surgical risks.

Algorithm	Accuracy (%)	Sensitivity (%)	Specificity (%)
Logistic regression	87	85	88
Random forest	90	89	92
Support vector machine	88	87	89

- Cox Proportional Hazards Model: This model is frequently utilized to analyze time-to-event outcomes, including survival rates after surgery.
- Decision trees and random forests: Decision trees are graphic representations of decision-making processes and can identify nonlinear associations between variables.
- Support vector machines : SVMs are effective in high-dimensional spaces and can assist in classifying intricate patterns in patient data.

In general, these models are validated by contrasting their predictions with the outcomes of a group of surgically treated patients. A model may be utilized in clinical practice to assist in decision-making if it can accurately and consistently predict a patient's risk factors (Table 12.1).

By using AI to identify patients at high risk for complications through predictive modeling, surgeons can improve preoperative strategies, optimize resource allocation, and improve patient safety.

12.2.2 Applications of predictive modeling and machine learning in preoperative risk evaluation

The application of predictive modeling and ML in preoperative risk analysis has yielded a number of advantages to clinical practice:

Individualized risk stratification: By using models to quantify the risk of each individual patient, a patient-specific risk profile might be created and applied to preoperative management. This can also help you prioritize interventions directed to risk, that is, tiers between very close follow-up patients at risk or delaying a procedure until the patient will have the most beneficial outcome. Determine high-risk patients for complications: Predictive algorithms could be used for identifying patients who have identified risks for complications derived from their diabetes, cardiac, surgical history, etc. Clinicians can then take action to use that information in preventative strategies, that is, not to proceed with an invasive intervention or more creatively interact with patients in the management of their disease. Resource utilization: Related to the predictive discrimination regarding complications, resources could be optimally used to intensively manage high-risk patients (Yenra, 2025).

Surgical advances in safety and success: Rather than using a predictor of some form, a prediction model assists with preparatory and preoperative decision-making, at least in some fashion, that enables the surgeon to plan for the type and timing of animals' veterinary surgery or the need for ongoing care with a surgical specialist. Therefore preoperative predictions improve the ability to make informed surgical decisions. Predictive modeling and ML represent a qualitative advancement to the era of preoperative risk assessment, which would be just dataset assessment and analysis that provides explanations to patterns and relationships that a human expert would never be able to pick up on. Overall, these technologies improve patient outcomes and efficiencies of safety by applying an observed probability of every outcome of surgery, including complications and the probability of death.

Predictive modeling means using statistical techniques to create models that will predict future events based on past behavior. In preoperative risk assessment, predictive models might be used to predict complications due to surgery (postoperative problems, hospitalization, or death). ML by the broader definition of AI, refers to algorithms that pick up knowledge from experience, identifying patterns and predicting outcomes without prior programming. In preoperative risk assessment, several ML techniques are employed:

- Supervised learning: This approach uses labeled data, which is data with the known outcome, to train the algorithm. This model gains knowledge of the relationship between the intended output (such as the risk of surgical complications) and patient demographics, comorbidities, and lab findings. SVMs, logistic regression, decision trees, and random forests are a few of the frequently used algorithms in supervised learning.

- Unsupervised learning: This method looks for hidden patterns in the data and is used when the target is unavailable. It is primarily applied to tasks that require dimensionality reduction, anomaly detection, and clustering. Although it is less used in preoperative risk assessment, learning can be used to gradually improve risk prediction models by learning from results and making adjustments.

12.3 Artificial intelligence in intraoperative applications

One of the most appealing uses of AI today, within the modernization context of healthcare, is intraoperative practice being totally revolutionized. Uses of artificial intelligence in surgery take advantage of the cutting-edge developments in technology to further enhance precision, safety, and efficiency in surgical practice. Surgeons can now use AI systems in real-time to help make decisions, improve real-time navigation, enhance patient tracking, and, most significantly, facilitate decision-making in robotic-assisted surgery (Zhang et al., 2025).

The introduction of technologies based on AI has allowed surgeons to utilize it to help make real-time decisions about them.

12.3.1 Artificial intelligence-driven surgical assistance

AI methodologies, either using robots and/or computer vision systems, are applied to real-time data streams obtained from surgical instruments or from the tissues/anatomical structures involved in the surgery. In addition to tracking surgical instruments, AI systems can also track the very motion of the instruments to ensure performance is sustained within safe and effective ranges of motion. Computer vision can also then assess landmarks and identify and alert the surgeon if they arrive at a vital risk zone. The operating room (OR) is a unique decision-making environment in healthcare, in real-time, with much of the assumed risk. In the context of a surgical case, there are many considerations, including advanced facelift movement, the actual precision, and the timing of the interventions, because the clinical picture of the patient may change. Technology and tradition can be valued and useful, but the OR will have difficulty bringing those together or acting on larger amounts of data that is moving quickly (Ingwersen et al., 2023).

On the other hand, AI can quickly ingest large amounts of data, such as medical imaging, patient vital signs, and surgical devices' motions, and provide instant feedback to the surgical team. AI systems are based on algorithms and continuous analysis of data and can improve optimal decision-making in the OR, reduce human error, and improve surgical outcomes. Possible AI applications in the context of procedures can be classified into the following categories:

- Surgical navigation
- Robotic-assisted surgery

- Intraoperative monitoring
- Decision support systems
- Automated instrumentation

Each of these tools has specific attributes that lend themselves to enhanced surgical accuracy, productivity, and safety.

12.3.2 Minimizing intraoperative complications

Intraoperative complications are important contributors to morbidity and mortality in the OR. Apart from issues relating to anesthesiology, perioperative management, and surgical skill, surgical complications are an important topic worth discussing. Intraoperative complications can present themselves as benign problems, for example, bleeding or infection, or serious and urgent problems, such as organ failure or cardiac arrest. Averting intraoperative complications decreases patient outcomes, minimizes the cost to the healthcare system, and ensures the most competent management of the surgical problem. Although increasing the amount of preparation provided to the patient can minimize the risk of intraoperative complications, they can arise at any point during the operation. Complications have a profound effect on patients' outcomes, convalescence, and healthcare costs, and they can vary from insignificant, curable pathology to lethal, even fatal, illness. With the progress in surgery, the primary issue for anesthetists, surgeons, and the whole surgical team nowadays is understanding the etiology of such complications and their policy to reduce their occurrence (Bonnett et al., 2019).

By using AI systems that make predictive recommendations, intraoperative complications like bleeding, nerve damage, or organ injury are reduced. AI-based systems can analyze the operating field through patterns observed in real-time and alert the surgeon to potential hazards. Intraoperative complications may arise due to a variety of reasons, including surgical error, patient-specific pathology, anesthesia-related issues, and unexpected environmental or technical issues. Amongst the most common intraoperative complications are:

1. Bleeding and hemorrhage: One of the most common side effects of surgery is excessive bleeding, especially when large vessels are being operated on, as in orthopedic, liver, or cardiovascular procedures. Uncontrolled bleeding can result in shock, organ failure, and even death.
2. Anesthesia-related complications: During surgery, an anesthesia reaction may result in respiratory distress, bradycardia, hypotension, or malignant hyperthermia. With preexisting comorbidities, complications from anesthesia may be particularly dangerous.
3. Infection: Self-sovereign identitys are the most frequent infection complication, especially in the gastrointestinal (GI) tract, orthopedic, or implant operations.

Infections significantly impact prolonging recovery duration, requiring additional therapy or inducing possibly life-threatening sepsis.

4. Cardiovascular and respiratory complications: The patient intraoperatively can develop cardiac arrhythmias, ischemia, or respiratory depression. The cause of these complications could be surgery stress, anesthesia effects, or existing cardiovascular or pulmonary disease.
5. Nerve injury: Nerve damage is a result of inappropriate use of surgical devices or nerve compression due to positioning. This can cause permanent impairment or loss of function.
6. Organ injury: Accidental injury to nearby organs, such as the bladder, liver, spleen, or bowel, can occur, which can cause further complications, for example, sepsis, hemorrhage, and further operations.
7. Vascular complications: Vascular trauma or thromboembolic events, such as deep vein thrombosis (DVT) or pulmonary embolism, are hazards, particularly in long operations or those with long immobilization (Fig. 12.2).

12.4 Postoperative monitoring and early complication detection

The postoperative period is an important phase in the patient's recovery process. After surgery, patients are at risk for several complications that may arise due to reasons such as surgical trauma, anesthesia, comorbid conditions, and individual responses to the operation. Appropriate observation after surgery and early detection of complications play a crucial role in improving patient outcomes, preventing morbidity, and allowing for early intervention. A most important phase in delivering maximum surgical outcomes is postoperative care. AI technologies improve patient outcomes as they allow real-time monitoring and the possibility of early diagnosis of postoperative complications.

The significance of postoperative observation

Postoperative monitoring refers to the continued or periodic observation of a patient's clinical status, laboratory findings, and vital signs following surgery. Postoperative monitoring begins as soon as the patient is transferred to the intensive care unit (ICU) or recovery room and extends until the patient is stable for discharge or transfer to a general ward. Appropriate monitoring facilitates:

1. Early complication identification: Early identification of complications enables healthcare professionals to act quickly, minimizing the risk of serious consequences, extended recovery, and the necessity for further interventions.
2. Tracking the progress of recovery: Continuous tracking aids in the evaluation of whether the patient is recovering as expected and allows for readjustments if treatment schedules are necessary.

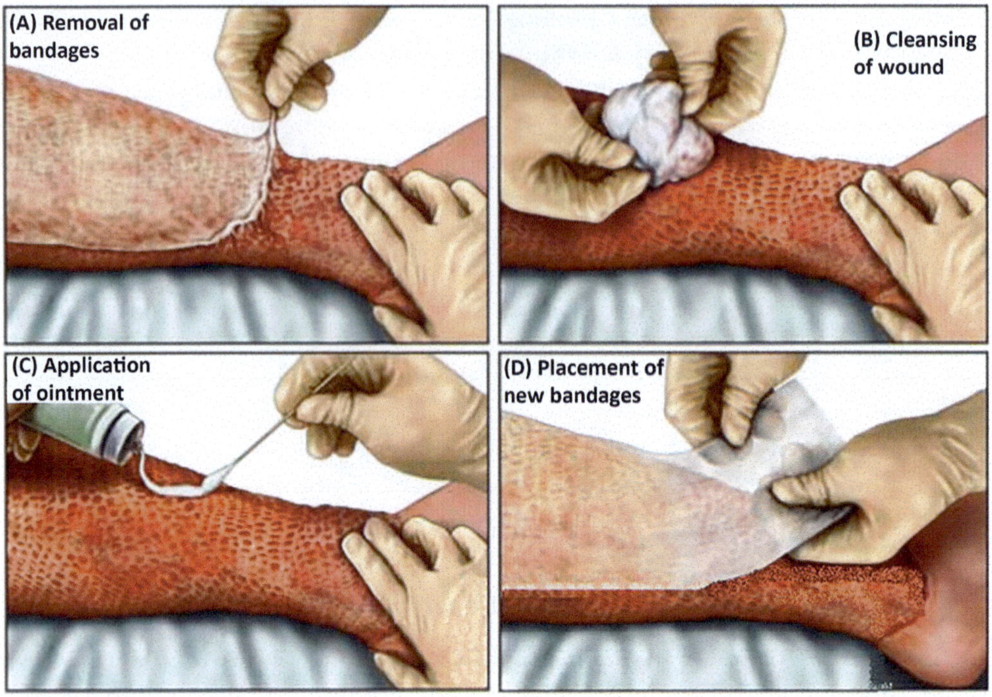

Figure 12.2 Wound care steps: (A) remove bandage, (B) clean wound, (C) apply ointment, and (D) place new bandage.

3. Maximization of pain and comfort: Frequent tracking of pain levels and other symptoms ensures that the patient's comfort is addressed, which can avoid immobility or wound breakdown complications.
4. Prevention of morbidity and mortality: Prompt identification of life-threatening circumstances like hemorrhage, respiratory distress, or infection can be life-saving and avert long-term effects.

Postoperative monitoring parameters

In monitoring a postoperative patient, there must be a systematic process. Important parameters to monitor are:

1. Vital signs: Rate and rhythm: Tachycardia, bradycardia, and arrhythmias can be indicators of cardiovascular instability, hemorrhage, or anesthetic complications. Blood pressure: Hypotension can be a sign of blood loss, dehydration, or sepsis, and hypertension may be due to pain, anxiety, or cardiovascular complications. Respiratory rate and oxygen saturation: Respiratory depression, hypoxia, and pulmonary conditions like atelectasis or pneumonia are usually seen postsurgery, especially if general anesthesia has been administered. Temperature: Fever is caused

by infection or inflammation, and hypothermia may be caused by anesthesia or exposure to cold for an abnormally long period of time during surgery.

2. Neurological status: Degree of consciousness: A sudden alteration in the mental status of a patient, that is, confusion, agitation, or somnolence, could be indicative of conditions such as hypoxia, drug effect, or neurological abnormalities. Pupil response: Pupil response and pupil size may provide data about neurologic status and also about drug-related issues (e.g., opiate overdose).
3. Fluid and electrolyte balance: Urine output: Carefully identifying the amount of urine produced is a vital sign that determines if fluids were received and how the kidneys are functioning. Anuria or oliguria suggests potential issues related to hypovolemia and/or renal failure or diseases of the urinary system. Fluid intake and output: Monitoring fluid and electrolyte status and ensuring appropriate hydration can prevent complications related to dehydration, enemas, and electrolyte imbalance.
4. Wound assessment: Wound site inspection: It is necessary to keep an eye out for symptoms of infection, severe bleeding, or dehiscence. Regular dressing changes are necessary, as is drainage monitoring for both quantity and consistency. Drainage tubes: Drains should be regularly checked for the amount and consistency of discharge, as these can indicate the possibility of infection or hemorrhage, if they are present.
5. Pain control: Levels of pain: The patient is more comfortable and recovers more quickly when pain is controlled, and appropriate intervention is used. Complications from poorly managed pain can include respiratory insufficiency, DVT, and immobility (Fig. 12.3).

12.4.1 Early detection of adverse events

The postoperative monitoring by AI algorithms of vital signs, laboratory findings, and images enables complications such as infections, hematomas, or organ failure to be detected early. Abnormalities within the norm frequently do not register with clinicians, whereas ML software is best suited to detect these. Healthcare adverse events, particularly those occurring during surgery, are unwanted outcomes that can assume a gargantuan negative burden on the patient's survival, quality of life, and recovery. These occurrences need to be identified in a timely manner to lessen their impact and prevent a subsequent complication. Adverse events may include a spectrum of potential harms ranging from relatively minor injuries or infections to genuinely severe complications, which may threaten life due to cardiovascular accidents, organ failures, or bleeding. By providing an early diagnosis, the physician can better manage adverse outcomes, avert unnecessary hospitalization, and avoid excessive costs. The merriment of the field lies in combining predictive modeling with ML techniques to sift through

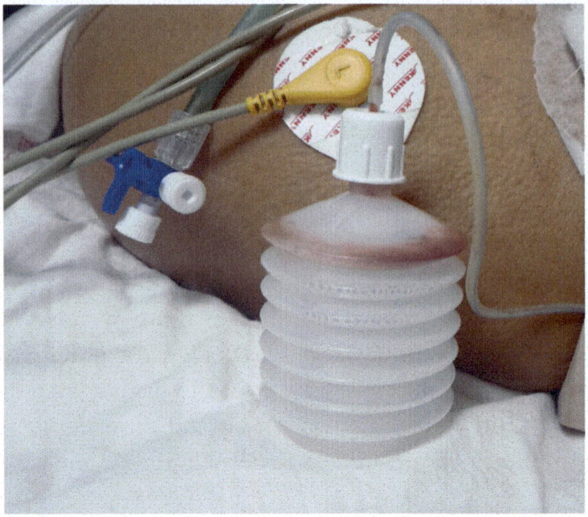

Figure 12.3 A surgical drain (closed suction system) is used for postoperative fluid collection.

petabytes of data to detect subtle patterns and predict adverse events just prior to their occurrence. In this paper, we present the use of predictive modeling and ML for early detection of adverse events in healthcare. The next steps involve summarizing the scientific bases for these technologies, analyzing their present applications in healthcare settings, and giving an overview of the challenges and potential future directions of this exciting field. Such econometric models for prediction are often defined in a statistical context as being a quantitative approach based on historical data to predict future events by building a model that estimates the likelihood of a future outcome happening. Within this domain, outcomes such as complications, readmittance, and even chronic illnesses have a parallel predictive element. While consulting a patient's clinical history, a more advanced predictive model is far more capable of identifying and predicting relationships and patterns that otherwise may be overlooked, assisting in the recognition of a clinically adverse event at an earlier time. There is a likelihood that postoperative complications can occur, and this possibility can be mapped by models that in the predictive analytics stage assess a patient's age, history, lab results, and vital signs, along with several other variables. When such risks are managed early, clinicians can alter any treatment plans or even take any approach in order to mitigate and decrease the chances of any adverse outcomes.

What makes ML a subclass of AI? One interesting form of AI is ML, which, with access to large datasets, is capable of improving its predictive abilities over time. Unlike other predictive models that need to be programmed for each step, ML models can autonomously discover and refine patterns within the data. In the healthcare industry, there is a broad range of data, including real-time monitoring, clinical data, genomic data, and patient demographics, that can be leveraged to train ML models. The developed models can now generate real-time predictions of risk for event-related adverse events, updating clinicians as to the patient's current condition. For instance, ML models have predicted sepsis, what about possible adverse effects of medication, and surgical complications, all beforehand, and triggering intervention in a timely manner.

Real-world examples

1. Predicting surgical complications: Graeßner et al. (2023) utilized a ML model to predict complications in patients who have undergone a surgical procedure, utilizing patient characteristics before surgery. The model was used to evaluate the characteristics that reliably represented whether patients were at risk for post-operative complications, such as the presence of comorbidities, laboratory findings, and the type of surgery. Additionally, medical care providers had the opportunity to intervene early based on this information once patients transferred from the hospital wards, which could result in a reduced likelihood of experiencing complications.
2. Sepsis prediction: Detecting sepsis, an infectious condition that could be fatal, can be challenging in its initial stages. To predict whether a patient will get sepsis hours before the clinical symptoms appear, researchers have developed ML algorithms that use real-time patient data, such as vital signs and blood tests. Early detection allows clinicians to initiate treatment sooner and achieve better patient outcomes.
3. Adverse drug reactions: Adverse drug reactions (ADRs) are one of the largest challenges for contemporary medicine. ML algorithms can predict ADRs risk by analyzing patient characteristics (age, sex, medical history, and currently prescribed drugs) and help clinicians adjust medications/doses to optimize risk/benefit ratio and safely provide drugs to the patient.

12.4.2 Continuous monitoring systems

Wearable technology using AI and monitoring from a distance collects and analyzes data on an ongoing basis, allowing us to act quickly when warranted (especially for high-risk patients or patients recovering from complex surgery). Continuous monitoring systems (CMS) monitor the patient's health status in real-time with sensors and devices. CMS allow for the ongoing collection of data more thoroughly and dynamically than older methods of merely taking snapshots of health on a designated schedule. Wearables, remote devices, and other sensors transmitting data points utilizing wireless technologies are generally part of these systems and can provide rapid intervention if necessary.

Major elements of continuous monitoring systems

1. Sensors are at the heart of CMS and ultimately measure physiological markers such as heart rate, blood pressure, temperature, oxygen saturation, and respiratory rate. They acquire continuous data and obtain information without interfering with a

patient's usual activities. Sensors can be developed into wearables, such as patches, bands, or garments, that measure physiological markers.

2. Data delivery: The doctors received the sensor data wirelessly. The experience of the real-time data transmission allows the physician to be able to access the patient data remotely so they can act quickly in the event an anomalous finding is discovered.
3. Data interpretation and analysis: Raw sensor data is processed and analyzed through algorithms and AI. Analysis provides healthcare workers with comparative data that can allow for spotting trends, predicting potential risk, and making more informed decisions in real-time—rather than simply observing this data manually.
4. Notifications and alerts: A primary benefit of CMS is its capability to notify medical professionals about an unexpected change in a patient's condition. If the clinician responds appropriately to the alert, they may be able to avert negative consequences such as infection, respiratory distress, or cardiac arrest.

Applications of continuous monitoring systems

CMS have applications in many healthcare environments, from postoperative care to chronic illness management. Here are a few examples of CMS use for improved patient care:

1. Chronic disease management*:* Chronic conditions such as diabetes, heart failure, and chronic obstructive pulmonary disease need a consistent way to monitor the condition to manage symptoms and identify issues early. For example, in diabetic patients, regular blood glucose monitoring lets medical practitioners follow blood glucose levels throughout the day and adjust medication or lifestyle alterations accordingly. Likewise, in heart failure patients, ongoing monitoring of vital signs (blood pressure, oxygen saturation levels, and heart rate) will enable practitioners to identify early indications of a decline (Harvey et al., 2023). Gomes et al. (2022) showed that continuous monitoring can be beneficial in heart failure care, where the use of real-time data led to real-time intervention, which resulted in reduced hospital readmission rates. The use of continuous monitoring solutions in the routine of care allowed clinicians to recognize symptoms and the worsening of heart failure many days/many visits earlier than it became medically serious, leading to better outcomes for patients and lower costs of healthcare.
2. Postoperative care: After surgery, the risk of complications occurs, including infection, blood clots, or an organ not working. Healthcare workers continuously monitor so that early action is possible, and any change in a patient's condition can be recognized immediately. A study by Sohn et al. (2021) reported an improvement in recovery rates, with early identification of complications such as sepsis and respiratory failure, due to the continuous monitoring postoperatively. CMS can provide warning alerts for conditions that would otherwise likely rely on a more invasive monitoring approach by continuously measuring heart rate, temperature,

and oxygen saturation. The monitoring will specifically help high-risk individuals and those recovering from major surgery.

3. Intensive care units and acute care: The ability to continuously monitor a multitude of variables for severely ill patients is required in ICUs. More recently, CMS added more detailed and predictive abilities to legacy monitors in ICUs that may monitor things such as oxygen saturation, blood pressure, and heart rate (Zhang et al., 2024) assessed a CMS in the ICU to improve monitoring of patients with acute respiratory distress syndrome (ARDS) and/or severe sepsis.

By using predictive algorithms to identify trends in their patients' conditions, the system was able to notify physicians of problems before they became life-threatening complications, leading to fewer adverse events in patients and better outcomes.

12.5 Robotic surgery: a leap toward precision and autonomy

The combination of robotics and AI ushered in an era of minimally invasive surgery. Robotic surgery has improved surgical accuracy, improved recovery, and reduced human error. Robotic surgery began in the 1980s as an experimental approach, and most systems were developed for either diagnosis or as adjuncts to minimally invasive surgery. The surgeon, still the de facto operator, frequently had direct manual control over the immature, expensive robots of this time. Over the decades, robotics systems have progressed through advances in engineering, AI, and data processing, allowing them to evolve into independent machines capable of contributing or executing certain actions with minimal human oversight.

The da Vinci Surgical System, which made its commercial debut in the late 1990s, was one of the first robotic surgery systems to demonstrate commercial success. It allowed for more precise and less invasive procedures through 3D visualization capabilities. Since then, advancements in AI and robotics have led to systems with increasingly advanced capabilities and levels of intelligence and autonomy. Published a study that emphasized the advantages of ongoing development of robotic-assisted surgeries and cited significant advances in technology that have made surgical procedures safer.

Key features of robotic surgery systems

A number of components make up robotic surgical systems, which are intended to help surgeons carry out delicate procedures. Among the primary characteristics are:

1. Robotic arms: These are the primary devices that hold surgical instruments, such as scalpels, scissors, or suturing devices. The robotic arms are very flexible and provide you with a larger range of motion than the human hand. You will be able to move more precisely and more controlled, as well as in confined, compromised, or difficult anatomical areas.
2. 3D visualization: 3D cameras with high-definition embedded in robotic systems provide real-time and precise visualization of the surgical field. Enhanced

visualization is especially beneficial in minimally invasive surgery when vision may be impaired due to internal structures and small incisions.

3. Precision tools: Robotic surgical instruments have a smaller size, improved precision, and higher flexibility in contrast to traditional instruments. The instruments can perform procedures with greater precision, which reduces trauma to surrounding tissues and organs.
4. Surgeon console: The surgeon typically operates the robot from a console located adjacent to the patient. The surgeon operates the robotic instruments through the console, with foot pedals and hand controls. The console offers real-time haptic feedback (or proprioception), allowing the surgeon to control robotic arms and instruments to an even greater degree of sensitivity and accuracy.
5. Artificial intelligence integration: Magnanimous robotic systems incorporate AI-backed algorithms to assist in surgery. They can streamline surgical processes, assess data gathered in the OR, and provide advice and professional suggestions in real-time. They can even offer some nondiscretionary functions in some cases.

12.5.1 Robotic surgical systems

Robot-assisted tools like the da Vinci Surgical System are now being used across several different specialties (e.g., urology, gynecology, and cardiothoracic surgery). Robotic instruments increase the dexterity for the surgeon and magnify the surgical field, enabling the surgeon to produce more precise incisions while keeping adjacent tissues intact. In addition to robotic instruments, AI programs can be employed for real-time decision-making during surgical procedures, and some components of the surgery can even be automated. For example, AI could automatically identify important landmarks during the procedure or navigate the robotic arms to avoid healthy tissues. Thiel et al. (2022) explored the possibility of using AI to assist in robotic-assisted laparoscopic surgery, where AI algorithms actively assisted the surgeon with real-time feedback on the motion of the instruments and the anatomy during the operation.

In addition, robotic surgery systems powered by AI are being developed with predictive analysis that has the capability to anticipate surgical complications depending on patient records before and during surgery. The predictability enhances preoperative planning and makes surgery safer and more precise.

Challenges and ethical issues

Although robotic surgery comes with many benefits, there are a number of challenges and ethical issues:

1. High cost: Robotic surgical systems are expensive to purchase, maintain in operation, and utilize. The initial capital requirements for hospitals may be too great, and certain providers cannot afford to do so without clear evidence of

improved patient results. Robotic surgery is controversial regarding cost-effectiveness.

2. Training and expertise: Robotic systems can be effectively utilized by surgeons only if their training is specialized. Because the learning curve for robotic-assisted surgery is steep, potential complications may arise from erroneous use of the robotic system. Adequate surgeon training is an essential factor in the safe performance of robotic surgery.
3. Ethical and legal issues: The increasing autonomy of robotic systems brings up accountability and initiative concerning errors. If a robotic surgery system that uses an AI system made an error, there may be some ambiguity in determining whether the surgeon, the manufacturer, or the technology itself is liable. This topic is nuanced and will need to be dealt with as robotic surgery becomes commonplace.

The future of robot-assisted surgery

The future of surgery with robotic assistance is full of potential. AI, ML, and data analysis will continue to advance and lead to robotic systems becoming increasingly intelligent and autonomous. Surgeons will increasingly collaborate with robotic systems through a hybrid model in which the robots not only provide real-time data feedback for surgeon decision-making but also autonomously complete specific procedures while clinician control is maintained in other areas contributing to the surgery process.

Additionally, there are vast opportunities for robotic surgery to extend outside the hospital and into remote and underserved settings. For example, telesurgery, or having a surgeon perform a procedure remotely using a robotic system, may provide improved surgical care in situations where surgical care is difficult to access, such as rural areas. Jenkins et al. (2023) suggested that tele-robotic surgery may address some inequities in access to surgical care by providing surgical care to people located in remote areas by referring to the surgical expertise elsewhere.

12.5.2 Artificial intelligence-enhanced robotics

The next generation of robotic systems utilizes AI to make decisions and readjust in real-time. AI can evaluate the most effective surgical plan and define a surgical route that accounts for subtle circumstances of evolving anatomy. AI-enabled robotics describes the assimilation of computer vision, ML, and numerous methodologies from the field of AI into robotic devices. AI systems, therefore, can allow robots to process vast amounts of data in real-time, access sufficient data for autonomy, and modify to account for real-time situations during surgery. AI can also optimize movement, anticipate complications, support precision work, and, when necessary, keep the neurosurgeon on a route through complex procedures.

Key elements of AI-powered robotics

1. Machine learning and data processing: Robot systems based on AI use ML algorithms that allow the system to "learn" using large datasets. Those datasets can include past surgical cases, medical images, patient histories, and sensor data collected during surgery. Over time, the algorithms of the robot will be able to optimize the processes they follow to help ensure superior surgical results and decision-making.
2. Computer vision and imaging: AI-programmed robots perform computer vision analysis of medical images (CT, MRI, and X-rays) to help identify key structures and abnormal signals. In conjunction with scanning real-time images, the robot can assist in targeting tumors, key blood vessels, and anatomy during the surgical procedure, resulting in improved accuracy.
3. Real-time decision-making: An important advantage of AI-infused robotic systems is the capacity to make decisions in real-time. During the surgical process, AI algorithms can analyze and process data from the patient in real-time and will detect changes or risks that will prompt the robot's action, whether that is to move or alert the surgical team. Predictive analytics will indicate the complications and recommend a deviation from the surgical plan.
4. Autonomous or semiautonomous functions: While a human surgeon will always maintain control, most robotic platforms with AI function semiautonomously. These functions may include automating repetitive movements, for example, suturing or making an incision, from real-time sensor data and algorithmic decision-making (Table 12.2).

Advantages of AI-aided robotics in surgery

The use of AI in robotic systems has several advantages for patients and healthcare practitioners. These are as follows:

1. Enhanced surgical accuracy: AI-driven robotic systems improve surgical movement precision, particularly in complex procedures requiring high dexterity and fine

Table 12.2 Key benefits of AI-assisted surgery.

Benefit	Description
Enhanced precision	AI assists in more accurate movements and adjustments.
Reduced surgical trauma	Minimizes damage to healthy tissue through minimally invasive techniques.
Shorter recovery time	Less trauma leads to faster healing and reduced hospital stays.
Real-time adaptation	AI systems adjust to changing surgical conditions automatically.

motor skills. For example, AI can improve implant placement, minimize the risk of cutting vital structures, and improve tissue handling in delicate surgery.

2. Less human error: The primary reason for most surgical complications is human error. Robotic systems with AI can aid in ridding errors stemming from fatigue, distraction, or poor judgment. AI systems can adapt their responses to best optimize surgical strategies and learn automatically from prior surgeries to deliver consistency that is very difficult for humans to provide.
3. Shorter recovery times: AI-enabled robotic systems facilitate minimally invasive surgical procedures that have reduced incision size, decreased blood loss, and shortened recovery times for patients. Nguyen et al. (2023) confirmed that patients who had AI-aided robotic surgery had shorter hospital stays and fewer postoperative complications than patients who underwent traditional open surgery.
4. Improved surgical planning: AI systems improve preoperative planning by analyzing patient data, images, and history to optimize surgical methods. The systems can identify risk and make recommendations to prevent risks and to improve the accuracy of the procedure and the safety of the patient.

12.5.3 The function of artificial intelligence in increasing precision and autonomy

The capability of AI to augment robotic surgery relates to its ability to process and analyze data rapidly and provide more accurate surgical procedures. AI allows for ongoing learning via the robotic systems getting better with time through processing larger datasets of surgical activity. The systems can identify potential problems that may arise, suggest other surgical procedures to consider, and also make clinical decisions independently.

A major advantage is that AI supports surgeon decision-making in real-time. For example, article describes how an AI-enabled robot used continuous information regarding patients to evaluate possible threats, such as renal bleeding or an organ injury, and modified surgical devices and instruments automatically.

AI's capability to enable semiautonomous surgeries is also the reason why AI-controlled robots can perform certain tasks, like suturing or stitching, under the supervision of a surgeon. These robots partially automate workflows, and are an option when greater efficiency and less fatigue of the surgeon are needed during long operations.

12.6 Clinical decision support systems

By leveraging data-driven recommendations for the patient's care, CDSS with AI will advance the future of surgery.

12.6.1 Decision support in preoperative planning

In a way that considers the patient's medical history, risks, and preferences, a CDSS can make recommendations for what surgery might be best to perform. Following through a database of groups of similar cases, a CDSS can then recommend what postoperative care and recovery would be best and appropriate for patients.

12.6.2 Intraoperative decision support

An AI-based CDSS can observe data during a surgical procedure in real-time and make recommendations based on that data, for example, alerting the surgeon to changes in vital signs or proposing an alternative technique if it becomes difficult to follow through with the original surgical technique.

12.7 Data-driven research and optimization of surgical practices

AI can advance surgical techniques in research by analyzing large datasets to determine if commonalities exist and what can be learned to optimize practices.

12.7.1 Predictive analytics in surgical research

AI will allow for surgical outcomes to be assessed across institutions and patient populations at scale. By assessing large surgical and patient outcomes datasets, AI can identify trends, best practices, and areas for improvement in surgical care.

12.7.2 Evidence-based guidelines development

AI-enabled evidence-based systems could combine science from clinical trials, observational studies, and real-life cases to support the updating of evidence-based surgery guidelines, with surgeons using the most current evidence to inform their clinical decision-making.

12.8 Artificial intelligence in surgical education

AI is revolutionizing surgical education by providing tools for simulation and performance monitoring of medical professionals.

12.8.1 Simulation-based learning

Surgeons can practice several procedures utilizing AI-based simulators in a safe, controlled environment. Such simulators can also simulate different surgical scenarios, providing feedback to the learner during performance.

12.8.2 Performance monitoring and feedback

AI tools can also monitor surgical residents' performance and perform those analyses. For example, AI can monitor the speed and precision of a resident's actions, capturing those actions to consolidate into constructive feedback that aids in residents' performance improvement.

12.9 Conclusion

The use of AI and data analytics within the surgical field is revolutionizing the practice of surgery. AI is improving risk stratification preoperatively, improving accuracy during intraoperatively, and supporting the early detection of postoperative complications, all of which will improve the safety of surgical patient care, and, overall, patient care will become more patient-centered. The evolution of robotic surgery, CDSS, and simulation-based education will also modify the way surgery is undertaken to become more accurate, patient-centered, and efficient. As we evolve, these technologies will ultimately improve patient safety and improvement in outcomes. Our approach to defining adverse events is progressing with the utilization of ML and predictive modeling in healthcare practice. These technologies will use large amounts of data to find detectable patterns that the human eye cannot see, to improve outcomes, and to allow preventative intervention. The potential advantages of these technologies are noteworthy, despite the challenges of explainability, data quality, and clinical implementation. With constant research and continued improvements to the system, earlier identification of adverse events through predictive modeling and ML will become standard practice within the clinic, optimizing and enhancing care safety and efficiency. CMS are revolutionizing the delivery of healthcare by offering a more proactive, data-driven model of patient management. CMS has proven its ability to improve outcomes, minimize complications, and reduce healthcare costs in a range of settings, including chronic disease management, postoperative care, and critical care. Before the majority of the potential can be realized, there are challenges like data overload, cost, and privacy that need to be addressed. As technology and data analytics continue to evolve, CMS is well-positioned to assume a larger role in healthcare in the next few years and improve the quality of care and more personalized care models. Robotic surgery represents a step toward precision, autonomy, and improved patient outcomes. With its ability to enhance operative precision, reduce recovery time, and improve patient safety, robotic surgery can potentially revolutionize the practice of surgery. Issues of cost, training, and ethics remain, but advances in robotic and AI technologies are likely to stimulate broader application and still more rapid evolution in surgical care. As research progresses in expanding the limits of what is possible, robotic surgery will inevitably be at the leading edge of defining the future of medicine.

References

Bonnett L, McDowell T, Montague L, et al. The role of clinical prediction models in healthcare: a guide for implementation. BMJ 2019;365:1737.

Gomes AM, Rodrigues AD, Silva PD. The effectiveness of continuous monitoring systems in heart failure management. Journal of Cardiac Monitoring 2022;29(3):225–33.

Graeßner M, Jungwirth B, Frank E, Schaller SJ, Kochs E, Ulm K, et al. Enabling personalized perioperative risk prediction by using a machine-learning model based on preoperative data. Scientific Reports 2023;13(1). Available from https://doi.org/10.1038/s41598-023-33981-8.

Harvey RM, Sun L, Wang X. Chronic disease management through continuous monitoring: a review of recent innovations. Journal of Telemedicine and Telecare 2023;29(2):89–97.

Ingwersen M, Ali M, Spijkerman JT, et al. Machine learning models in clinical practice for the prediction of postoperative complications. World Journal of Surgery 2023;47(2):439–48. Available from https://link.springer.com/article/10.1007/s00595-023-02662-4 (Accessed 24 February 2025).

Jenkins R, Kumar P, Chang Y. Tele-surgery and the future of remote robotic surgery. Telemedicine Journal and E-Health 2023;29(4):217–23.

Nguyen T, Lin H, Zhang C. AI-assisted robotic surgery: faster recovery and fewer complications. Journal of Surgical Robotics 2023;10(4):212–20.

Sohn CH, Lee SM, Kwon HW. Postoperative complications detection using continuous monitoring systems in high-risk patients. Journal of Surgical Research 2021;47(1):43–50.

Yenra, AI-powered autonomous surgical robots: advancements in surgery, 2025. Available from: https://yenra.com/ai-tech/autonomous-surgical-robots/ (Accessed 24.02.25).

Zhang T, Lin Y, Yang J. Continuous monitoring in ICU: a comprehensive review of patient management improvements. Critical Care Journal 2024;52(4):301–10.

Zhang, Y, Li, H, Wei, L, Spatiotemporal AI model for analysing surgical skills in open surgical videos, 2025. Available from: https://arxiv.org/abs/2112.07219 (Accessed 24 February 2025).

PART IV

Economic, social, and cost considerations

CHAPTER 13

Cost-effectiveness analyses of artificial intelligence based medical robotics

Shiji Valsan[1], Daria A. Nefedova[2], and Donald S. Shepard[2]
[1]Global Health, Boston Children's Hospital, Interventional Radiology, Boston, MA, United States,
[2]Heller School for Social Policy and Management, Brandeis University, Waltham, MA, United States

13.1 Introduction

Artificial intelligence (AI) is the study and development of algorithms that allow machines to reason and perform cognitive functions, such as problem-solving and decision-making. The most impactful adoption of AI on human life is in healthcare, where it is used to help make more precise decisions and predict patient outcomes with more certainty. Surgery has experienced one of the largest impacts with the adoption of AI, as more surgeries are performed using robotic assistance (Knudsen et al., 2024).

13.1.1 Artificial intelligence integrated robotic assisted surgery: enhancing process and outcomes

Robotic-assisted surgery (RAS) has revolutionized procedures involving the abdominal and pelvic cavities, as well as synovial joints, by enabling precision through minimally invasive approaches. However, the dynamic intraoperative environment, shifting anatomy, smoke from electrocautery, and varying tissue textures pos challenges to consistent visualization (Knudsen et al., 2024). Modern robotic systems increasingly leverage AI-driven image processing tools for real-time denoising, deblurring, and color correction. These enhancements improve intraoperative visibility, particularly in procedures like knee arthroscopy, where surgical smoke can otherwise obscure the field (Knudsen et al., 2024). AI also aids in identifying safe surgical planes, helping surgeons avoid critical structures such as arteries and nerves, reducing intraoperative risks and long-term complications. In surgical oncology, especially during complex procedures like radical prostatectomy, AI can assist in navigating challenging anatomy, such as previous surgeries or enlarged median lobes, enhancing precision and reducing recurrence (Kumazu et al., 2021). A transformative application of AI lies in surgical workflow recognition. Surgeries consist of distinct phases and tasks; training AI to recognize these has traditionally required large, annotated datasets. Today, deep learning

AI-Powered Developments in Medical Robotics
DOI: https://doi.org/10.1016/B978-0-443-27783-2.00010-7

models are streamlining this process, enabling real-time task recognition, anticipatory guidance, and workflow optimization (Shi et al., 2020).

In summary, AI integration in robotic surgery improves visualization, enhances intraoperative decision-making, standardizes workflows, and supports safer, more efficient procedures with better patient outcomes. The continued adoption of robot-assisted surgery is multifactorial, driven in part by aggressive marketing strategies from manufacturers, hospitals, and surgeons aiming to stimulate patient demand. As the number of elderly persons in the United States and globally continues to rise, the number of individuals requiring surgical intervention, including those eligible for robotic procedures, is also increasing. This shift toward more technologically advanced and consequently more expensive treatment modalities has substantial public health and economic implications. The rising cost of medical care is often driven by technological innovation in clinical management. With increasing healthcare expenditures and constrained resources, economic considerations are instrumental in guiding resource allocation in healthcare (Neumann et al., 2017). Decision-makers must evaluate the capital and operational costs of robotic platforms against alternative surgical modalities while simultaneously considering clinical outcomes and quality metrics to determine overall value. Laviana et al. critically examined the economic and systemic drivers behind the rapid adoption of robotic surgery. They highlight how competitive pressures and the pursuit of technological prestige among hospitals have fueled growth, despite substantial upfront and maintenance costs. The authors draw attention to the opportunity costs, such as the diversion of resources from other essential areas, such as staff training or facility upgrades. Although robotic surgery may reduce postoperative complications and length of stay, these potential benefits must be weighed against higher procedural costs and extended operating times, particularly during the learning phase. Nonclinical factors, such as patient demand and surgeon preference, often outweigh evidence-based outcomes in influencing adoption. The authors suggested meticulous cost-effectiveness evaluations to ensure clinical gains justify long-term financial investments (Laviana and Hu, 2017). The cost-effectiveness of robotic surgery remains a timely and critical issue as it increasingly competes with and in some cases replaces the gold-standard open and conventional laparoscopic approaches. The use of robotic surgery for all general surgery procedures increased from 1.8% to 15.1% from 2012 to 2018 (Sheetz et al., 2020). Robotic-assisted hysterectomy became one of the most performed minimally invasive gynecologic surgeries in the United States by early 2010 after the widespread usage of the da Vinci system (Bouquet de Joliniere et al., 2016).

A 2023 systematic review and meta-analysis found that an estimated 65% of hysterectomies in the United States are performed using a minimally invasive approach, with up to 43% of laparoscopic surgeries done with robotic assistance (Randell et al., n.d.). The adoption of robotic surgery for general procedures rose from 15.1% in 2018 to 22% by 2022. The number of robotic surgeries performed annually in the United States has shown a steep upward trend, increasing from 205,000 in 2010 to an estimated 2.68 million in 2024, reflecting the rapid adoption and integration of robotic technology in surgical practice (Fig. 13.1, Source 9). Also, 42.5% of small hospitals (<200 beds) adopted robotics, and 84.9% of teaching hospitals use robotic systems (Deb, 2025; Glick et al., 2014).

Robotic-assisted radical prostatectomy (RARP) has progressively integrated AI features, blurring the line between conventional RARP and AI-enhanced RARP (AI-RARP). These features include augmented reality overlays, real-time tissue recognition, and automated movement refinement, all designed to support intraoperative decision-making and enhance surgical precision. As such, the AI-RARP described in this case study represents a series of points along a continuum of technological evolution, with varying degrees of AI implementation across platforms and studies (Hung et al., 2018). Since the introduction of robotic-assisted prostatectomy in 2000, the adoption of robotic technology, gradually augmented by AI, has significantly reduced complications such as biochemical recurrence and nerve-sparing deficits by enhancing surgical precision and improving operative performance (Westerman et al., 2024). These gains

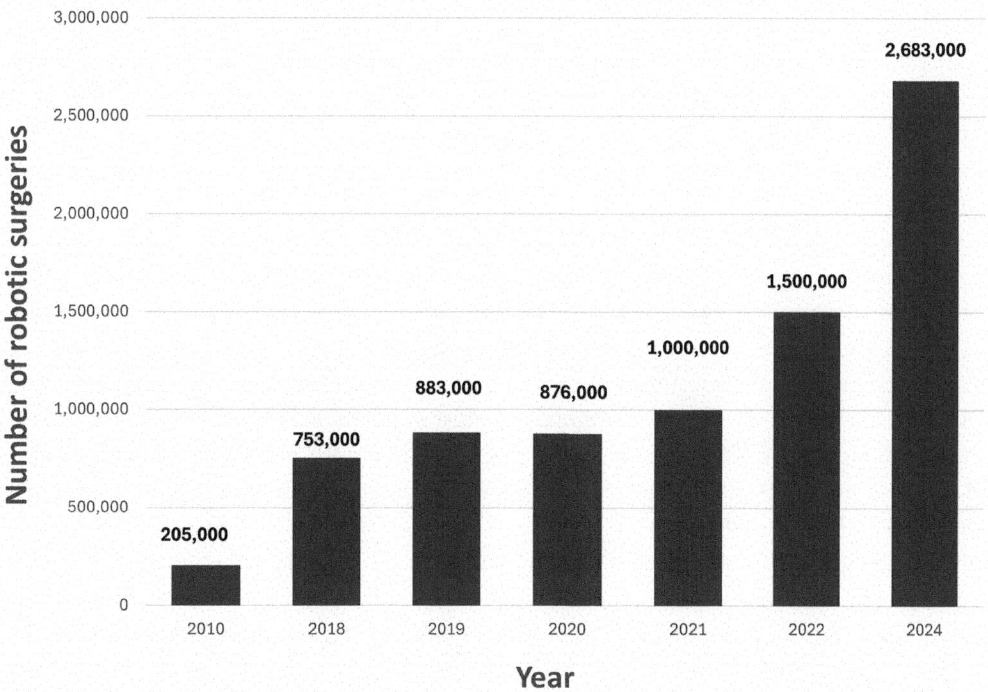

Figure 13.1 Number of robotic surgeries in the US by year, 2010–24. *Data taken from a published or online source without its original presentation.*

point to the importance of considering AI capabilities as an evolving attribute within the broader context of robotic surgery.

Accordingly, there is an urgent need for rigorous cost-effectiveness analyses and comparative effectiveness research to evaluate the economic impact and clinical utility of newly introduced surgical technologies. However, such evidence remains sparse in the current literature, highlighting a critical gap in informing evidence-based policy and practice.

13.1.2 Theoretical foundations of cost-effectiveness analysis

Cost-effectiveness analysis (CEA) is a methodological framework intended to guide decision-makers on how best to allocate limited healthcare resources to achieve improved outcomes. Rooted in economic theory, CEA assumes that when individuals or institutions face constraints, typically financial, they must make trade-offs between competing alternatives. These exchanges can be examined using quantitative frameworks that assess both costs and consequences of potential interventions (Neumann et al., 2017). Meltzer et al. (2016) emphasize that in real-world healthcare systems, decision-makers (patients, providers, and insurers) often do not bear the full costs or receive the full benefits of their decisions. This market imperfection results in inefficient allocation of resources. The CEA helps bridge this gap by providing a summary measure of value, typically in the form of cost per unit of health outcome. These are the costs per quality-adjusted life year (QALY) gained or per disability-adjusted life year (DALY) averted. By aggregating both costs and benefits across different health interventions, CEA facilitates comparisons and supports rational policy choices in the presence of resource scarcity.

However, applying CEA in practice involves important normative and contextual considerations. Decisions must be made about which objectives to prioritize, whose preferences to consider, and what constraints (e.g., budgets, ethics, and equity) should guide allocation. These value-laden choices influence how outcomes and costs are defined and measured. For instance, sometimes productivity losses or caregiver time should be included, or equity weighted against efficiency (Neumann et al., 2017). Context also shapes CEA's role in policy. In countries with centralized health systems and defined budgets, CEA often informs formal coverage decisions. In more fragmented systems like the United States, CEA plays a more variable and indirect role, often filtered through private payers and institutional policies. Standardized frameworks such as "reference case" guidelines have been developed to improve comparability across studies, but local factors may still justify methodological variation (Neumann et al., 2017). In this conceptual framework, CEA is considered a decision-support tool that combines economic reasoning with empirical evidence to guide health policy under resource constraints. It does not dictate ethical or political choices but offers structured insight into how individual health outcomes translate into population-level value, supporting more informed and balanced decision-making.

In this chapter, we explore CEA principles with a focus on AI-enhanced medical robotics. We first present an overview of CEA methodology, explaining how costs and health impacts are categorized and quantified. We then explain CEA in the context of AI-assisted robotic surgery through a detailed case study that compares conventional radical prostate cancer surgery to an AI-assisted robotic surgery. Radical prostatectomy was selected for the case study because it is one of the most performed procedures using robotic assistance with high clinical volume and well-documented outcomes, making it a representative model for analyzing cost-effectiveness in surgical innovation (Shin and Lee, 2022). The case study illustrates the cost components, health outcomes (complications, recovery), QALY calculations, and the incremental cost-effectiveness ratio (ICER). While AI-enhanced RARP might have been compared with standard RARP, we chose open radical prostatectomy (ORP) as the comparator because it remains a widely used reference in cost-effectiveness studies, is still prevalent in many health systems, and has more complete cost-outcome data available. Future work could explore the marginal impact of AI by comparing it to standard RARP.

13.1.3 Principles of cost-effectiveness analysis in healthcare

Building on its role as a decision-support tool, CEA systematically compares two or more health interventions in terms of both their costs and their health outcomes. The core analytic output is the ICER, which represents the additional cost required to achieve one additional unit of health benefit (e.g., per QALY gained or DALY averted) (Glick et al., 2014).

Mathematically, the ICER is expressed as:

$$\text{ICER} = (\text{Cost}_{\text{new}} - \text{Cost}_{\text{usual}})/(\text{Effect}_{\text{new}} - \text{Effect}_{\text{usual}})$$

This ratio yields a cost per unit of health gained (e.g., cost per QALY gained or per DALY averted). A lower ICER indicates more favorable cost-effectiveness (lower cost per health benefit), while an extremely high ICER suggests the added benefits might not be worth the extra costs.

Decision-makers often compare the ICER to the willingness to pay threshold. This benchmark reflects the maximum amount a health system or society is willing to pay for one additional unit of health benefit (e.g., per QALY gained or DALY averted). A widely used rule of thumb in global health suggested that an intervention is considered cost-effective if its cost per DALY averted is less than thresholds of one to three times the country's GDP per capita (Macroeconomics and health: Investing in health for economic development / report of the Commission on Macroeconomics and Health,

2001). However, this approach has limitations: it fails to account for cross-country variation in factors such as life expectancy, health-adjusted life expectancy, health expenditure per capita, and economic capacity. More refined methods, such as those proposed by Pichon-Rivière et al., construct country-specific cost-effectiveness thresholds based on empirical relationships between changes in health expenditure and gains in life expectancy, offering a more contextualized measure of value (Pichon-Riviere et al., 2023).

In our model, the ICER for RARP should be compared against cost-effectiveness thresholds both currently and in the coming years. RARP offers potential clinical advantages, such as reduced complication rates and faster recovery. At the same time, the additional costs associated with robotic equipment, disposable instruments, and, currently, longer operating room time raise the total cost per patient. With increased surgical experience in robotic procedures, surgeons can optimize instrument utilization by employing fewer tools more efficiently. This leads to reduced instrument exchanges, contributing to lower procedural costs and shorter operating room durations (Feldstein and Coussons, 2020).

13.1.4 Cost components: direct and indirect costs

When conducting a CEA, it is essential to enumerate all relevant costs of each intervention. Costs are divided into direct and indirect categories. Direct costs are those tied to the medical intervention itself and related healthcare utilization (Feldstein and Coussons, 2020). These include costs of hospital stays, surgical procedures, medical devices, medications, diagnostic tests, and healthcare personnel. For example, in the context of surgery, direct costs would encompass the operating room time, surgical instruments (and, in the case of robotics, the cost of robotic equipment and its maintenance and disposables), anesthesia, postoperative hospital bed days, and any treatment of complications (Yousefi et al., 2014). By contrast, indirect costs are the broader economic consequences of illness and care, often borne outside the healthcare sector (Neumann et al., 2017). Indirect costs include lost productivity (for instance, inability to work during recovery or a family member's time off to provide care), transportation costs for treatment visits, and home care needs. In a prostate surgery scenario, if one technique allows faster return to work than another, the productivity gains can be considered a reduction in indirect costs. Intangible elements refer to pain, suffering, or decreased quality of life due to illness or side effects (Drummond et al., 2015). These are hard to quantify in monetary terms and are captured on the "effectiveness" side of CEA via quality-of-life metrics rather than as direct financial costs. In this case study, we have not considered intangible costs for the CEA. For a robust CEA, the perspective must be defined—for example, a health system perspective includes only direct medical costs, whereas a societal perspective includes indirect costs as well. In our analysis, we will primarily consider direct healthcare costs of surgery (since hospitals, insurers, and governments focus on these), but we will also comment on indirect costs, such as lost work time for completeness. These conceptual distinctions between direct and indirect costs and between different analytic perspectives become particularly salient when evaluating surgical innovations. Procedures incorporating advanced technologies, such as robotic assistance, often substantially shift the cost-effectiveness results. While advanced technologies may increase certain direct costs (e.g., equipment and operating time), they can also lead to downstream savings by reducing complication rates or shortening recovery time. This adjustment is central to the evaluation of AI-integrated RARP versus ORP, an area where the nature of the technology used influences both clinical outcomes and cost dynamics. Robotic surgeries tend to have higher direct costs compared to open surgeries, primarily due to the expensive technology. The robotic system can cost on the order of \$1–\$2 million to acquire, with additional annual maintenance contracts, and each case incurs costs for disposable instrument parts. In contrast, open surgery has no high-tech disposable costs, though it may involve slightly longer hospital stays. A 2021 U.S. study of over 11,000 patients found that the unadjusted costs for RARP (\$26,554) were about \$2630 higher than for ORP (\$23,924) (Okhawere et al., 2021).

However, costs after surgery (postdischarge) favor the robotic approach. The same U.S. study found that within 6 months, the cost gap narrowed, and by 12 months, there was no significant difference in total cost between RARP and ORP. The reason is that RARP patients utilized fewer healthcare resources postoperatively: fewer readmissions, less rehabilitation, and management of complications. In fact, the RARP group had significantly lower outpatient and follow-up costs, translating to an average saving of about \$2929 in the year after surgery due to reduced complications and faster recovery. This saving offset the higher initial surgical cost. From a societal perspective, faster recovery also meant RARP patients missed fewer days of work (1.7 days fewer, on average, in that study), highlighting an indirect cost advantage of the robotic method (Okhawere et al., 2021).

13.1.5 Effectiveness metrics: quality-adjusted life years and disability-adjusted life years

Effectiveness in CEA is measured in terms of health outcomes. Common metrics include lives saved, disease cases prevented, deaths averted, or composite measures such as QALYs gained and DALYs averted, which capture both the length and quality of life. While QALYs are used in high-income countries, DALYs are applied in low- and middle-income settings and in global health evaluations due to their emphasis on population-level disease burden and standardization across diverse conditions (Drummond et al., 2015). In this chapter, we emphasize QALYs as the primary outcome metric, consistent with most health economic evaluations in the United States. One QALY represents one year of life in perfect health. If a patient lives for a year in a healthy state with a utility weight of 0.8 (reflecting moderate impairment), that year contributes 0.8 QALYs. QALYs are calculated by multiplying the time spent in a health state by a utility weight ranging from 0 (equivalent to death) to 1 (perfect health) (Zeckhauser and Shepard, 1976). This approach allows for incorporating both life extension and quality-of-life improvements into a single outcome measure.

As an example, we calculate QALYs in postoperative recovery. Consider a patient who experiences a postoperative complication following prostatectomy that reduces their quality of life for one year. Suppose life with this complication is assigned a utility weight of 0.8, meaning the patient experiences 80% of full health during that year. The QALYs accrued during this period would be calculated as:

$$\text{QALYs} = \text{Utility Weight} \times \text{Duration} = 0.8 \times 1 = 0.8$$

In this scenario, compared to a year lived in perfect health, the patient loses 0.2 QALYs (i.e., 1.0–0.8). If a robotic surgical technique reduces the frequency or severity of such complications, it would lead to QALYs gained relative to the conventional approach, capturing health improvements and quality-of-life preservation. For comparison, many global health and low-income country evaluations use DALYs, which measure health loss instead of health gain (Drummond et al., 2015). A DALY represents one year of healthy life lost due to illness or premature death (The Global Health Observatory: Mortality and burden of disease. Disability-adjusted life years (DALYs), 2025). DALYs are calculated as the sum of:

Years of Life Lost (YLL) due to early death, and
Years Lived with Disability (YLD) due to nonfatal health conditions compared to no disability during those years.

YLL is calculated by multiplying the number of deaths by the remaining life expectancy at the average age of death (The Global Health Observatory: Mortality and burden of disease. Disability-adjusted life years (DALYs), 2025). For instance, if a patient dies at 60 years and their life expectancy would otherwise be 80 years, that death represents 20 YLL (undiscounted). YLD is calculated as the duration of the health condition multiplied by a disability weight (ranging from 0 to 1) that reflects its severity. For example, if a complication lasts one year with a disability weight of 0.2, it yields 0.2 YLD. Whether using QALYs or DALYs, the goal is the same: to express both fatal and nonfatal health impacts in a unified metric that enables meaningful comparisons across interventions. In CEA, the difference in QALYs gained or DALYs averted between two strategies serves as the measure of effectiveness.

13.2 Cost-effectiveness of artificial intelligence enhanced robotic radical prostatectomy : A Case study

13.2.1 Background

Radical prostatectomy is a widely accepted treatment for localized prostate cancer and has traditionally been performed as an open surgery (ORP). However, RARP has gained traction globally, particularly in high-resource hospital settings, due to its potential to improve surgical precision, reduce complications, accelerate recovery, and reduce absenteeism. Despite higher upfront costs, RARP's integration with AI-enhanced robotics may provide long-term value through reduced disability and improved quality of life. This case study uses a decision-analytic framework to evaluate the cost-effectiveness of AI-integrated RARP compared to ORP. The analysis incorporates both direct and indirect costs, as well as modeled health outcomes. The primary outcome metric is the QALY, derived from modeled (DALYs) through established conversions.

13.2.2 Methods

Our CEA used a decision-analytic model. The model included common complications associated with prostatectomy, such as urinary incontinence, erectile dysfunction, urethral stricture, and surgical complications, each assigned a disability weight and duration derived from Global Burden of Disease (GBD) studies (Salomon et al., 2015). The model estimated the YLD for each complication and calculated total DALYs per patient for each intervention arm. To align with cost-effectiveness conventions in high-income country settings (e.g., the United States), DALYs were converted to QALYs for use in ICER calculation.

13.2.3 Cost-effectiveness of open radical prostatectomy versus AI enhanced robotic assisted radical prostatectomy

To evaluate the cost-effectiveness of AI-RARP compared to ORP, we constructed a health outcome model using standardized disability weights and published estimates for complication risks and durations (Table 13.1).

The model compares both procedures across nine health states relevant to prostate cancer surgery, including common complications (e.g., urinary incontinence, erectile dysfunction, and urethral stricture), acute recovery, and long-term outcomes such as biochemical recurrence (BCR), metastatic progression, and perioperative mortality. Disability weights and durations of conditions were sourced from the GBD studies and peer-reviewed clinical literature (Disability weights for diseases and conditions, 2025).

Each state is defined by: (Gupta et al., 2024; Guarino et al., 2024; Maan et al., 2012; Park et al., 2013; Boys et al., 2016)

Table 13.1 Model parameters.

Row (A)	Health state and source (B)	Disability weight (C)	Duration with ORP (discounted years) (D)	Duration with AI-RARP (discounted years) (E)	Risk with ORP (F)	Risk with AI-RAR-P (G)	DALY or QALY (ORP) (H)	DALY or QALY (AI-RARP) (I)	DALYs averted or QALYs gained (J)
(1)	Urinary incontinence (DALYs) (Salomon et al., 2015; Carbin et al., 2024; Ficarra et al., 2012)	0.139	1.00	1.00	12.00%	7.50%	0.0167	0.0104	0.0063
(2)	Erectile dysfunction (Impotence), (DALYs) (Salomon et al., 2015; Carbin et al., 2024; Moretti et al., 2024)	0.017	1.00	1.00	59.00%	35.00%	0.0100	0.0059	0.0041
(3)	Urethral stricture, (DALYs) (Altıntaş et al., 2024; Sujenthiran et al., 2018; Disability weights for diseases and conditions, 2025)	0.151	2.00	2.00	6.90%	2.51%	0.0208	0.0076	0.0133

Table 13.1 (Continued)

Row (A)	Health state and source (B)	Disability weight (C)	Duration with ORP (discounted years) (D)	Duration with AI-RARP (discounted years) (E)	Risk with ORP (F)	Risk with AI-RAR-P (G)	DALY or QALY (ORP) (H)	DALY or QALY (AI-RARP) (I)	DALYs averted or QALYs gained (J)
(4)	Major surgical complications (e.g., infection, reoperation), (DALYs) (Salomon et al., 2015; Carbin et al., 2024; Moretti et al., 2022; Shah et al., 2024)	0.300	0.25	0.25	7.01%	2.83%	0.0053	0.0021	0.0031
(5)	Blood transfusion complications (DALYs) (Salomon et al., 2015; Pichon-Riviere et al., 2023)	0.100	0.02	0.02	19.77%	1.70%	0.0004	0.0000	0.0003
(6)	Biochemical recurrence before metastasis (DALYs) (Salomon et al., 2015; Bravi et al., 2023; Freedland et al., 2005)	0.300	8.00	8.00	16.00%	12.04%	0.3840	0.2890	0.0950
(7)	Metastatic disease phase (DALYs) (Salomon et al., 2015; Freedland et al., 2005)	0.500	5.00	5.00	0.41%	0.19%	0.0103	0.0048	0.0055

(Continued)

Table 13.1 (Continued)

Row (A)	Health state and source (B)	Disability weight (C)	Duration with ORP (discounted years) (D)	Duration with AI-RARP (discounted years) (E)	Risk with ORP (F)	Risk with AI-RAR-P (G)	DALY or QALY (ORP) (H)	DALY or QALY (AI-RARP) (I)	DALYs averted or QALYs gained (J)
(8)	Postoperative acute recovery phase (DALYs) (Salomon et al., 2015; Moretti et al., 2024; Porcaro et al., 2022)	0.200	0.12	0.08	100%	100%	0.0231	0.0154	0.0077
(9)	Perioperative mortality (death), (DALYs) (Salomon et al., 2015)	1.000	13.46	13.46	0.24%	0.04%	0.0323	0.0054	0.0269
(10)	**Total (DALYs)**						**0.5029**	**0.3406**	**0.1622**
(11)	**QALYs (1-DALYs)**						**0.4971**	**0.6594**	**0.1622**

AI-RARP, AI-enhanced robotic-assisted radical prostatectomy; *ORP*, open radical prostatectomy; *QALY*, quality-adjusted life year; *DALYs*, derived from modeled Disability-adjusted life years.
Source: Data from a published or online source without its original presentation.

- Disability weight: quantifies the severity of health loss on a scale from 0 (full health) to 1 (death).
- Duration: represents the average time (in discounted life years) a patient remains in that state.
- Risk: the probability of the condition occurring, specific to each surgical approach (ORP vs AI-RARP).

Risk estimates in the AI-RARP arm were adjusted downward from standard RARP values based on empirical evidence to reflect the benefits of AI integration. The risk of urinary incontinence was reduced from 9% to 7.5%, and erectile dysfunction from 41% to 35%, due to enhanced nerve-sparing techniques and improved anatomical recognition supported by AI-guided surgical planning (Carbin et al., 2024; Ficarra et al., 2012; Moretti et al., 2024). The incidence of urethral stricture was adjusted from 3.3% to 2.5%, and major surgical complications, such as infection or reoperation, from 3.54% to 2.8%, based on evidence that AI-enhanced robotic systems provide superior real-time visualization and intraoperative control (Altıntaş et al., 2024; Sujenthiran et al., 2018; Moretti et al., 2022; Shah et al., 2024). Additionally, the probability of biochemical recurrence was reduced from 14% to 12%, informed by studies showing that machine learning can improve operative strategy by better predicting oncologic margins and recurrence risk (Bravi et al., 2023; Freedland et al., 2005). These adjustments in the model demonstrate how AI may enhance surgical precision, reduce complications, and improve long-term outcomes.

We applied a 3% annual discount rate when modeling the impacts of health states in future years. Specifically, the discounted life expectancy for a 65-year-old male in the United States was calculated as 13.46 years, based on an undiscounted life expectancy of 17.5 years according to actuarial tables (Life expectancy at age 65 by sex: United States, 2022). This discounted value was used in the calculation of YLL for mortality-related DALY estimates.

To estimate the disability burden for each health state, we calculated YLD by multiplying the disability weight by the duration of the condition and then by the risk of occurrence under each intervention arm:

$$\text{YLD} = \text{Disability Weight} \times \text{Duration} \times \text{Risk}$$

This product was used as the DALY estimate for each nonfatal outcome, under the assumption that YLL was negligible for these complications. For mortality, DALYs were calculated by multiplying the risk of perioperative death by the discounted life expectancy (13.46 years) for a 65-year-old male in the United States:

$\text{DALY for Death} = \text{Mortality Risk} \times \text{Discounted Life Expectancy}$

This structure allowed DALYs to be calculated separately for ORP and AI-RARP by incorporating arm-specific risk values for each health outcome. Summing across all

Table 13.2 Stratified metastatic risk calculation (RARP vs ORP).

Row	Item and sources	Low	Intermediate	High	Very high	Total
(1)	% of ORP patients*	40%	30%	20%	10%	100%
(2)	% of RARP patients**	50%	30%	15%	5%	100%
(3)	Risk of BCR under ORP*	4.0%	6.0%	8.0%	7.0%	
(4)	Risk of BCR under RARP**	5.0%	6.0%	6.0%	3.5%	
(5)	Metastasis risk after BCR	10%	20%	40%	70%	
(6)	Risk of metastasis given BCR under ORP*	25%	25%	25%	25%	
(7)	Risk of metastasis given BCR under RARP***	15%	15%	15%	15%	
(8)	Risk of metastasis under ORP,* (3) × (5) × (6)	0.100%	0.300%	0.800%	1.225%	
(9)	Risk of metastasis under RARP,** (4) × (5) × (7)	0.075%	0.180%	0.360%	0.368%	
(10)	Crude risk of metastasis under ORP,* (1) × (8)	0.040%	0.090%	0.160%	0.123%	0.413%
(11)	Crude risk of metastasis under RARP,** (2) × (9)	0.038%	0.054%	0.054%	0.018%	0.164%
(12)	Standardized risk of metastasis under RARP,** (1) × (9)	0.030%	0.054%	0.072%	0.037%	0.193%

BCR, biological cancer risk; *ORP*, open radical prostatectomy; *RARP*, robotic-assisted radical prostatectomy. *Pound et al. (1999); **Seo et al. (2016); ***Bravi et al. (2023).
Note: Standardized risk uses the percentage of ORP patients by stratum (the reference treatment) as the population to which RARP was standardized. The risks of BCR in rows (3) and (4) are the authors' estimates based on literature on overall risk (subtracting survival from 100%) and the advantages of RARP over ORP (Bravi et al., 2023), (Pound et al., 1999), (Seo et al., 2016). Using literature-informed assumptions (e.g., metastasis risk of 25% following BCR), we calculated crude and standardized metastatic risks across groups. For example, low-risk patients under AI-RARP were assumed to have a higher BCR rate than under ORP (5% vs 4%), but a lower probability of metastasis after recurrence (15% vs 25%) (Pound et al., 1999; Seo et al., 2016). Conversely, very high-risk patients were less frequently observed in RARP cohorts and contributed proportionally less to the population-level metastatic burden. By multiplying recurrence rates by conditional metastasis risks and standardizing for patient distribution, we derived a weighted metastatic progression rate of 6.25% under ORP and 3.75% under RARP. These values were then used in the main DALY model to compute the health loss associated with metastatic disease (row 7 in Table 13.1).

Source: Data from a published or online source without its original presentation.

health states, the total DALYs per patient were 0.5029 for ORP and 0.3406 for AI-RARP, resulting in an improvement of 0.1622 DALYs averted by AI-RARP. This translates to 0.1622 QALYs gained from switching to robotic surgery.

13.2.4 Stratified risk modeling for metastatic disease progression

To estimate DALYs associated with metastatic disease, we applied a stratified risk model based on four clinically relevant patient risk groups: low, intermediate, high, and very high risk. This approach reflects real-world heterogeneity in prostate cancer prognosis. Each group was characterized by its prevalence among patients receiving either ORP or RARP, its baseline risk of biochemical recurrence (BCR), and the conditional risk of developing metastases after BCR (Table 13.2).

This stratified structure allowed the model to capture both the direct impact of robotic surgery on recurrence/metastasis risk and the indirect benefit of better patient selection and case distribution in real-world practice. The risk stratification was included to account for the differing baseline risks between groups and is a valid reflection of the current robotic surgery system.

13.2.5 Cost inputs

The anticipated direct and indirect medical costs per patient for ORP and AI-RARP are expressed in 2025 USD (Table 13.3). The analysis is conducted from a health system perspective, with select indirect costs noted only qualitatively or minimally modeled (Table 13.3).

The surgical and OR costs for ORP are estimated at $8000 per procedure, comprising $2500 for surgeon and operating room anesthesiologist fees, $3000 in OR time, and $2500 in disposables (Bolenz et al., 2010; Tomaszewski et al., 2012). In contrast, baseline AI-RARP incurs $13,500 in these categories due to longer OR time (typically 30–60 minutes longer) and more expensive robotic instruments and disposables (Bolenz et al., 2010; Tomaszewski et al., 2012). OR time in AI-RARP is billed at higher per-minute rates (usually $50–$100/min). Additionally, $2500 per case is allocated to AI-RARP for robotic system purchase and maintenance with high surgical volumes (e.g., at least 200 cases/year). In lower-volume settings, this fixed per-case cost would be higher. Hospitalization costs are modeled at $6000 for ORP and $3500 for AI-RARP, reflecting average lengths of stay of approximately 3 days and 1.5 days, respectively, consistent with findings from systematic reviews of prostatectomy recovery times (Okhawere et al., 2021; Wang et al., 2023). Okhawere et al. estimate recovery and follow-up care costs (including catheter management, outpatient visits, and rehabilitation) at $1200 for ORP and $1000 for AI-RARP (Okhawere et al., 2021). Costs of complication management were estimated at $2000 for ORP and $1000

Table 13.3 Anticipated medical and indirect costs per patient (2025 USD).

Row	Cost category and source	Conventional surgery (ORP)	AI robotic surgery (RARP) now	AI robotic surgery (RARP) in 5 years	AI robotic surgery (RARP) in 5 years, breakeven	Cost reduction for breakeven (%)
(1)	Fixed costs (robot purchase and annual maintenance amortized per patient) (Bolenz et al., 2010)	$0	$2500	$1875	$1875	25%
(2)	Surgeon and anesthesiologist fees (Tomaszewski et al., 2012)	$2500	$2500	$2500	$2500	0%
(3)	Operating room time (staff, facility use) (Bolenz et al., 2010; Tomaszewski et al., 2012)	$3000	$4500	$4500	$3525	22%
(4)	Surgical supplies and disposables (Bolenz et al., 2010; Tomaszewski et al., 2012)	$2500	$4000	$4000	$4000	0%
(5)	Hospitalization costs (avg. hospital stay) (Okhawere et al., 2021; Wang et al., 2023)	$6000	$3500	$3500	$3500	0%
(6)	Recovery and follow-up (clinic visits, catheter care, rehab) (Okhawere et al., 2021)	$1200	$1000	$1000	$1000	0%
(7)	Complication management costs (infection, strictures, reoperation) (Wang et al., 2023)	$2000	$1000	$800	$800	20%
(8)	**Subtotal, direct costs per patient**	**$17,200**	**$19,000**	**$18,175**	**$17,200**	9%
(9)	Transportation and **indirect** costs (travel, productivity loss) (Okhawere et al., 2021)	$1,000	$500	$500	$500	0%
(10)	**Total health sector costs per patient**	**$18,200**	**$19,500**	**$18,675**	**$17,700**	9%

AI, artificial intelligence; *ORP*, open radical prostatectomy; *RARP*, robotic-assisted radical prostatectomy.
Source: Data from a published or online source without its original presentation.

for AI-RARP, based on rates of infection, stricture, transfusion, and reoperation found in published meta-analyses (Wang et al., 2023). AI-RARP consistently shows a lower complication burden, contributing to these cost differences (Tomaszewski et al., 2012).

We also included a simplified estimate of short-term indirect costs such as transportation, missed workdays, and caregiver support during recovery—\$1000 for ORP and \$500 for AI-RARP. Although most prostatectomy patients are near or past retirement age, an earlier return to functional independence offers economic and quality-of-life benefits. For example, on average, AI-RARP patients have been shown to miss 1.7 fewer workdays compared to ORP patients (Okhawere et al., 2021). The total estimated cost per patient was \$18,200 for ORP and \$19,500 for AI-RARP in 2025.

13.2.6 Cost-effectiveness assessment

Using the cost and health outcome estimates from Tables 13.1 to 13.3, we calculated the ICER for AI-RARP compared to ORP (Table 13.4). Direct costs per patient were estimated at \$17,200 for ORP and \$19,000 for AI-RARP, yielding an incremental cost of \$1800. AI-RARP patients gained an average of 0.2192 additional QALYs compared to ORP. The resulting ICER is:

$$ICER = \frac{1,800}{0.1622} \approx 11,095 \; per \; QALY \; gained$$

To assess whether this ICER represents a cost-effective use of health resources, we compared it to a country-specific threshold based on past health spending. For the United States, this threshold was 1.33 × GDP per capita (Pichon-Riviere et al., 2023). With the U.S. GDP per capita in 2024 estimated at \$86,601, the threshold equals \$115,180 per QALY gained (GDP per capita (current US\$) – United States, 2024). Since the ICER \$11,095 < \$115,180, and RARP delivers positive health gains at a modest additional cost. We conclude that RARP is not cost-saving, but it is cost-effective under this threshold.

13.2.7 Projected evolutions

Although Moore's Law is frequently cited to project exponential cost declines of 50% every two years in digital technologies. However, its assumptions are not directly applicable to the domain of RAS, which depends heavily on specialized hardware, regulated environments, and constrained production volumes. Instead, we base our projections on market competition, workflow optimization, and learning curve gains. By 2030 we project that the fixed-cost per patient for robotic hardware and maintenance will decline by 25%, from \$2500 to \$1875, as market entrants increase competition and platforms become more standardized. The plausibility of this trajectory

Table 13.4 Assessment of cost-effectiveness from a health perspective (in 2025 USD).

Row	Parameter	Source	Value now	AI robotic surgery (RARP) in 5 years	AI robotic surgery (RARP) in 5 years, breakeven
(1)	Health sector cost of conventional surgery, ORP	Table 13.3	$17,200	$17,200	$17,200
(2)	Health sector costs of AI-RARP surgery	Table 13.3	$19,000	$18,175	$17,200
(3)	Difference in cost	(2)–(1)	$1800	$975	$0
(4)	QALY, ORP	Table 13.1	0.4971	0.4971	0.4971
(5)	QALY, AI-RARP	Table 13.1	0.6594	0.6594	0.6594
(6)	Net difference in QALY gained	(5)–(4)	0.1622	0.1622	0.1622
(7)	ICER	(3)/(6)	$11,095	$6010	$0
(8)	GDP per capita 2024	(GDP per capita (current US$) – United States, 2024)	$86,601	$86,601	$86,601
(9)	Threshold as proportion of GDP per capita	(Pichon-Riviere et al., 2023)	1.33	1.33	1.33
(10)	Cost-Effectiveness threshold	(8) × (9)	$115,180	$115,180	$115,180
(11)	Is AI-RARP cost-saving?	(3) <0	No	No	Yes
(12)	Is AI-RARP cost-effective?	(7) < (10)	Yes	Yes	Yes

GDP, gross domestic product; *ICER*, incremental cost-effectiveness ratio; *ORP*, open radical prostatectomy (Conventional); *QALY*, quality-adjusted life year; *RARP*, robotic-assisted radical prostatectomy (Intervention); *USD*, United States Dollars.

Source: Data from a published or online source without its original presentation.

is supported by trends in robotic system manufacturing and wider global adoption (Eckhoff et al., 2023). In parallel, OR time costs are projected to fall by 22%, from $4500 to $3525 per patient. This estimate is based on evidence that AI-guided workflow recognition, predictive case duration models, and simulator-enhanced training can reduce procedure length and improve scheduling accuracy. Using machine learning to predict case duration in robot-assisted surgery significantly improved OR efficiency and reduced time variance (Zhao et al., 2019). These time savings can translate into measurable cost reductions on a larger scale.

Furthermore, complication management costs are expected to decline by 20% (from $1000 to $800 per patient) as AI improves intraoperative decision support and early risk detection, reducing the likelihood of adverse events such as structures or transfusions. As summarized in Table 13.4, the total cost per AI-RARP patient should decline from $19,500 in 2025 to $18,675 under fixed-cost reductions alone, and further to $17,700 when combined with reductions in OR time and complication-related savings. This would bring AI-RARP to cost parity with ORP, reaching an economic breakeven point. In this scenario, AI-RARP becomes not just cost-effective but potentially cost-saving, given its superior clinical outcomes and comparable or lower total costs. These conservative, evidence-informed projections suggest that AI-enhanced robotic surgery is on a pathway toward long-term sustainability. Even in the absence of radical technological breakthroughs, incremental gains in system efficiency and declining per-unit costs position AI-RARP as an increasingly attractive strategy for high-volume health systems seeking value-based care delivery. Under a conservative 5-year scenario, where fixed costs decline by 25% and complication management improves, the incremental cost falls to $975, and the ICER drops to $6010 per QALY, strengthening its economic value. In a breakeven scenario, which assumes both capital cost reductions and enhanced OR time efficiency (22% improvement), the incremental cost approaches zero, thereby making AI-RARP cost-saving (Table 13.4).

13.2.8 Strengths and limitations of the case study

This case study offers an early, structured attempt to evaluate the cost-effectiveness of AI-RARP using established health economic modeling techniques. A key strength lies in integrating clinical outcome data, DALY and QALY modeling, and projected cost dynamics, including plausible future scenarios informed by Moore's Law and surgeon proficiency gains. By incorporating real-world parameters (e.g., disability weights, complication rates, and U.S.-based costs), the model aligns well with decision-making needs in high-resource health systems.

However, several limitations must be acknowledged. First, published empirical data on AI-enhanced robotic prostatectomy remain sparse. Most studies examine conventional RARP rather than platforms fully augmented by AI (e.g., workflow recognition, real-time image processing, or predictive analytics). As such, the model relies partly on extrapolations from early-stage research or general trends in AI optimization rather than AI-RARP-specific clinical trial data.

Second, outcome assumptions (e.g., risk reduction in biochemical recurrence or urethral stricture with AI integration) are guided by broader literature on AI in urology or expert projections, rather than condition-specific randomized evidence. This assumption introduces uncertainty into the magnitude of predicted QALY gains.

Third, cost projections, particularly future reductions in capital expenditure and OR time, are based on plausible but unvalidated assumptions, such as Moore's Law and improved learning curves. While grounded in historical trends, these forecasts remain speculative until confirmed in larger multiinstitutional analyses.

Finally, indirect costs (e.g., societal productivity losses or caregiver time) are only partially modeled, potentially underestimating the full economic benefits of faster recovery associated with AI-RARP. Despite these limitations, the study provides a transparent, reproducible framework to simulate emerging innovations in robotic surgery and can be adapted as stronger real-world data become available.

13.3 Policy projections

13.3.1 Urology

Robotic surgery has been widely adopted in urology for radical and partial prostatectomy and kidney cancer with robotic-assisted laparoscopic radical and partial nephrectomy, lymphadenectomy, robotic-assisted kidney transplantation, and bladder and urethral reconstructive techniques for congenital malformations. In 2000 the use was extended to benign gynecological disease conditions, hysterectomy and myomectomy being the most common (Rivero-Moreno et al., 2023). Even though robotics-assisted surgery allows easy dissection and fine suturing and has a faster recovery rate, lack of tactile feedback limits its extensive use. The use of robotic surgery in gynecological malignancies is yet to be explored due to a lack of studies (Chandrakar et al., 2022).

13.3.2 Cardiovascular disease

Cardiovascular disease conditions are another major field where coronary artery bypass grafting (CABG) is adopted in robotic surgery in the United States (Cardiac Intervention: An Update, 2025; Giambruno et al., 2018). The application of robotics in cardiac surgery is associated with low mortality and morbidity when compared with traditional sternotomy. Hence, the utilization of robotic technology in CABG, mitral valve (MV) replacement, tumor resection, and atrial septal defect repair has continued to grow over time (Cardiac Intervention: An Update, 2025). Although the cost-effectiveness of robotic MV surgery has been extensively studied, findings often suggest that comparable value to conventional approaches is achievable primarily in high-volume centers, and the issue remains controversial (Mihaljevic et al., 2014; Moss and Halkos, 2017; Tomšič and Palmen, 2023).

The Cleveland Clinic has compared the cost of robotic MV surgery to complete sternotomy, partial sternotomy, and anterolateral thoracotomy. They evaluated the costs of capital investment and fixed yearly maintenance costs to implement the robotic MV surgery program. In robotic surgery, the higher technical operating, capital investment, and maintenance costs were balanced by lower costs of recovery from surgery. Yet, the costs for robotic surgery remained 15% higher than those of conventional cardiovascular interventional surgeries (Mihaljevic et al., 2014). The cost of robotic surgery is related to the volume of surgeries performed, so the additional costs of robotic surgery get reduced with higher yearly case volumes. Robotic MV surgery reduced the hospital admission length and early readmission rate, which balanced the additional costs for robotic surgeries, specifying the need for optimized robotics to make it cost-effective (Tomšič and Palmen, 2023).

13.3.3 Oral and oropharyngeal surgery

Robotic platforms have expanded into oral and oropharyngeal surgery through transoral robotic surgery (TORS), which offers improved visualization, expertise in anatomically constrained regions of the head and neck, and access to pharyngeal tumors compared to conservative open surgery, with low morbidity, faster postoperative recovery, and better functional outcomes. Early experience shows that TORS can be safely completed with neck dissection, and functional outcomes, such as swallowing and speech, have generally been good, although more in-depth evaluations of functional outcomes after synchronous neck dissections are necessary. Decisions about the need for adjuvant therapy for radiotherapy and chemotherapy are based not only on pathological risk factors but also on the surgical approach. Systematic reviews now show that TORS is a safe, minimally invasive approach for select cases of oropharyngeal squamous cell carcinoma, with promising functional and oncologic outcomes, and while some cohort analyses suggest that quality of life following TORS for early-stage disease may be similar to that following definitive radiotherapy, direct comparisons are limited by the high rate of adjuvant therapy in many of the TORS patients (Park et al., 2013).

A systematic review on robotic applications in otolaryngology identified that a hub-and-spoke approach might be important for centralizing robotic services in high-volume settings, thereby enhancing cost efficiency through economies of scale. High capital costs and specialized training requirements have concentrated TORS in tertiary cancer centers, while smaller community hospitals face barriers due to lower case volumes and limited reimbursement differentials. Further formal cost-benefit analyses on the application of surgical robotics in head and neck surgery settings are needed, especially on the evaluation of perioperative and postoperative outcomes (Maan et al., 2012). A recent systematic review confirms that TORS is now considered a valuable minimally invasive intervention for oropharyngeal squamous cell carcinoma, associated with favorable functional outcomes in speech and swallowing and good oncologic control in selected patients when compared with historical open surgery and chemoradiotherapy series (Guarino et al., 2024). Moreover, meta-analyses of recent cohorts suggest that, in selected early-stage disease, speech and swallowing quality of life after TORS is comparable to radiotherapy, although definitive comparisons remain challenging due to frequent use of adjuvant therapy in many patients treated with TORS (Gupta et al., 2024).

13.3.4 Future changes

The case study we adopted in this chapter illustrates the multifaceted nature of cost-effectiveness analyses for AI-enhanced robotics in radical prostatectomy for prostate cancer to enhance the outcomes. AI-assisted robotic prostatectomy offers clinical advantages—less blood loss, quicker recovery, fewer complications in a few domains, and comparable cancer control—which translate into health gains (DALYs averted) for patients (Wang et al., 2023; Labban et al., 2022). However, technology comes with higher upfront costs, and the cost-effectiveness depends on the magnitude of benefit relative to the extra cost, as well as the healthcare setting's ability to pay.

From a policy perspective, our analysis suggests several important considerations:

In wealthier healthcare systems, even small improvements in patient outcomes may justify higher costs, especially if the ICER falls below accepted thresholds (Labban et al., 2022). The United Kingdom and other high-income countries have embraced robotic surgery for prostate cancer, often finding it cost-effective in formal analyses supporting its use in high-volume centers (Labban et al., 2022; Close et al., 2013). In contrast, low- and middle-income countries must be more selective due to tighter resource constraints. An ICER of approximately $11,000 per QALY, while acceptable in wealthier settings, exceeds the cost-effectiveness thresholds typically used in countries with a GDP per capita under $5000, where context-specific thresholds often range from $500 to $2000 per QALY (Pichon-Riviere et al., 2023).

Cost-effectiveness improves significantly with higher surgical volumes. The capital cost of robotic platforms is fixed, so centers performing a higher volume of robotic surgeries per year will have a far lower per-surgery cost than those performing fewer. Consolidating complex surgeries, such as prostatectomy, into high-volume tertiary centers can thus make robotic surgery more economically viable. Additionally, training surgeons to operate more efficiently can reduce operative time, saving on operating room (OR) costs, and improve outcomes, translating into more QALYs gained or

DALYs averted by avoiding complications. Such consolidation need not imply a single tertiary center; however, for example, different centers could become hubs for oral and prostatic robotic surgery, as the equipment and expertise likely differ between these sites.

Emerging advances in AI may further amplify this efficiency. In complex domains such as heart or liver transplantation, AI is increasingly being used to optimize surgical planning, predict organ rejection, assist in intraoperative decision-making, and support postoperative risk stratification. These tools can shorten decision times, standardize quality across institutions, and even guide resource allocation for better donor–recipient matching (Chowdhury et al., 2025). If AI enables such procedures to be performed faster, more accurately, and consistently, it could unlock the capacity to perform a greater volume of operations, spreading fixed infrastructure costs across more patients and improving population-level outcomes.

In the long-term, integrating AI with robotic systems may also contribute to learning about health systems, where performance improves over time through feedback loops. For high-cost interventions such as organ transplantation or robotic prostatectomy, this synergy between volume, technology, and precision may shift the cost-effectiveness threshold, making formerly expensive innovations more justifiable from a health system perspective.

The "AI-enhanced" aspect of medical robotics is evolving. Currently, surgical robots like the da Vinci provide excellent mechanics and visualization, but the "intelligence" is human-controlled. However, AI is increasingly being integrated—for example, machine learning algorithms that can guide surgeons by highlighting anatomical structures (nerve pathways, blood vessels) in real-time or robotic automation for certain standardized tasks such as suturing and camera control. As these AI features advance, the field of surgery may see further improvements in outcomes: even fewer complications, shorter operative times, and better oncological precision. Such improvements would increase the effectiveness side of the CEA equation (more DALYs averted). If, at the same time, competition in the robotic market increases (potentially driving costs down), the cost side might also improve. Together, these trends could make future AI-enhanced robotics more cost-effective than the current generation. The field of urology is already witnessing multiple companies developing new robotic systems, which may bring capital and maintenance costs down.

13.3.5 Intangible impacts

Robotic surgery has beneficial intangible effects on patients, caregivers, and health systems beyond the direct clinical outcomes. These intangible benefits are particularly noticeable in specialties such as urology, gynecology, and head and neck surgery, where the preservation of function (e.g., continence, sexual function, speech, or swallowing) may influence psychological well-being and an earlier return to daily life activities. While these outcomes are not easily monetized, they contribute substantially to perceived value within patient populations. In radical prostatectomy, reduced erectile dysfunction or urinary incontinence leads to better emotional well-being and reduced stigma, while in TORS, functional preservation of swallowing and speech reduces reliance on feeding tubes and speech rehabilitation, facilitating an earlier return to daily life activities. When recovery is accelerated and disability minimized, caregivers face fewer physical, emotional, and financial demands. Patient satisfaction, faster return to normal daily activities, and less scarring or pain are valuable outcomes. Studies have found that members of the public view RAS as safer, faster, less painful, or yielding better results than conventional methods, although this perception does not always translate into a clear preference when actual surgical choices are considered (Boys et al., 2016). A 2014 survey cited in a 2022 needs assessment study found that 72% of respondents believed RAS was safer, faster, less painful, and offered better results compared to conventional methods. Gaps in the patient information should also be addressed: 25% of respondents did not fully grasp how RAS differed from open or laparoscopic surgery, and over 20% incorrectly believed the robotic systems operated with some degree of autonomy (Boys et al., 2016). While quality of life considerations may not override cost considerations, they contribute to the adoption of robotic systems. If the cost is not prohibitive, hospitals may invest in robotics partly for these qualitative reasons and the attractive factor for both patients and surgeons.

13.3.6 Ethical and regulatory considerations

The expanded use of AI-enhanced robotic surgery raises several ethical and regulatory questions that define adoption strategies and long-term policy decisions. The high cost of AI-enhanced robotics raises concerns about equitable access. Not all hospitals can afford a robot, and not all patients can travel to a center that has one. If robotic surgery truly improves outcomes in meaningful ways, it could create a disparity where only certain populations benefit from better surgery. Health systems might consider this in deciding how widely to disseminate the technology. If an intervention is only marginally better but much more expensive, sticking with the conventional method could free resources for other healthcare services that yield larger population health gains. Our analysis showed modest health differences, suggesting that in a constrained system, one might question investing heavily in robots. Health systems should consider alternative uses of the resources such as funding more cancer screenings or other treatments that prevent more DALYs overall. Addressing these disparities will require coordinated efforts involving reimbursement reform, regionalization strategies, and targeted capital investment in underserved areas. While AI-RARP incurs higher upfront costs than ORP, the AI-integrated system results in measurable reductions in patient disability, especially regarding postsurgical complications. These benefits translate into health gains and potential downstream cost savings through fewer complications and reduced rehabilitation needs.

In pediatric and academic hospitals such as Boston Children's, robotic surgery programs also serve as platforms for surgical innovation, training, and research, offering broader institutional value beyond pure cost-effectiveness metrics. The US Policy on Safe, Secure, and Trustworthy Development and Use of Artificial Intelligence put into effect by an executive order in 2023 is promoting AI-automated robotics and declares that the United States should be a global leader in AI development and adoption by engaging with international allies and partners, leading efforts to develop common AI regulatory and accountability principles, and advancing responsible global technical standards for AI (Executive Office of the President. Safe, Secure, and Trustworthy Development and Use of Artificial Intelligence, 2023). The integration of AI into robotic surgery represents a change in basic assumptions in surgical care, promising improved precision, reduced complication rates, and data-informed decision-making. However, this evolution also introduces significant ethical and regulatory challenges that must be addressed to ensure responsible and equitable adoption. AI systems depend heavily on large-scale, high-resolution datasets, such as surgical videos, electronic health records, and intraoperative imaging, for training and real-time performance. These datasets inherently include sensitive patient information, raising serious concerns about privacy, data ownership, and cybersecurity. Ensuring compliance with established regulations such as the Health Insurance Portability and Accountability Act in the United States and the General Data Protection Regulation in the European Union is critical. Furthermore, as AI systems are integrated into real-time surgical workflows, the need for robust data encryption, access controls, and secure cloud-based storage becomes paramount to mitigate the risk of breaches or unauthorized use of patient data (Topol, 2019), challenges that must be addressed to ensure responsible and equitable adoption. Prompt regulations should be enforced to monitor the practices in the field of robotic surgery in healthcare. The Massachusetts Department of Public Health has constituted a commission to (1) investigate and review the use of robotic surgery in the Commonwealth, (2) work to develop a training protocol for each application of robotic surgery currently being performed in the Commonwealth, (3) develop an application process for hospitals to propose the certification of any new applications of robotic surgery, and (4) establish guidelines for surgeon training and experience to perform such an operation (An Act establishing a special commission on robotic surgery. Senate Bill No. 1069, 2013). Training and credentialing present additional ethical and regulatory considerations. Robotic systems require specialized technical proficiency, structured learning curves, and ongoing competency assessment. Absent standardized national credentialing frameworks, hospitals rely on internal privileging standards, creating variability in training oversight and potential patient safety concerns. Professional societies have begun to articulate competency standards and simulation-based curricula, yet regulatory harmonization remains incomplete.

One of the most pressing ethical concerns in AI-assisted surgery is the potential for algorithmic bias, which can arise from imbalanced training datasets or nonrepresentative clinical inputs. Studies have demonstrated that AI systems trained on data from certain racial or socioeconomic groups may produce less accurate or beneficial outcomes for underrepresented populations (Obermeyer et al., 2019). In surgical settings, this could result in disparities in complication prediction, operative planning, or even patient selection for advanced robotic procedures. Addressing such inequities requires deliberate strategies for bias detection and mitigation, including the use of diverse, high-quality datasets, bias audits during development, and transparent performance reporting across demographic subgroups. The regulatory framework for AI in surgical applications is still evolving and varies significantly across jurisdictions. In the United States, the Food and Drug Administration (FDA) currently evaluates AI-powered robotic systems under its Software as a Medical Device framework. However, the dynamic and adaptive nature of AI algorithms, especially those capable of real-time learning, challenges traditional regulatory models. Recent FDA initiatives, such as the proposed total product life cycle approach, aim to address these complexities by emphasizing continuous monitoring, postmarket surveillance, and the need for algorithmic transparency and human oversight (Artificial Intelligence and Machine Learning in Software, 2025). In the European Union, the Medical Device Regulation outlines similar pathways but requires additional clinical evidence and postmarket reporting (Regdesk, 2025). Globally, organizations such as the International Medical Device Regulators Forum are working toward harmonizing standards to ensure consistent safety and efficacy criteria (Group, 2020). As AI technologies continue to evolve, the establishment of adaptive regulatory models that balance innovation with accountability is essential to protect patient welfare and public trust.

13.3.7 Imapct on healthcare costs

The cost of healthcare in the United States and globally is a critical question for policymakers and the public. As we found AI-RARP to be an effective and cost-effective procedure for prostate cancer, it is useful to discuss how it might affect healthcare costs in five years. In Table 13.4, we projected that the healthcare costs of AI-RARP and ORP would each be \$17,200 (in 2025 prices). Since these numbers are identical, we might be tempted to infer that there would be no impact on healthcare costs. This inference needs some clarification, however. The more precise meaning of Table 13.4 is that for an average patient with prostate cancer who is definitely being treated, the two approaches cost the same to the healthcare system. To further unpack this technology, we need to understand how the technology may change the number of patients with prostate cancer who are being treated. To address that question, let us recall that each patient with prostate cancer is unique. Table 13.2 showed that these patients with prostate cancer have a spectrum of risks of metastasis. Similarly, rows 1 through 9 in Table 13.1 show that patients face a variety of risks of complications from each treatment modality. For our population-level analysis, we rightly used the mean values in column C. However, each patient has their own preferences, or weights, for each of these complications. Prior to RARP, the only treatment option for a patient with prostate cancer was ORP. In this case, some patients at low metastatic risk and with a high aversion to the potential complications might have rationally chosen "watchful waiting" and, in some cases, never received any surgical treatment. Except for row 8, where the values are identical, the risks of each complication are lower under AI-RARP than under ORP. This reasoning indicates that the availability of a lower-risk treatment option would be expected to shift some patients from "watchful waiting" to active treatment. In conclusion, a lower-risk treatment will likely increase utilization. Thus we hypothesize that a highly cost-effective treatment will increase utilization and services. A similar analysis would apply to other treatments that are cost-effective and highly effective. They will likely encourage some patients who would have delayed or foregone treatment to seek care.

13.3.8 Future directions

As AI and robotics continue to reshape surgical practice, several future directions emerge in technology, research, policy, and clinical integration. Understanding these trajectories is essential for stakeholders, including clinicians, researchers, policymakers, and industry partners, to harness innovation responsibly and effectively. Advances in machine learning, computer vision, and real-time data processing are expected to significantly extend the capabilities of AI-enhanced robotic surgery. One promising development is automated surgical task execution, such as AI-guided suturing, which has been reported to reduce surgeons' perceived physical demand and task duration (Lim et al., 2023). Future systems may integrate multimodal inputs, including visual, haptic, and auditory data, for better decision support. However, critical challenges remain. Current robotic systems lack tactile feedback, limiting the surgeon's perception of tissue texture and tension. Additionally, instrument delineation during dynamic movements in constrained surgical fields remains a technical hurdle, affecting precision and safety in real-time AI guidance. Despite progress, substantial gaps in knowledge persist regarding the reliability, generalizability, and contextual adaptability of AI algorithms. For instance, most AI systems are trained on nonrepresentative datasets, limiting performance across diverse populations. Moreover, extensive data sharing, essential for algorithmic improvement, raises substantial concerns about patient privacy and cybersecurity. Unchecked, these risks may result in data breaches and loss of public trust. Another research frontier involves the physical and cognitive burden on surgeons operating robotic systems. While robotic assistance improves precision and reduces invasiveness, it imposes demands such as constrained surgical field visibility, limited access to deep anatomical structures, and the need for simultaneous multiinstrument control, demanding the multitasking skills of the surgeon to operate the mechanical hands of the robot system, manage probes for cameras, and lead associated medical staff for better workflow (Kiyasseh et al., 2023; Schuler et al., 2023). Lim et al. demonstrated that surgeons operating under high workload conditions achieved 79% accuracy in task execution, highlighting the system's promise and the need for ergonomic and cognitive workflow improvements (Lim et al., 2023).

Despite its technological promise, AI-enhanced robotic surgery faces critical institutional barriers that could slow its widespread adoption. Resistance from clinicians, especially surgeons accustomed to traditional methods, may stem from concerns over automation encroaching on clinical autonomy or unfamiliarity with AI systems. In parallel, the shortage of personnel trained in both robotics and AI presents a serious constraint, particularly in low-resource or decentralized health systems. Administrative and bureaucratic inertia also plays a role. Hospitals and insurers may be slow to approve or reimburse innovative technologies, especially when short-term cost savings are unclear. As robotic surgery evolves from the niche to mainstream, healthcare policies must adapt. Evidence-based policy frameworks should support regulatory oversight of adaptive AI systems, ensuring safety across diverse populations, and funding for multidisciplinary research combining surgical science, computer engineering, ethics, and incentive mechanisms for AI developers and robotic manufacturers to prioritize automation that enhances clinical outcomes while minimizing disparities. Financial incentives and transparent evaluation criteria can accelerate the development of next-generation robotic platforms that are affordable, scalable, and inclusive. Policymakers should also consider value-based reimbursement models that reward cost-effective use of robotic technology, especially in high-burden surgical domains. AI-enhanced robotic surgery, despite its promise, is not without risk. AI prediction errors can lead to adverse surgical outcomes, especially in high-stakes procedures. These systems must be treated as decision-support tools, not replacements for clinical judgment. Moreover, robust fail-safes, audit trails, and explainability mechanisms must be incorporated to prevent and investigate mishaps.

13.3.9 Conclusion

AI-integrated robotic surgery, while not yet cost- saving, offers substantial clinical and strategic value. These systems contribute to enhanced surgical precision, reduced complication rates, faster patient recovery, and long-term institutional capacity-building. Although the current implementation remains costly, due to expensive equipment, prolonged operating room time, and a steep learning curve, these barriers are expected to diminish over time. Guided by Moore's Law, the cost of robotic hardware is projected to decline within the next five years, while surgical proficiency and workflow efficiency are likely to improve, enhancing cost-effectiveness. However, the ethical use of AI is crucial. Algorithms trained on limited or homogenous clinical datasets risk embedding structural biases, thus jeopardizing equitable care. Ethnicity-based disparities in training data can lead to inaccurate recommendations or inconsistent performance across diverse patient groups. This indicates the importance of transparency, inclusive data practices, and continuous bias auditing in AI development. Furthermore, regulatory oversight must safeguard against the misuse of AI for financial exploitation, such as recommending unnecessary tests, drugs, or procedures contrary to established clinical guidelines. AI-integrated robotic surgery, while cost-effective based on our case study, is not cost-saving. However, the advantages of AI-enhanced robotic surgery are expected to improve over time, delivering clinical and strategic value through improved health outcomes and institutional capability-building. Future research should include more diverse populations, longer-term cost offsets, and comprehensive metrics such as patient-reported outcome measures to better capture the full value proposition of AI-enhanced surgical systems in modern healthcare.

Acknowledgment

The authors thank Clare L. Hurley, Heller School for Social Policy Management, Brandeis University, Waltham, MA, for editorial assistance and insights.

References

Altıntaş E, Şahin A, Babayev H, Gül M, Batur AF, Kaynar M, et al. Machine learning algorithm predicts urethral stricture following transurethral prostate resection. World Journal of Urology 2024;42(1). Available from https://doi.org/10.1007/s00345-024-05017-x.

An act establishing a special commission on robotic surgery. Senate Bill No. 1069. Massachusetts General Court, 2013.

Artificial Intelligence and Machine Learning in Software. US Food and Drug Administration, 2025.

Bolenz C, Gupta A, Hotze T, Ho R, Cadeddu JA, Roehrborn CG, et al. Cost comparison of robotic, laparoscopic, and open radical prostatectomy for prostate cancer. European Urology 2010;57(3):453–8. Available from https://doi.org/10.1016/j.eururo.2009.11.008.

Bouquet de Joliniere J, Librino A, Dubuisson JB, Khomsi F, Ben Ali N, Fadhlaoui A, et al. Robotic surgery in gynecology. Frontiers in Surgery 2016;3 Available from https://doi.org/10.3389/fsurg.2016.00026, journal.frontiersin.org/journal/surgery.

Boys JA, Alicuben ET, DeMeester MJ, Worrell SG, Oh DS, Hagen JA, et al. Public perceptions on robotic surgery, hospitals with robots, and surgeons that use them. Surgical Endoscopy 2016;30(4):1310–6. Available from https://doi.org/10.1007/s00464-015-4368-6, https://rd.springer.com/journal/464.

Bravi CA, Dell'Oglio P, Mazzone E, Moschovas MC, Falagario U, Piazza P, et al. The surgical learning curve for biochemical recurrence after robot-assisted radical prostatectomy. European Urology Oncology 2023;6(4):414–21. Available from https://doi.org/10.1016/j.euo.2022.06.010.

Carbin DD, Shah A, Kusuma VRM. Artificial intelligence in robot-assisted radical prostatectomy: where do we stand today? Journal of Robotic Surgery 2024;18(1) Available from https://doi.org/10.1007/s11701-024-02143-x, https://www.springer.com/journal/11701.

Cardiac Intervention: An Update, 2025.

Chandrakar I, Pajai S, Toshniwal S. Robotic surgery: the future of gynaecology. Cureus 2022;14(10). Available from https://doi.org/10.7759/cureus.30569.

Chowdhury MA, Rizk R, Chiu C, Zhang JJ, Scholl JL, Bosch TJ, et al. The heart of transformation: exploring artificial intelligence in cardiovascular disease. Biomedicines 2025;13(2):427. Available from https://doi.org/10.3390/biomedicines13020427.

Close A, Robertson C, Rushton S, Shirley M, Vale L, Ramsay C, et al. Comparative cost-effectiveness of robot-assisted and standard laparoscopic prostatectomy as alternatives to open radical prostatectomy for treatment of men with localised prostate cancer: a health technology assessment from the perspective of the UK national health service. European Urology 2013;64(3):361–9. Available from https://doi.org/10.1016/j.eururo.2013.02.040.

Deb, Robotic surgery statistics 2025 by–procedures, safety, complications, 2025.

Disability weights for diseases and conditions, 2025.

Drummond MF, Sculpher MJ, Claxton K. Methods for the Economic Evaluation of Health Care Programmes. Oxford University Press; 2015.

Eckhoff JA, Müller DT, Brunner SN, Fuchs HF, Meireles OR. Do the costs of robotic surgery present an insurmountable obstacle? A narrative review. International Journal of Abdominal Wall and Hernia Surgery 2023;6(2):71–6. Available from https://doi.org/10.4103/ijawhs.ijawhs_62_22.

Executive Office of the President. Safe, secure, and trustworthy development and use of artificial intelligence. Federal Register 2023;75191.

Feldstein J, Coussons H. Achieving robotic program best practice performance and cost versus laparoscopy: two case studies define a framework for optimization. The International Journal of Medical Robotics and Computer Assisted Surgery 2020;16(3). Available from https://doi.org/10.1002/rcs.2098.

Ficarra V, Novara G, Rosen RC, Artibani W, Carroll PR, Costello A, et al. Systematic review and meta-analysis of studies reporting urinary continence recovery after robot-assisted radical prostatectomy. European Urology 2012;62(3):405–17. Available from https://doi.org/10.1016/j.eururo.2012.05.045.

Freedland SJ, Humphreys EB, Mangold LA, Eisenberger M, Dorey FJ, Walsh PC, et al. Risk of prostate cancer-specific mortality following biochemical recurrence after radical prostatectomy. JAMA: the Journal of the American Medical Association 2005;294(4):433–9. Available from https://doi.org/10.1001/jama.294.4.433, http://jama.ama-assn.org/cgi/reprint/294/4/433.

GDP per capita (current US$) – United States. World Bank Data, 2024.

Giambruno V, Chu MW, Fox S, Swinamer SA, Rayman R, Markova Z, et al. Robotic-assisted coronary artery bypass surgery: an 18-year single-centre experience. International Journal of Medical Robotics and Computer Assisted Surgery 2018;14(3) Available from https://doi.org/10.1002/rcs.1891, https://onlinelibrary.wiley.com/journal/1478596x.

Glick HA, Doshi JA, Sonnad SS, Polsky D. Comparing Cost and Effect: Point Estimates for Cost-Effectiveness Ratios and Net Monetary Benefit. Oxford University Press (OUP); 2014. p. 140–53. Available from http://doi.org/10.1093/med/9780199685028.003.0007.

Group, MDCG 2020-1. Guidance on Clinical Evaluation (MDR) / Performance Evaluation (IVDR) of Medical Device Software, 2020.

Guarino P, Chiari F, Cordeschi S, D'Alessio P, Ingelido C, Motta G, et al. A comprehensive systematic review on functional results, speech and swallowing outcomes after trans-oral robotic surgery for oropharyngeal squamous cell cancer. Journal of Clinical Medicine 2024;13(20) Available from https://doi.org/10.3390/jcm13206039, https://www.mdpi.com/journal/jcm.

Gupta KK, De M, Athanasiou T, Georgalas C, Garas G. Quality of life outcomes comparing primary Transoral Robotic Surgery (TORS) with primary radiotherapy for early-stage oropharyngeal squamous cell carcinoma: a systematic review and meta-analysis. European Journal of Surgical Oncology 2024;50(7) Available from https://doi.org/10.1016/j.ejso.2024.108434, https://www.sciencedirect.com/science/journal/07487983.

Hung AJ, Chen J, Che Z, Nilanon T, Jarc A, Titus M, et al. Utilizing machine learning and automated performance metrics to evaluate robot-assisted radical prostatectomy performance and predict outcomes. Journal of Endourology 2018;32(5):438–44. Available from https://doi.org/10.1089/end.2018.0035, www.liebertonline.com/end.

Kiyasseh D, Laca J, Haque TF, Otiato M, Miles BJ, Wagner C, et al. Human visual explanations mitigate bias in AI-based assessment of surgeon skills. npj Digital Medicine 2023;6(1). Available from https://doi.org/10.1038/s41746-023-00766-2 Available from: https://www.nature.com/npjdigitalmed/.

Knudsen JE, Ghaffar U, Ma R, Hung AJ. Clinical applications of artificial intelligence in robotic surgery. Journal of Robotic Surgery 2024;18(1). Available from https://doi.org/10.1007/s11701-024-01867-0.

Kumazu Y, Kobayashi N, Kitamura N, Rayan E, Neculoiu P, Misumi T, et al. Automated segmentation by deep learning of loose connective tissue fibers to define safe dissection planes in robot-assisted gastrectomy. Scientific Reports 2021;11(1). Available from https://doi.org/10.1038/s41598-021-00557-3.

Labban M, Dasgupta P, Song C, Becker R, Li Y, Kreaden US, et al. Cost-effectiveness of robotic-assisted radical prostatectomy for localized prostate cancer in the UK. JAMA Network Open 2022;5(4):e225740. Available from https://doi.org/10.1001/jamanetworkopen.2022.5740.

Laviana AA, Hu JC. Understanding the market forces and opportunity costs of robotic surgery. Perioperative Management in Robotic Surgery. United States: Cambridge University Press; 2017. p. 241–8. Available from http://doi.org/10.1017/9781316534229, http://doi.org/10.1017/9781316534229.023.

Life expectancy at age 65 by sex: United States. (2022)

Lim C, Barragan JA, Farrow JM, Wachs JP, Sundaram CP, Yu D. Physiological metrics of surgical difficulty and multi-task requirement during robotic surgery skills. Sensors 2023;23(9):4354. Available from https://doi.org/10.3390/s23094354.

Maan ZN, Gibbins N, Al-Jabri T, D'Souza AR. The use of robotics in otolaryngology–head and neck surgery: a systematic review. American Journal of Otolaryngology 2012;33(1):137–46. Available from https://doi.org/10.1016/j.amjoto.2011.04.003.

Macroeconomics and health: Investing in health for economic development / report of the Commission on Macroeconomics and Health. WHO Commission on Macroeconomics and Health, 2001.

Meltzer DO, Basu A, Sculpher MJ. Theoretical Foundations of Cost-Effectiveness Analysis in Health and Medicine. Oxford University Press (OUP); 2016. p. 39–66. Available from http://doi.org/10.1093/acprof:oso/9780190492939.003.0002.

Mihaljevic T, Koprivanac M, Kelava M, Goodman A, Jarrett C, Williams SJ, et al. Value of robotically assisted surgery for mitral valve disease. JAMA Surgery 2014;149(7):679–86. Available from https://doi.org/10.1001/jamasurg.2013.5680, http://archsurg.jamanetwork.com/data/Journals/SURG/930571/soi130143.pdf.

Moretti TBC, Magna LA, Reis LO. Erectile dysfunction criteria of 131,350 patients after open, laparoscopic, and robotic radical prostatectomy. Andrology 2024;12(8):1865–71. Available from https://doi.org/10.1111/andr.13634, http://onlinelibrary.wiley.com/journal/10.1111/(ISSN)2047-2927.

Moretti TBC, Magna LA, Reis LO. Surgical results and complications for open, laparoscopic, and robot-assisted radical prostatectomy: a reverse systematic review. European Urology Open Science 2022;44:150–61. Available from https://doi.org/10.1016/j.euros.2022.08.015, https://www.journals.elsevier.com/european-urology-open-science.

Moss E, Halkos ME. Cost effectiveness of robotic mitral valve surgery. Annals of Cardiothoracic Surgery 2017;6(1):33–7. Available from https://doi.org/10.21037/acs.2017.01.03, http://www.annalscts.com/article/view/13277/13634.

Neumann PJ, Saunders GD, Russell. Cost-Effectiveness in Health and Medicine. Oxford University Press; 2017.

Obermeyer Z, Powers B, Vogeli C, Mullainathan S. Dissecting racial bias in an algorithm used to manage the health of populations. Science (New York, N.Y.) 2019;366(6464):447–53. Available from https://doi.org/10.1126/science.aax2342.

Okhawere KE, Shih I-F, Lee S-H, Li Y, Wong JA, Badani KK. Comparison of 1-year health care costs and use associated with open vs robotic-assisted radical prostatectomy. JAMA Network Open 2021;4(3):e212265. Available from https://doi.org/10.1001/jamanetworkopen.2021.2265.

Park ES, Shum JW, Bui TG, Bell RB, Dierks EJ. Robotic surgery. A new approach to tumors of the tongue base, oropharynx, and hypopharynx. Oral and Maxillofacial Surgery Clinics of North America 2013;25(1):49–59. Available from https://doi.org/10.1016/j.coms.2012.11.002.

Pichon-Riviere A, Drummond M, Palacios A, Garcia-Marti S, Augustovski F. Determining the efficiency path to universal health coverage: cost-effectiveness thresholds for 174 countries based on growth in life expectancy and health expenditures. The Lancet Global Health 2023;11(6):e833. Available from https://doi.org/10.1016/s2214-109x(23)00162-6.

Pound CR, Partin AW, Eisenberger MA, Chan DW, Pearson JD, Walsh PC. Natural history of progression after PSA elevation following radical prostatectomy. Journal of the American Medical Association 1999;281(17):1591–7. Available from https://doi.org/10.1001/jama.281.17.1591.

Porcaro AB, Rizzetto R, Amigoni N, Tafuri A, Shakir A, Tiso L, et al. Severe intraoperative bleeding predicts the risk of perioperative blood transfusion after robot-assisted radical prostatectomy. Journal of Robotic Surgery 2022;16(2):463–71. Available from https://doi.org/10.1007/s11701-021-01262-z.

Randell R, Alvarado N, Honey S, Impact of Robotic Surgery on Decision Making: Perspectives of Surgical Teams AMIA Annual Symposium Proceedings. AMIA Symposium, n.d.

Regdesk, European Union Medical Device Regulations, 2025.

Rivero-Moreno Y, Echevarria S, Vidal-Valderrama C, Stefano-Pianetti L, Cordova-Guilarte J, Navarro-Gonzalez J, et al. Robotic surgery: a comprehensive review of the literature and current trends. Cureus 2023;15(7). Available from https://doi.org/10.7759/cureus.42370.

Salomon JA, Haagsma JA, Davis A, de Noordhout CM, Polinder S, Havelaar AH, et al. Disability weights for the Global Burden of Disease 2013 study. The Lancet Global Health 2015;3(11):e712. Available from https://doi.org/10.1016/S2214-109X(15)00069-8, http://www.elsevier.com/journals/the-lancet-global-health/2214-109x.

Schuler N, Shepard L, Saxton A, Russo J, Johnston D, Saba P, et al. Predicting surgical experience after robotic nerve-sparing radical prostatectomy simulation using a machine learning–based multimodal analysis of objective performance metrics. Urology Practice 2023;10(5):447–55. Available from https://doi.org/10.1097/upj.0000000000000426.

Seo HJ, Lee NR, Son SK, Kim DK, Rha KH, Lee SH. Comparison of robot-assisted radical prostatectomy and open radical prostatectomy outcomes: a systematic review and meta-analysis. Yonsei Medical Journal 2016;57(5):1165–77. Available from https://doi.org/10.3349/ymj.2016.57.5.1165, http://www.eymj.org/Synapse/Data/PDFData/0069YMJ/ymj-57-1165.pdf.

Shah N, Khalid U, Kavia R, Batura D. Current advances in the use of artificial intelligence in predicting and managing urological complications. International Urology and Nephrology 2024;56(11):3427–35. Available from https://doi.org/10.1007/s11255-024-04149-8.

Sheetz KH, Claflin J, Dimick JB. Trends in the adoption of robotic surgery for common surgical procedures. JAMA Network Open 2020;3(1):e1918911. Available from https://doi.org/10.1001/jamanetworkopen.2019.18911.

Shi X, Jin Y, Dou Q, Heng P-A. LRTD: long-range temporal dependency based active learning for surgical workflow recognition. International Journal of Computer Assisted Radiology and Surgery 2020;15(9):1573–84. Available from https://doi.org/10.1007/s11548-020-02198-9.

Shin TY, Lee YS. Robot-assisted radical prostatectomy in the treatment of patients with clinically high-risk localized and locally advanced prostate cancer: single surgeons functional and oncologic outcomes. BMC Urology 2022;22(1) Available from https://doi.org/10.1186/s12894-022-00998-6, http://www.biomedcentral.com/bmcurol.

Sujenthiran A, Nossiter J, Parry M, Charman SC, Aggarwal A, Payne H, et al. National cohort study comparing severe medium-term urinary complications after robot-assisted vs laparoscopic vs retropubic open radical prostatectomy. BJU International 2018;121(3):445–52. Available from https://doi.org/10.1111/bju.14054 Available from: http://onlinelibrary.wiley.com/journal/10.1111/(ISSN)1464-410X.

The Global Health Observatory: mortality and burden of disease. Disability-adjusted life years (DALYs). World Health Organization, 2025.
Tomaszewski JJ, Matchett JC, Davies BJ, Jackman SV, Hrebinko RL, Nelson JB. Comparative hospital cost-analysis of open and robotic-assisted radical prostatectomy. Urology 2012;80(1):126–9. Available from https://doi.org/10.1016/j.urology.2012.03.020, www.elsevier.com/locate/urology.
Tomšič A, Palmen M. Robotic mitral valve repair surgery: where do we go from here? Frontiers in Cardiovascular Medicine 2023;10. Available from https://doi.org/10.3389/fcvm.2023.1156495.
Topol EJ, Deep Medicine How Artificial Intelligence Can Make Healthcare Human Again. scirp, 2019.
Wang J, Hu K, Wang Y, Wu Y, Bao E, Wang J, et al. Robot-assisted versus open radical prostatectomy: a systematic review and meta-analysis of prospective studies. Journal of Robotic Surgery 2023;17(6):2617–31. Available from https://doi.org/10.1007/s11701-023-01714-8.
Westerman M, R. Robotic, Prostatectomy, A Game-Changer in Prostate Cancer Treatment. American Urological Association, 2024.
Yousefi M, Assari Arani A, Sahabi B, Kazemnejad A, Fazaeli S. Household health costs: direct, indirect and intangible. Iranian Journal of Public Health, Iran Iranian Journal of Public Health 2014;43(2):202–9. Available from http://ijph.tums.ac.ir/index.php/IJPH/article/download/5256/2324.
Zeckhauser R, Shepard D. Where now for saving lives? Law and Contemporary Problems 1976;40(4):5. Available from https://doi.org/10.2307/1191310.
Zhao B, Waterman RS, Urman RD, Gabriel RA. A machine learning approach to predicting case duration for robot-assisted surgery. Journal of Medical Systems 2019;43(2) Available from https://doi.org/10.1007/s10916-018-1151-y, https://link.springer.com/journal/10916.

CHAPTER 14

Real-world case studies in artificial intelligence-driven medical robotics

Vishal Kumar Sharma[1], Palak Dwivedi[1], and Al Ahmed[2]
[1]AI Research Centre, Woxsen University, Hyderabad, Telangana, India,
[2]Harbourside Institute of Australia, Pyrmont, NSW, Australia

14.1 Introduction

The evolution of artificial intelligence (AI) alongside medical robotics has opened a new way of delivering healthcare, especially in surgical, orthopedic, and rehabilitative settings. This combination is shifting the environment from human-operated mechanical devices to intelligent, semiautonomous systems with capabilities to learn, perceive, make decisions, and have flexibility of use. These technologies are changing the surgical workflow, improving accuracy, decreasing complications, and enabling personalized treatment pathways across numerous clinical environments (Esteva et al., 2019; Yang et al., 2017). The original development of medical robotic systems sought to build on the physical capacity of human surgeons to provide better dexterity, tremor cancellation, and minimally invasive access. Early platforms like the da Vinci Surgical System served as a prototype to allow surgeons to translate movements of their hands into exact movements in the body of the patient. It is important to highlight that for the most part, these systems worked as sophisticated telemanipulators with little to no real-world decision-making capabilities (Hashimoto et al., 2018). The recent advancements in machine learning, computer vision, and real-time data processing have allowed designers to incorporate cognitive resource use into robotic platforms. Now, AI-enabled operating room robots can analyze intraoperative pictures, safely predict the best paths for incisions, and specify instrument motions for guidance on payload trajectories inside the body, all in real-time. These advances are apparent in the da Vinci Xi, MAKO SmartRobotics, and ROSA Knee platforms. The ROSA platform, for example, uses 3D modeling and AI prediction capabilities to enable total knee arthroplasty (TKA) planning without the need for a preoperative CT scan, allowing less cost and less radiation use (Shademan et al., 2016). Robots capable of would accurately cut bone and secure screws, use AI as segmentation to identify anatomy and assess to determine risk while providing intraoperative analysis to elevate a surgeon's confidence and accuracy in areas—and tissues—where traditional comparisons are limited. The use of AI in preoperative planning, intraoperative navigation, and

AI-Powered Developments in Medical Robotics
DOI: https://doi.org/10.1016/B978-0-443-27783-2.00006-5

postoperative outcome prediction is supporting a transition to data-driven and evidence-based surgery (Kassahun et al., 2016b; Kassahun et al., 2016a).

Rehabilitation robotics has also seen tremendous advancements thanks to the use of reinforcement learning and adaptive control algorithms. The Lokomat and MIT-Manus systems modify the assistance and robotic support depending on real-time observations of patient performance (gait asymmetry, joint torque, neuromuscular feedback, etc.). These AI-based rehabilitation technologies may promote neuroplasticity, greater patient adherence, and quicker motor recovery in patients who have experienced a stroke or spinal cord injury (Reinkensmeyer et al., 2004). Macro-level factors are shifting many health systems toward adopting robotic systems with AI-enabled capabilities. These drivers include the increased global burden of chronic diseases, an aging population with highly complex surgeries, an increase in the volume of surgical procedures, and a shortage of qualified surgeons in many locations. The COVID-19 pandemic has also demonstrated an even greater demand for intelligent robotic systems that permit less human exposure while still achieving efficiency and safety of procedure (Putra et al., 2024; Topol, 2019).

There are four functional tiles for AI at the center of these systems:

- Perception: The real-time perception of multimodal sensory stimuli: visual data (computer vision to follow tissue margins, detect bleeding, ensure that tissue is viable during surgery), haptic data, and auditory data (Hashimoto et al., 2018).
- Decision-making: AI-based recommending systems for surgeons in selecting optimal pathways, identifying high-risk zones, and predicting surgical complications by using predictive models (Khan et al., 2023).
- Control: Includes motion scaling, tremor suppression, and corrections of preset trajectories for unanticipated disturbances. AI control models can even perform shared autonomy, with the robot and human sharing control during execution (Feng et al., 2021a).
- Learning: Indicates time-based learning from the data collection through reinforcement or supervised learning. Robots learn from the surgical videos the procedural success or failure and from the annotations dataset to improve their performance in subsequent procedures (Chen et al., 2020).

Even with its transformative possibilities, there are many technical and ethical hurdles that are limiting the application of AI–robotics in clinical practice. The cost of purchase and upkeep, especially for complex systems with AI and machine learning capabilities, is another factor holding things back. There is also the challenge of training and certifying clinical staff to operate robotic systems; this is even more provably challenging in resource-poor settings. Another misunderstanding about these systems is that there is often no haptic feedback, which is important in many operations requiring the sense of touch, such as palpation of a tumor or dissection of a vessel. Inappropriately, regulatory frameworks have failed to keep up with the pace of technology. The use of semiautonomous systems carries its own implications in terms of responsibilities for failure or mistaking a misjudgment by the system, and protocol deviations also have implications. The challenge of assuring AI decisions is an added fear. Of even greater importance is when an AI system is providing intraoperative training decisions. Some existing regulatory frameworks in the public space request transparency not only of decisions, but also of the reasons why these decisions were made (Li et al., 2023). The field of explainable AI (XAI) in medical robotics has become an emergent area of research focusing on trust, accountability, and interpretability benefits of the research (Li et al., 2023; Wang et al., 2023a). Local innovation is allowing emerging markets to address these challenges. The SSI Mantra system, created in India from the ground up, is a multispecialty robotic platform that was designed for value, modularity, and teleoperation, and it uses AI-based navigation and imaging to execute complex surgeries in resource-constrained environments. It is an embodiment of how design that was influenced by context and AI can democratize access to high-quality surgical care.

Robotic systems are evolving, which is motivating the development of novel architectures emphasizing cloud-edge computing with federated learning so that models can be trained in different institutions, and multimodal sensory integrations of data for enhanced situational awareness. While full autonomy is not possible or warranted in high-hazard procedures such as neurosurgery and cardiovascular interventions, human-in-the-loop autonomy is usually our desired configuration when using robotics. AI-based decision support selected designs can be used to support human capability rather than replace it, which agrees with the studies in constrained environments referenced by responsible AI healthcare (Shah et al., 2022; Mahmud et al., 2021). This chapter examines five evidence-based AI-powered medical robotic systems, including da Vinci, MAKO, SSI Mantra, Versius, and ROSA Knee. Each system will be assessed against architecture, level of AI-integration, clinical effectiveness, and user experience. By comparing the robotic systems by use cases, technical functionalities, and deployment contexts, each evidence-based robotic system will help clarify the state of current and future possibilities with AI-powered medical robotics.

14.2 Methodological framework

This chapter takes a formal case study approach to investigate five leading AI-enabled surgical robotics systems—da Vinci, MAKO, SSI Mantra, Versius, and ROSA Knee. The rationale for these systems included a triad of important evaluation criteria: technological innovation, clinical integration, and demonstrated patient impact. The selection of these frameworks is considered the result of both contemporary literature reviews and evaluations of comparative impacts of AI in surgery on the need for additional real-world clinical impact and translational capacity for consideration as the criteria for choosing robotic systems frameworks (Holzinger et al., 2017). Technological innovation signifies the level of incorporation of AI capabilities of the robot systems, such as real-time image analysis, intraoperative assistance, autonomous control capabilities, and adaptive cognition. Clinical integration is the level of adoption and embedding of the system in a real-world hospital environment and its integration with existing clinical workflow patterns across surgical specialties using the system. Demonstrated patient impact is based on clinically validated endpoints such as accuracy of intervention, complication rates, speed of recovery, and patient satisfaction.

The robotic systems selected for assessment will be evaluated based on four operational performance criteria, providing context and understanding as to how AI adds value to surgical systems:

- Accuracy: This includes assessing robotic interventions' accuracy, which can be measured in millimetric deviation when making an incision, suturing, or placing an implant, for example. AI algorithms provide measurable enhancements that impact accuracy through preoperative imaging and intraoperative sensing that guide tools with submillimeter accuracy. The ROSA Knee system provided some examples that show <1.0 degree for alignment, which ultimately led to improved joint function (Li et al., 2023).
- Efficiency: The efficiency paradigm assesses robotic system efficiencies in time of the procedure, intraoperative movement, and, to a lesser extent, how a robotic system would affect the overall surgical process. Generally, robotic systems with AI capabilities around preoperative planning and intraoperative guidance can reduce the operative time and the actual length of stay. For example, the MAKO robotic system has been shown to reduce variability of procedures and operating time through real-time AI planning and cutting guides (Attanasio et al., 2020).
- Safety: Safety includes measurements such as intraoperative complications, conversions from minimally invasive to open surgery, and postoperative outcomes based on function. AI-enhanced perception and feedback provide more rigid redundancy for human errors, such as unintentional variability, and built-in redundancies provide fail-safes to important steps. For example, reviews of AI-enabled robotic surgery demonstrated a statistically significant reduction in adverse events and surgical revision (Innovations, no date).
- Acceptance: Surgeon and staff acceptance is critical for long-term integration. Metrics include ergonomic design, interface intuitiveness, learning curve duration, and training infrastructure. Systems such as da Vinci and Versius have reported high surgeon satisfaction due to intuitive interfaces and modular designs. AI-based simulation tools are also facilitating faster onboarding and skill acquisition (Biomet, no date).

To systematically analyze AI integration, each robotic system is deconstructed along four functional layers of AI capability:

- Perception: This is the system's ability to use visual and sensor data in real-time. Computer vision algorithms are able to help in the identification of anatomical landmarks, classification of tissue types, and complications such as bleeding or signs of inflammation; describe convolutional neural networks and surgical field analysis systems for autonomous tool navigation analysis (Fasoli et al., 2021).

- Decision-making: AI utilization at this level allows for organizations to make recommendations or select surgical actions based on real-time inputs. Central to this is path planning, risk identification, and predictive analytics. These systems use algorithms that include Bayesian inference, deep reinforcement learning, and supervised learning models to plan intraoperative strategies (Intuitive Surgical, no date).
- Control: The control level AI recognize the motion of the tools based on algorithms for tremor filtering of motion, as well as motion scaling or optimizing motion force. This affords additional stability and precision when engaging in high-risk tasks. In the da Vinci and Versius systems, AI-controlled manipulators were employed, and they enabled a consistent tool trajectory, and physiological tremor was filtered out of tool trajectories (Feng et al., 2021b).
- Learning: This level allows robotic systems to learn through experience. Learning systems apply learning algorithms and previous exposure to data. Learning algorithms encompass surgical video archives, annotated datasets, and real-time feedback from procedure/s. Systems such as MAKO and ROSA learn continually; they import outcome data into their part-planning module, facilitating their ability to learn to improve surgical personalization over the long term.

Together, these evaluation parameters and AI capability layers form the analytical framework used throughout this chapter. This structure ensures a consistent, evidence-based comparison of how AI is embedded in robotic platforms to augment clinical precision, operational safety, and workflow adaptability.

14.3 Case sudy 1: robotic-assisted surgery with artificial intelligence-based decision support (da Vinci Surgical System) system overview

The da Vinci Surgical System, from Intuitive Surgical, is currently the most widely used AI-enabled robotic platform with application to minimally invasive surgical procedures. The system encompasses the surgeon console, robotic arms, and high-definition 3D scope, thus providing the surgeon incredible dexterity, visualization, and control (Haidar et al., 2021). The da Vinci Surgical System is used for a wide variety of procedures, including urology, gynecology, cardiothoracic surgery, and general surgery.

14.3.1 Role of artificial intelligence

AI in the da Vinci System supports various intraoperative functions:

- Perception: The robot utilizes enhanced imaging algorithms and computer vision to support tissue differentiation and anatomical boundary detection.

- Decision Support: Real-time AI suggestions for instrument trajectory optimization and camera navigation reduce surgeon fatigue and improve efficiency.
- Control: Motion scaling and tremor filtration algorithms powered by AI allow for micrometer-scale precision, which is particularly important in delicate procedures such as prostatectomy or mitral valve repair.

Emerging research is also integrating machine learning into the da Vinci platform to support performance analytics, such as motion economy and tool path prediction. These functions are designed to assess surgical skill objectively and to customize real-time assistance during procedures (Feng et al., 2021a).

14.3.2 Clinical outcomes

Clinical studies comparing da Vinci-assisted surgeries to conventional laparoscopic techniques have reported:

- Reduced intraoperative blood loss,
- Lower conversion rates to open surgery,
- Fewer complications,
- Quicker patient recovery,
- Improved postoperative functional outcomes (Ramsay et al., 2012).

These advances are due mainly to the robot's mechanical accuracy and support AI modules that support user control and visualize the surgery task.

14.3.3 Takeaways

- The da Vinci System is representative of collaborative intelligence as it enhances the surgeon's skill with AI-guided visualization.
- Limitations include high cost, no haptic feedback, and a steep learning curve.
- Still considered a gold-standard system globally, it is setting the company and research community in directions to produce next-generation intelligent robot systems.

14.4 Case study 2: artificial intelligence-enabled rehabilitation robotics

14.4.1 System overview

AI-enabled rehabilitation robots such as Lokomat (Hocoma, Switzerland) and MIT-Manus have significantly advanced motor recovery protocols in poststroke and spinal-injury therapy. These platforms support task-specific and repetitive gait training using robotic exoskeletons. Unlike static or therapist-dependent systems, these robots incorporate AI algorithms to dynamically tailor resistance and support levels based on patient-specific motor status and recovery trajectories.

14.4.2 Role of artificial intelligence

The core AI functions in these systems lie in adaptive resistance control, trajectory planning, and real-time feedback modulation. By monitoring physiological metrics such as joint torque, limb velocity, and gait symmetry, the robot adjusts the assistance provided in each cycle. Reinforcement learning models are employed to optimize performance by rewarding active participation and penalizing passivity, facilitating neuroplastic adaptation in stroke patients (Fasoli et al., 2004).

These systems use closed-loop control, where AI algorithms continuously refine behavior based on patient feedback and therapeutic goals. For example, MIT-Manus uses error-based adaptive strategies to encourage engagement while avoiding over-assistance known to hinder long-term recovery (Marchal-Crespo and Reinkensmeyer, 2009; Frisoli et al., 2009).

14.4.3 Clinical outcomes

Studies consistently show that AI-driven rehabilitation robotics outperform conventional physical therapy in terms of:

- Faster improvement in gait speed and stride symmetry,
- Greater neuromuscular coordination and motor relearning,
- Enhanced patient motivation and adherence due to dynamic personalization (Reinkensmeyer et al., 2004).

These outcomes are particularly significant in neurorehabilitation, where recovery is highly individualized. Robots with AI capabilities reduce therapy time, improve long-term retention of motor skills, and minimize relapse rates.

14.4.4 Key takeaways

- AI-enhanced rehabilitation robots embody continuous learning, adjusting assistance across sessions.
- They offer quantitative, real-time data for clinicians, enabling evidence-based decisions.
- These systems democratize high-quality rehabilitation through consistency and scalability.

14.5 Case study 3: versatile surgical robotics—the SSI Mantra system

14.5.1 System overview

The SSI Mantra Surgical System, from SS Innovations (India), is a cost-effective, modular solution designed for AI-enabled surgery. The ability to almost universally utilize the platform in support of general, pediatric, gynecological, and cardiac surgeries requires a minimally priced marketplace when compared to other international providers. The 3D 4K camera system, open-console interface, and collective stabilized-control framework offer participants ergonomic and collaborative benefits (Fasoli et al., 2021).

14.5.2 Artificial intelligence role

The system has features with machine learning–enabled functionality to allow real-time image recognition, trajectory prediction, and automated warnings for threats. Though the system is not fully autonomous, its intelligent modules add precision to surgical procedures and safety support during the intraoperative phase. The remote telesurgery and telementoring functions are highly impactful in low-resource environments (Fasoli et al., 2021).

14.5.3 Clinical outcomes

Early deployment studies across hospitals in India report the following:

- Significant reduction in operative time and intraoperative bleeding,
- High visual clarity due to its advanced optics,
- Smooth adaptation across surgical specialties,
- Successful implementation in pediatric nephrectomies and cardiac valve repairs (Gamazeliuc et al., 2024).

14.5.4 Key takeaways

- SSI Mantra provides a cost-effective solution (~70%–80% less than the da Vinci).
- Its AI-assisted vision and control features make it ideal for emerging economies.
- Its modularity and scalability ensure adaptability across various hospital infrastructures.

14.6 Case study 4: artificial intelligence-driven orthopedic robotics—the ROSA Knee system

14.6.1 System overview

The ROSA Knee system by Zimmer Biomet is tailored for TKA procedures. The system integrates intraoperative sensors and 2D-to-3D image synthesis to facilitate

CT-free alignment planning using X-Atlas technology. This platform is particularly noted for its ability to improve reproducibility in joint alignment (Kaur, 2024).

14.6.2 Artificial intelligence role

ROSA applies AI-enhanced models for the following:

- Intraoperative soft-tissue balancing,
- Predictive analytics for implant positioning,
- Learning from past surgical data to refine procedural recommendations (Kaur, 2024).

These AI functions allow the robot to adapt intraoperatively, guiding surgeons based on real-time measurements and historical outcome data. This increases both the precision and personalization of TKA procedures.

14.6.3 Clinical outcomes

Clinical trials and retrospective studies have shown:

- Alignment deviations reduced to <1.0 degree,
- Significant decreases in hospital stay durations,
- Shortened rehabilitation timelines,
- Reduction in radiation exposure due to CT-free workflows.

14.6.4 Key takeaways

- ROSA improves both accuracy and personalization in knee replacements.
- The system has a short learning curve, with most surgeons getting comfortable within 10–15 procedures.
- This system is a good fit for institutions looking to manage costs while improving accuracy and consistency within procedures.

14.7 Case study 5: smart tissue autonomous robot for autonomous surgical suturing

14.7.1 System overview

The smart tissue autonomous robot (STAR) is an advanced surgical platform developed through a collaboration between researchers at Carnegie Mellon University and the Children's National Hospital. STAR's primary aim is to perform soft-tissue surgery autonomously, tackling one of the greatest challenges in surgical robotics, safely interacting with deformable, dynamic tissues. Unlike traditional robotic systems that rely heavily on direct human control, STAR integrates real-time intraoperative imaging, precision robotic manipulators, and machine learning-driven decision-making to execute delicate procedures such as intestinal anastomosis (sewing together sections of bowel) with minimal human intervention. The platform combines stereovision cameras, force sensors, and preoperative imaging data to create a dynamic model of the surgical field. STAR's motion control is driven by AI algorithms trained on expert surgeon behavior and optimized through reinforcement learning techniques.

14.7.2 Artificial intelligence role

AI is at the heart of STAR's capability. Key roles include:

- Deep learning models process live imaging to segment tissues and track deformation in real-time, an enormously complex task due to tissue movement, bleeding, and variable lighting.
- Reinforcement learning policies determine optimal suture placement and instrument trajectories, adapting to variability during surgery.
- AI refines instrument movement at submillimeter scales, smoothing motion and compensating for tremor or unexpected shifts in tissues.
- A supervisory AI layer constantly evaluates risk thresholds and can pause or adjust actions if unsafe conditions are detected.

14.7.3 Clinical outcomes

In preclinical and early clinical investigations:

- STAR completed autonomous suturing of intestinal tissues with higher consistency and fewer leaks than human surgeons in animal models.
- Benchmarks showed reduced tissue trauma, lower variability in suture spacing, and a notably lower rate of postoperative complications in comparison to conventional laparoscopic suturing.
- Surgeons acting in oversight roles reported that STAR's performance was predictable and reliable and that AI-driven action reduced procedural time variability.

While full human clinical trials remain limited, STAR's use in controlled environments has demonstrated that autonomous assistance can enhance precision and free up surgical attention for strategic decision-making rather than repetitive manual tasks.

14.7.4 Key takeaways

- Autonomy beyond preprogrammed paths: STAR represents a shift from teleoperated systems to *intelligent autonomy* capable of adjusting to dynamic surgical conditions.

- AI as a safety amplifier: Real-time perception and control help maintain safe operation even when tissues move unpredictably.
- Human–AI collaboration model: STAR is not replacing surgeons but augmenting them; the surgeon oversees and intervenes, while the system handles repetitive yet critical tasks.
- Translational barriers: Regulatory approval, rigorous validation in diverse human anatomies, and integration with hospital workflows remain key hurdles for clinical adoption.

14.8 Case study 6: cyberknife with artificial intelligence-enhanced treatment planning in radiosurgery

14.8.1 System overview

The CyberKnife System (Accuray Incorporated) is a robotic radiosurgery platform used to treat tumors throughout the body with focused radiation beams. It features a six-axis robotic arm that can deliver high-dose radiation with submillimeter precision. Unlike traditional linear accelerators, CyberKnife does not require rigid frames and can track tumors that move during respiration or patient movement. Traditionally, treatment planning in radiosurgery has been a laborious process requiring significant clinician input. With the incorporation of AI tools, CyberKnife workflows have evolved to include automated segmentation, optimized dose mapping, and real-time motion prediction.

14.8.2 Artificial intelligence role

AI enhances CyberKnife performance in several critical areas:

- Automated segmentation: Deep learning models automatically delineate tumor boundaries and organs-at-risk from CT/MRI scans, reducing planning time and interoperator variability.
- Dose optimization: Machine learning algorithms propose optimal radiation dose distributions, balancing tumor coverage with normal tissue sparing.
- Motion management: Predictive AI models analyze respiratory patterns and adjust beam delivery in real-time, improving accuracy during treatments of lung, liver, and pancreatic tumors.

14.8.3 Clinical outcomes

Clinical studies and retrospective evaluations have highlighted several benefits of CyberKnife when combined with AI-driven planning:

- Improved target accuracy: Automated segmentation and dynamic tracking resulted in *higher conformity indices* and *lower radiation exposure to healthy tissues* compared with manual planning in complex anatomical regions.

- Reduced planning time: AI tools have cut preparation time from several hours to under an hour in many cases, enabling faster treatment initiation, which is crucial for aggressive cancers.
- Outcomes in tumor control: For certain indications (e.g., early-stage lung cancer and spinal metastases), local control rates and pain relief outcomes using AI-enhanced CyberKnife were at least on par with traditional techniques and sometimes superior due to more precise dose delivery.
- Patient throughput: Efficiency gains helped clinics treat more patients without compromising quality.

14.8.4 Key takeaways

- Automating labor-intensive planning steps without sacrificing quality has significant operational benefits.
- Predictive models that compensate for patient breathing represent a major advancement over static-beam approaches.
- AI adoption hinges on tools that fit seamlessly into clinician workflows and maintain transparency so physicians trust automated suggestions.
- The gains have been seen across a range of tumor types, demonstrating the generalizability of AI in treatment planning.

14.9 Case study 7: EksoGT exoskeleton for neurological rehabilitation

14.9.1 System overview

The EksoGT exoskeleton (by Ekso Bionics) is a wearable robotic device designed to assist gait rehabilitation for patients with neurological impairments such as spinal cord injury, stroke, or multiple sclerosis. The system supports ambulation by providing motorized assistance to the hip and knee joints while the user engages in intentional movement. Initially, exoskeletons were purely mechanical or provided fixed levels of assistance. The addition of AI has transformed EksoGT into a responsive, adaptive rehabilitation partner.

14.9.2 Artificial intelligence role

AI contributes in several dimensions:

- Adaptive assistance algorithms: Machine learning analyzes user gait patterns in real-time, adjusting support levels based on strength, balance, and fatigue.
- Personalized progression: AI tracks longitudinal performance data and tailors progression goals, offering clinicians actionable biomarkers to inform therapy plans.

- Predictive analytics: AI forecasts potential loss of balance or user fatigue moments, prompting preemptive adjustments to reduce fall risk.

14.9.3 Clinical outcomes

Multiple clinical evaluations have shown:

- Improved functional gains: Patients using EksoGT with AI-assisted adaptive control showed greater improvements in walking speed, stride symmetry, and independence compared to conventional therapy.
- Engagement and motivation: Real-time feedback and adaptable challenge levels increased patient engagement, which correlated with better adherence and outcomes.
- Objective metrics for clinicians: The system's data outputs enabled therapists to monitor progress quantitatively, leading to more informed therapy adjustments and better long-term outcomes.

14.9.4 Key takeaways

- Real-time adaptation enhances rehab: AI-driven adjustment to assistance levels helps match therapy to patient capability.
- Data-driven rehabilitation: Continuous monitoring and analytics offer a deeper understanding of recovery trajectories.
- Human–robot teamwork: The exoskeleton augments but does not replace physical therapists; it enhances their ability to deliver tailored therapy.
- Scaling potential: AI allows the system to be effective across a wide range of impairments and recovery stages.

14.10 Comparative insights across case studies

An in-depth comparative study, such as the one discussed regarding the five AI-based robots, da Vinci, MAKO, SSI Mantra, Versius, and ROSA, is a useful tool to learn how AI is changing the surgical and rehabilitation workflow today. The goal of the comparison was to determine what enablers for success exist across systems, what barriers prevent scale, and what advancements will formulate the future.

14.10.1 Key success factors

A principal commonality in these systems is the quality and quantity of clinical data that is available to train and test AI models. We know that over years of utilization, the da Vinci and ROSA platforms have combined their years of surgical video, kinematics, and surgical outcomes into a deep learning model for instrument navigation and for tissue segmentation (Irfan et al., 2023). This is an example of using high-quality annotated datasets as the foundation for machine learning models capable of providing real-time decision support. Regulatory compliance has emerged as a critical success factor in gaining institutional trust and accelerating deployment. All five systems analyzed have either secured or are undergoing approval processes with regulatory bodies such as the FDA (U.S.), Conformité Européenne (CE) (Europe), and Central Drugs Standard Control Organisation (CDSCO) (India). These certifications not only validate safety but also ensure interoperability with hospital protocols and standards of care (Irfan et al., 2023).

Also, these platforms capitalize on structured user training and straightforward interfaces, minimizing the cognitive load faced by clinicians. Training approaches that incorporate real-life simulation, virtual reality, and a pathway for a graduate certificate reduced learning curves and increased surgeon proficiency. The MAKO and SSI Mantra systems reinforce user capability and safety, helping diminish the chances of iatrogenic errors.

14.10.2 Issues related to across platforms

Although there are positive developments, there are still important technical and ethical challenges. One issue of concern is AI explainability and transparency. There still remains a desire among many surgeons and hospital administrators to elevate surgical decisions that could lead to life-critical outcomes, with AI limited to a black-box model for real-time decision-making. Additionally, the inability to follow how the AI algorithm developed a trajectory or surgical decision is certainly seen as an edge against utilizing the very technology we designed to assist us. Trust in automation is also a barrier for some. Trust in automation is challenging; this can be fragile even when validated performance exists, especially when it is not well understood or respected (if this goes wrong, it's not our fault??). ROSA and MAKO technologies have adopted a framework that keeps the surgeon in the loop while reducing biomechanical workload, but semiautonomy offers few efficiency improvements compared to purely automated solutions (De Rossi et al., 2021).

Data privacy and security are additional pain points, particularly in the context of shared surgical data and collaborative machine learning. Systems must adhere to HIPAA, GDPR, and other regulations to ensure that patient data remains anonymized and protected.

14.10.3 Emerging trends

Several trends are coming together to expand the scope of what AI's role in augmented surgical systems can accomplish. For example, the use of federated learning allows many hospitals to build a pool of common models from decentralized data without compromising patient privacy. By doing so, not only is the diversity of data increased, but the models are also able to learn from the world's surgical practices (Ur Rehman et al., 2024). The migration of AI from cloud to edge computing is a further trend as we see applications in latency-sensitive areas such as real-time feedback in robotic surgery. Systems like SSI Mantra are including AI accelerators onboard to keep some level of independence from external server infrastructures, ensuring the system is more resilient at decisive moments in surgery. Interoperability is gaining traction too, as robotic systems are developed to connect with electronic medical records, picture archiving and communication systems, and preoperative planning systems. This should lead to an ecosystem in which a single user interface can give real-time combined use of analytics on the spot, previous histories, and predictive algorithms. Human–machine symbiosis is becoming the preferred paradigm. Rather than supplanting the surgeon, AI is being optimized to provide decision support, to assist in tedious or high-precision procedures, or to augment human cognition. For instance, Versius employs AI to determine suggested instrument positioning based on the patient's anatomy; ROSA collects and shares real-time monitoring data about placement for enhanced implant fit.

In conclusion, this comparative analysis suggests that while algorithmic sophistication might be a factor in the success of AI robotics in surgery and rehabilitation, there is more to successful adoption than that. For instance, there is data quality, regulatory considerations, clinician education, and the resilience of the platform itself, whereas transparency, trust, and data security must be fully solved; this innovation is moving forward to a more extensive degree of adaptability, intelligence, and cooperation, which can benefit patient care (Wang et al., 2023b; Zhang et al., 2024). As technology, it will be important for all stakeholders, surgeons, data scientists, regulators, and patients, to collaborate in ensuring that AI can be used ethically, transparently, and safely within a clinical environment.

14.11 Future directions and research gaps

The five AI-integrated robotic systems reviewed in this work demonstrate innovative implementations, as well as significant gaps that need to be addressed in the future. These proposed areas are important to investigate systematically in order to produce medical robotics with its maximum potential capacity and accessibility.

14.11.1 Personalized artificial intelligence–robotics platforms

One of the most exciting areas for research is the development of a customizable robotic platform. The platform would be able to be personalized in regard to anatomy, surgical risk profile, motor limitations, and recovery. Current systems are exact and specific;

however, all systems except Da Vinci utilize static algorithms and some form of standardized protocol template. Future systems need to incorporate dynamic learning, where algorithms and performance plans can adapt to and learn from specific patient data, such as muscle tone, response to therapy, and intraoperative variability. Systems that learn would rely on ongoing data collection and an AI model that can support reinforcement learning and real-time neural change (Zhang et al., 2024). For example, orthopedic robotic systems could optimize implant positioning while using real-time kinematic feedback to ensure the patient's functional recovery is optimal.

14.11.2 Multimodal artificial intelligence sensing

Current robotic platforms primarily use visual modalities—such as video and x-ray—for input. However, surgical and rehabilitative environments need greater modalities, and future robotic systems will need to use multimodal sensing, integrating visual, haptic, auditory, thermal, and biomechanical signals. This would broaden the sense-making of the robot's surroundings and enable surgeons to receive complete environmental data as procedures unfold (Wang et al., 2023b). For example, including haptic feedback in surgical arms will allow the robot to recognize tissue characteristics, and bio-acoustic information would facilitate the localization of unusual physiological occurrences such as an aneurysm rupture or airway obstruction.

14.11.3 Regulatory and legal frameworks for artificial intelligence autonomy

Robotic systems are developing autonomous features, and existing regulatory standards have to quickly change to reflect this. Guidelines from the FDA—and other international bodies—are unfit to deal with accountability challenges related to AI autonomy. Issues of legal liability, explainability of AI decisions, and validation of self-learning algorithms remain unanswered at this time. Regulation scholars and roboticists must work together to create a dynamic regulatory framework that can adjust to future innovations while ensuring patient safety (Zhang et al., 2024). The framework(s) should also establish provisions for postmarket monitoring, tracking algorithmic decision-making, and indicating the respective human and AI roles in situations.

14.11.4 Human-in-the-loop versus fully autonomous

There is a philosophical and functional tension between fully autonomous AI systems and human-in-the-loop systems. While full autonomy may have an opportunity to streamline low-risk, repetitive procedures such as laparoscopic hernia repairs, higher-risk surgeries such as open-heart surgery or neurosurgery still require oversight and control from a clinician. The future may involve hybrid systems wherein the robot can perform standard measures autonomously but will rely on a shared decision-making model when faced with a nonlinear decision or complication (De Rossi et al., 2021). Research should center on optimally creating this collaboration, incorporating real-time intent recognition, interfaces for shared control, and calibrating trust.

14.11.5 Ethical and socioeconomic considerations

As AI-enhanced surgical robotics grows, so do the responsibilities of addressing ethical factors and health equity. Many low- and middle-income countries still find advanced systems like da Vinci or MAKO to be prohibitively expensive. The economic inequality already present in global health can evolve into a larger concern within autonomous surgical robotics. In autonomous surgical robotics, ethical factors such as AI biases in data ownership and data used to develop consent in automated interventions also need to be addressed. Researchers should prioritize ethical aspects of design for AI, ensuring accessibility, transparency, and cultural responsiveness.

14.11.6 Real-world evaluation metrics

Developing standardized metrics for assessing the performance of AI-enabled robotic systems in real-world situations should be another priority in the research agenda. Generally, assessments are based on surrogate endpoints (e.g., incision accuracy, blood loss); however, it will be essential to measure long-term outcome measures after surgery, for example, quality of life, number of revisions, and cost-effectiveness, over periods of 10–15 years in future studies. This will require collaboration between clinical researchers, data scientists, and health services researchers with the aim of developing these benchmarks.

14.11.7 Cloud and edge computing collaboration in robotics

The combination of cloud-based collective learning and on-device edge computing will be fundamental to building strong, context-aware surgical robotics. Whereas cloud-based systems allow robots to learn from every other institution that interfaces with them around the world, edge computation provides low-latency input that is critical to making real-time decisions. Future research should incorporate a safe and encrypted AI structure that encourages bi-directional learning while preserving patient confidentiality and system resilience (Zhang et al., 2024).

14.12 Conclusion

This chapter provided a detailed analysis of five AI-driven robotic systems that are transforming surgery and rehabilitation. Each case study, from the precision multiarm configurations of the da Vinci system to the adaptive orthopedic capabilities of the ROSA Knee platform, demonstrates the transformative potential of AI through physiological bodies.

Despite differences in base design principles, technical platform, and affordability in economic accessibility, all reviewed systems have in common a trend toward intelligent, personalized, explainable automation. AI is no longer a technology of the future, but it enables excellence in delivering clinical practice. These systems will become more scalable, affordable, and inclusive as innovations in edge computing, federated learning, and cloud robotic integration develop.

In the near term, translational research should prioritize:

- Extending regulatory pathways that find a place for AI autonomy,
- Creating global communities of collaboration involving learning networks and open robotic architectures,
- Development of low-cost, high-impact solutions for developing and underserved areas,
- Embedding ethical guardrails in system design.

At the end, we can conclude that AI-enabled robotic technology is not just a means of enhancing surgical performance but a means to redefine what is possible surgically and therapeutically. By meeting clinical need and placing collaborative innovation on the terms of inclusion, transparency, and bottom-up clinical need, we can usher in a new age of medical robotics that is safer, smarter, and more humane.

References

Attanasio A, Grasso RF, Rea S, et al. Artificial intelligence in surgery: current state and future perspectives. Artificial Intelligence in Medicine 2020;108:101917. Available from https://doi.org/10.1016/j.artmed.2020.101917.

Biomet, Z, ROSA knee system: surgical platform summary and data brief, n.d., Available from: https://www.zimmerbiomet.com (accessed 2024).

Chen C, Chen X, Liu C, et al. Reinforcement learning in robotic gait rehabilitation: a review of the state-of-the-art. Biomedical Signal Processing and Control 2020;60:101965. Available from https://doi.org/10.1016/j.bspc.2020.101965.

De Rossi G, Minelli M, Roin S, Falezza F, Sozzi A, Ferraguti F, et al. A first evaluation of a multi-modal learning system to control surgical assistant robots via action segmentation. IEEE Transactions on Medical Robotics and Bionics 2021;3(3):714–24. Available from https://doi.org/10.1109/tmrb.2021.3082210.

Esteva A, Robicquet A, Ramsundar B, Kuleshov V, DePristo M, Chou K, et al. A guide to deep learning in healthcare. Nature Medicine 2019;25(1):24–9. Available from https://doi.org/10.1038/s41591-018-0316-z.

Fasoli SE, Krebs HI, Stein J, Frontera WR, Hughes R, Hogan N. Robotic therapy for chronic motor impairments after stroke: follow-up results. Archives of Physical Medicine and Rehabilitation 2004;85(7):1106–11. Available from https://doi.org/10.1016/j.apmr.2003.11.028.

Fasoli SE, Iversen MD, Krebs HI, Hogan N. Distributed machine learning in collaborative rehabilitation robotics. Computer Methods and Programs in Biomedicine 2021;199:105946. Available from https://doi.org/10.1016/j.cmpb.2020.105946.

Feng Y, Li Y, Gao L. Robotic-assisted surgery: a review of recent advancements and future directions. Biomedical Signal Processing and Control 2021;68:102707. Available from https://doi.org/10.1016/j.bspc.2021.102707.

Feng Y, Gao B, Liu Y, Zhang Y, Liu Y. Bridging the gap between technology and clinical practice in robotic surgery. IEEE Reviews in Biomedical Engineering 2021;14:112–24. Available from https://doi.org/10.1109/RBME.2020.3021885.

Frisoli A, Salsedo F, Bergamasco M, Rossi B, Carboncini MC. A force-feedback exoskeleton for upper-limb rehabilitation in virtual reality. Applied Bionics and Biomechanics 2009;6(2):115–26. Available from https://doi.org/10.1080/11762320902959250.

Gamazeliuc G, Nechita RM, Ulerich O, Bădiță-Voicu LL, Cristoiu C. Fusion of AI and robotics for transforming neurosurgery. International Journal of Mechatronics and Applied Mechanics 2024;2024(18):175–83. Available from https://doi.org/10.17683/ijomam/issue18.21, https://ijomam.com/wp-content/uploads/2024/12/pag.-175-183_FUSION-OF-AI-AND-ROBOTICS-FOR-TRANSFORMING-NEUROSURGERY.pdf.

Haidar D, Charara J, Louati H, et al. Surgical data science: enabling next-generation surgery using AI and big data. Computer Methods and Programs in Biomedicine 2021;200:105897. Available from https://doi.org/10.1016/j.cmpb.2021.105897.

Hashimoto DA, Rosman G, Rus D, Meireles OR. Artificial intelligence in surgery: promises and perils. Annals of Surgery 2018;268(1):70 6. Available from https://doi.org/10.1097/SLA.0000000000002693, http://journals.lww.com/annalsofsurgery/pages/default.aspx.

Holzinger A, Biemann C, Pattichis CS, Kell DB. What do we need to build explainable AI systems for the medical domain? Revue d'Intelligence Artificielle en Médecine 2017;67:1–18. Available from https://doi.org/10.1016/j.artmed.2016.11.004.

S.S. Innovations, Clinical evaluation report on the SSI Mantra surgical robotic system, n.d. Available from: https://www.ssinnovations.com (accessed 2024).

Intuitive Surgical, The da Vinci surgical system: technology overview, n.d. Available from: https://www.intuitive.com/en-us/products-and-services/da-vinci/systems (accessed 2024).

Irfan B, Céspedes N, Casas J, Senft E, Gutiérrez LF, Rincon-Roncancio M, et al. Personalised socially assistive robot for cardiac rehabilitation: critical reflections on long-term interactions in the real world. User Modeling and User-adapted Interaction 2023;33(2):497–544. Available from https://doi.org/10.1007/s11257-022-09323-0, https://www.springer.com/journal/11257.

Kassahun Y, Yu B, Tibebu AT, Stoyanov D, Giannarou S, Metzen JH, et al. Surgical robotics beyond enhanced dexterity instrumentation: a survey of machine learning techniques and their role in intelligent and autonomous surgical actions. International Journal of Computer Assisted Radiology and Surgery 2016;11(4):553–68. Available from https://doi.org/10.1007/s11548-015-1305-z, http://www.springer.com/dal/home?SGWID=1-102-70-132298836-0&SHORTCUT=www.springer.com/11548.

Kassahun Y, Yu B, Tibebu AT, Stoyanov D, Giannarou S, Metzen JH, et al. Surgical robotics beyond enhanced dexterity instrumentation: a review of machine learning and augmented reality. Annals of Biomedical Engineering 2016;44(12):3575–89. Available from https://doi.org/10.1007/s10439-016-1632-7.

Kaur, J., Synergistic evolution: pioneering frontiers of artificial intelligence and robotics in healthcare. In: International Conference on Artificial Intelligence and Smart Energy, Springer Nature Switzerland, Cham, 2024, March, pp. 147-157. https://doi.org/10.1007/978-3-031-61475-0_12.

Khan M, Gupta R, Singh P. Comparative outcomes of CT-based vs. CT-free robotic knee surgeries. Journal of Orthopaedic Robotics 2023;9(1):112–21. Available from https://doi.org/10.1111/os.14055.

Li Z, Tan Y, Bai X, et al. Smart surgical systems: progress, applications, and challenges. Information Fusion 2023;91:1–15. Available from https://doi.org/10.1016/j.inffus.2023.03.012.

Mahadevan B, Jain A. Principles of responsible AI in healthcare robotics. The International Journal of Medical Robotics + Computer Assisted Surgery MRCAS 2023;19(4):e2502. Available from https://doi.org/10.1002/rcs.2502.

Mahmud N, Khan S, Ahmed K. Human-in-the-loop AI in surgical robotics: a hybrid learning approach. Computers in Biology and Medicine 2021;138:104931. Available from https://doi.org/10.1016/j.compbiomed.2021.104931.

Marchal-Crespo L, Reinkensmeyer DJ. Review of control strategies for robotic movement training after neurologic injury. Journal of Neuroengineering and Rehabilitation 2009;6(1). Available from https://doi.org/10.1186/1743-0003-6-20.

Putra KT, Arrayyan AZ, Hayati N, Firdaus, Damarjati C, Bakar A, et al. A review on the application of Internet of Medical Things in wearable personal health monitoring: a cloud-edge artificial intelligence approach. IEEE Access 2024;12:21437–52. Available from https://doi.org/10.1109/ACCESS.2024.3358827, http://ieeexplore.ieee.org/xpl/RecentIssue.jsp?punumber=6287639.

Ramsay C, Pickard R, Robertson C, et al. Systematic review and economic modelling of the clinical effectiveness and cost-effectiveness of minimal access surgery for prostate cancer. Health Technology Assessment (Winchester, England) 2012;16(41):1–313. Available from https://doi.org/10.3310/hta16410.

Reinkensmeyer DJ, Emken JL, Cramer SC. Robotics, motor learning, and neurologic recovery. Annual Review of Biomedical Engineering 2004;6:497–525. Available from https://doi.org/10.1146/annurev.bioeng.6.040803.140223.

Shademan A, Decker RS, Opfermann JD, Leonard S, Krieger A, Kim PCW. Supervised autonomous robotic soft tissue surgery. Science Translational Medicine 2016;8(337) Available from https://doi.org/10.1126/scitranslmed.aad9398, http://stm.sciencemag.org/content/scitransmed/8/337/337ra64.full.pdf.

Shah S, Ravi V, Dey D. Regulatory considerations for AI-driven autonomous surgical systems. Journal of Health Policy and Technology 2022;11(2):100636. Available from https://doi.org/10.1016/j.hlpt.2022.100636.

Topol EJ. High-performance medicine: the convergence of human and artificial intelligence. Nature Medicine 2019;25(1):44–56. Available from https://doi.org/10.1038/s41591-018-0300-7, http://www.nature.com/nm/index.html.

Ur Rehman M, Zoha A, Ali Jamshed M, Ramzan N. Advances in Multi-Modal Intelligent Sensing Multimodal Intelligent Sensing in Modern Applications. Wiley; 2024. p. 1–28. Available from http://doi.org/10.1002/9781394257744.ch1.

Wang J, Liu H, Zhou Z, et al. Personalized robot learning in surgical automation. Engineering Applications of Artificial Intelligence 2023;120:105881. Available from https://doi.org/10.1016/j.engappai.2023.105881.

Wang S, Liu Z, Yang W, Cao Y, Zhao L, Xie L. Learning-based multimodal information fusion and behavior recognition of vascular interventionists' operating skills. IEEE Journal of Biomedical and Health Informatics 2023;27(9):4536–47. Available from https://doi.org/10.1109/jbhi.2023.3289548.

Yang GZ, Cambias J, Cleary K, et al. Medical robotics—regulatory, ethical, and legal considerations. Science Robotics 2017;2(4):eaam8638. Available from https://doi.org/10.1126/scirobotics.aam8638.

Zhang R, Zhou Y, Zhang J, Zhao J. Cloud-integrated robotics: transforming healthcare and rehabilitation for individuals with disabilities. Proceedings of the Indian National Science Academy 2024;90(3):752–63. Available from https://doi.org/10.1007/s43538-024-00264-3.

PART V

Challenges and future directions

CHAPTER 15

Challenges in artificial intelligence driven robotic surgery

Sachin Sharma[1] and Seshadri Mohan[2]
[1]Amity School of Engineering and Technology, Amity University Punjab, Mohali, Punjab, India,
[2]Systems Engineering Department, University of Arkansas at Little Rock, Little Rock, AR, United States

15.1 Introduction

Artificial intelligence (AI)-based robotic surgery is revolutionizing the restorative field by combining AI, machine learning (ML), and mechanical technology to upgrade exactness, productivity, and quiet results in surgical strategies. The equipment comprises robotics arms that are prepared with surgical apparatuses for exact and steady developments. Sensors that accumulate real-time information on constraint, position, and tissue reaction. End-effectors that are disobedient after mechanical arms are custom-made for particular surgical errands. AI encourages decision-making, where calculations analyze persistent information and recommend ideal surgical techniques. Design an acknowledgment that joins AI frameworks to translate medical images (e.g., CT, MRI) to distinguish tumors, organs, or basic structures. Supervised learning makes a difference in AI to learn from explained surgical datasets to imitate master activities. Fortification learning is where the robots learn ideal activities by collaborating with recreated or genuine situations. The specialists control mechanical frameworks using haptic input that gives material sensations, upgrading exactness. Real-time imaging offers augmented reality (AR) or 3D visuals to make strides in situational mindfulness. The key innovations have computer vision, where AI forms high-resolution surgical pictures to distinguish anomalies, direct disobedience with millimeter accuracy and improve the route in negligibly obtrusive surgeries. Natural language processing empowers voice-activated mechanical commands. It helps in real-time documentation amid methods. Predictive analytics, where AI predicts potential complications and offers preemptive arrangements based on past history and live information. Cloud frameworks empower information sharing, preparing models, and upgrading robot capabilities. Edge computing guarantees real-time reactions amid surgery without latency. Exactness incorporates robots to dispose of hand tremors and guarantee reliable execution. Negligible intrusive procedures empower smaller cuts to lead to quicker recuperation and diminished disease dangers. Customization offers AI that tailors surgical plans to a person's understanding of life systems. The real-time adjustment empowers AI to alter automated activities based on live input. There are numerous challenges, that is, information reliance, where AI frameworks require tremendous, high-quality datasets for preparation. Ethical concerns like responsibility and

AI-Powered Developments in Medical Robotics
DOI: https://doi.org/10.1016/B978-0-443-27783-2.00012-0

educated assent require tending to. Fetched and openness, where Tall is starting a venture and operational costs constrain the broad selection. Direction rigid endorsement forms for AI frameworks in healthcare. The number of applications empowered, that is, common surgery of hernia repairs, gallbladder evacuation. Cardiovascular surgery, that is, valve repairs, bypass surgeries. Oncology, that is, tumor resections with exactness. Orthopedics, that is, joint substitutions and spine surgeries. Neurosurgery, that is, tiny and negligibly invasive methods. Independent surgery, that is, AI frameworks performing assignments with negligible human mediation. Upgraded preparation on virtual reality (VR) recreations for specialists utilizing AI direction. Global accessibility empowers AI, democratizing surgical mastery in farther or underserved districts.

15.2 Literature review

In Knudsen et al. (2024), clinical applications of AI in robotic surgery, the survey discusses AI's contributions to robotic surgery, centering on intraoperative improvements such as progressed measurements, detection of positive surgical margins, and computerization of surgical steps. It also addresses AI's part in surgical instruction and moral contemplations. In Iftikhar et al. (2024), the authors examine the current landscape of AI-driven robotic surgical systems, specifying their benefits, restrictions, and future prospects. In Brian et al. (2024), the authors investigate how AI can address educational challenges in robotic surgery, proposing solutions and identifying future directions. In Vasey et al. (2023), the authors review intraoperative AI applications currently under development and highlight their potential to improve surgical techniques. In Byrd IV and Tignanelli (2024), examine AI's role in surgery, discussing current applications, future directions, and addressing moral, legal, and ethical challenges. In Moglia et al. (2021), the authors analyze AI in robot-assisted surgery, recognizing current confinements and challenges. In Boal et al. (2024), the authors evaluate objective, manual, computerized, and AI-based methods used in robotic surgery skill assessment, providing insights into their effectiveness. In Qian and Ren (2025), the authors audit the application of deep reinforcement learning in surgical mechanical autonomy, highlighting its potential to make strides in computerization and exactness in surgical assignments. In Thai et al. (2020), the authors examined the advanced intelligent systems in surgical mechanical autonomy, centering on detection, incitation, and brilliantl control to upgrade surgical methods. In Ali et al. (2022), the authors survey recent deep learning-based strategies applied to surgical data, including instrument localization, tracking, and 3D scene reconstruction.

15.3 Importance of emerging technologies in artificial intelligence-based robotic surgery

The integration of emerging technologies into AI-based robotic surgery plays an essential part in changing the scene of healthcare and surgery. These headways offer improved exactness, improved patient outcomes, and more prominent effectiveness in surgical strategies. The significance of rising advances in AI-based mechanical surgery cannot be exaggerated. These advancements are reshaping the field of surgery, driving advances in accuracy, safety, quiet recovery, and overall healthcare efficiency. AI systems' capacity to back specialists in real-time, personalize care, and give prescient analytics is changing how surgeries are performed and observed. As these advances proceed to advance, their integration into healthcare will not, as it were, improve surgical results but, moreover, contribute to a more evenhanded and open healthcare framework, all-inclusive. Here are the key reasons why rising advances are vital in AI-based mechanical surgery:

- AI algorithms: Advances in AI algorithms improve the accuracy of robotic systems. For example, AI can analyze preoperative imaging, such as CT scans or MRIs, to make detailed 3D models of a patient's life structures. This permits more exact arranging and real-time alterations amid surgery.
- Mechanical arms and AI integration: The combination of robotic arms with AI-driven direction enables micro-level precision, particularly in fragile methods like neurosurgery, cardiothoracic surgery, and ophthalmology. These advances diminish the hazard of human blunder, especially in complex surgeries, guaranteeing more noteworthy security for patients.
- Faster recovery: With negligible surgical disturbance, patients encounter diminished postoperative torment and can recuperate more rapidly, driving to shorter clinic stays and quicker return to typical exercises. The general healthcare burden is diminished as patients can be released sooner. AI-based robotic frameworks empower negligibly intrusive surgeries by making smaller entry points, which lead to less injury to encompassing tissues. This not only decreases blood misfortune but also abbreviates recuperation time and diminishes the chance of complications like contaminations.
- Intraoperative AI support: During surgery, AI systems can provide real-time data and recommendations, supporting intraoperative decision-making. AI can analyze the continuous method, identify potential complications, and recommend adjustments to guarantee an ideal outcome. This steady input makes a difference in accomplishing better outcomes.
- Personalized surgical plans: Emerging technologies permit AI to form individualized surgical plans custom-fitted to a patient's special life structures and well-being conditions. This personalized approach guarantees that the method is optimized for the particular needs of the understanding, improving the general adequacy and lessening dangers.
- Simulation and testing: AI can recreate different surgical approaches and test diverse scenarios before actual surgery. This capacity to test and refine strategies in a virtual

environment can diminish the time that goes through within the working room and move forward with results.

- Postoperative observing: After surgery, AI can help monitor patient recovery by analyzing information from wearable gadgets or sensors. This continuous data stream permits healthcare experts to track crucial signs, identify complications early, and alter treatment plans in like manner.
- Long-term predictive analytics: AI can be utilized to anticipate long-term results and survey the adequacy of the surgery over time. For example, AI frameworks can track the movement of a tumor after surgery or screen the victory of a joint substitution.
- Reduced operating time: With more prominent exactness and real-time AI direction, mechanical surgeries often take less time compared to conventional strategies. Shorter surgeries cruel less working room time, less assets used, and decreased costs for healthcare suppliers. By reducing operating time, AI also enables more efficient resource allocation, lowering costs for hospitals.
- Tele-surgery and remote help: Rising advances such as 5G networks and telecommunication tools permit AI-based automated surgery to be performed remotely. Specialists can direct mechanical frameworks from all over the world, advertising high-quality care to underserved zones and low-resource settings where master specialists may not be accessible.
- Standardized quality of care: AI systems ensure greater consistency in surgical outcomes, reducing variability and promoting standardized quality of care. This could be particularly useful in districts with constrained access to exceedingly specialized surgical skills.
- Intelligent robotic arms: The most recent progressions in robotics have made automated arms more adaptable, responsive, and able to replicate human dexterity with more noteworthy accuracy. AI upgrades these frameworks by permitting them to adjust in real-time to a surgeon's inputs, advertising a much more extensive extent of movement and exactness than conventional surgical instruments.
- Haptic feedback: Rising innovations have empowered mechanical frameworks to supply specialists with haptic feedback—realistic touch sensations that replicate the tactile sensation of tissues amid surgery. This permits specialists to perform assignments with improved affectability, moving forward with their decision-making and execution.
- Information analytics: AI-based frameworks can analyze large volumes of surgical data, giving insights into the execution of different procedures, specialist abilities, and quiet results. These bits of knowledge can offer assistance in refining surgical conventions and progress in preparing hones.
- Enhanced surgeon training: Developing AI advances, such as VR and AR, can be utilized for simulation-based preparation, allowing surgeons to practice in a

controlled environment. AI can evaluate a surgeon's execution while preparing and provide personalized feedback for skill improvement.

15.4 Overview of artificial intelligence-based robotic surgery

AI-based robotic surgery combines advanced robotic systems with AI to perform complex surgical methods with enhanced precision, safety, and efficiency. This development coordinates mechanical technology, ML, and real-time information investigation to help robotize surgical assignments, changing conventional surgery as noted by Bhatt et al. (2024). Sharma et al. (2024) explore how integration transforms traditional surgical practices. Robotic arms enhance precision by eliminating human tremors, enabling controlled movements. AI facilitates accurate targeting of tissues and structures. AI ensures accurate targeting of tissues and anatomical structures. AI provides real-time decision support by analyzing patient data, imaging, and surgical conditions to guide surgeons. Predictive analytics help anticipate complications and recommend corrective actions. Minimally invasive techniques use smaller incisions, reducing trauma, blood loss, and recovery time for patients (Sharma et al., 2023). AI improves navigation in narrow or delicate anatomical regions. Core technologies include ML, which enables robots to learn from surgical data and improve performance (Singh et al., 2023). Computer vision processes medical images to provide accurate navigation and identification of anatomical structures. Haptic input provides tactile sensations that enhance control and safety. The benefits of AI-based robotic surgery include improved outcomes, reduced risk of complications and errors, consistent quality across procedures, greater efficiency through shorter operation times and optimized workflows, and faster patient recovery due to minimally invasive techniques (Nautiyal et al., 2023). Reliable quality over strategies and upgraded proficiency that has shorter operation times with optimized workflows. Faster recovery for patients due to negligibly obtrusive procedures. Personalized surgery, where AI tailors procedures to each patient's anatomy and condition, further enhances effectiveness.

15.5 Types of artificial intelligence-based robotic surgery

AI-based robotic surgery encompasses a range of systems and approaches designed for specific medical applications. These sorts can be categorized based on the innovation utilized and the nature of the surgical methods. Fig. 15.1 illustrates the main types of AI-based robotic surgery.

- Teleoperated robotic surgery: In teleoperated robotic surgery, the surgeon controls robotic arms remotely using a console. Although the surgeon provides input for each movement, AI enhances accuracy and reduces errors.

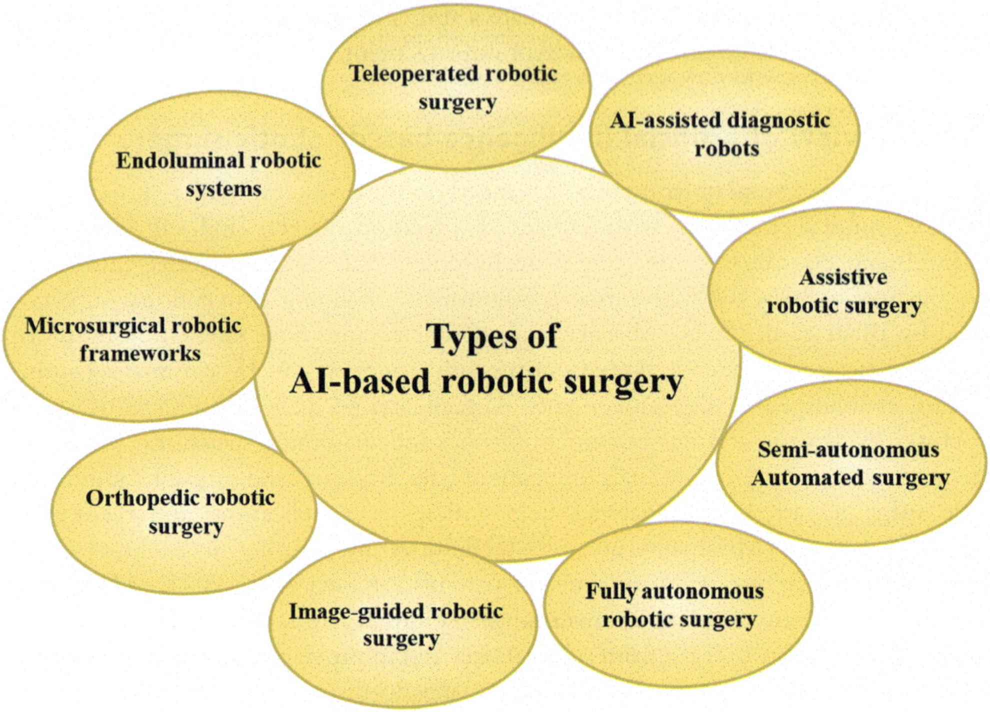

Figure 15.1 Types of AI-based robotic surgery.

- Assistive robotic surgery: Assistive robotic systems support surgeons by holding instruments, stabilizing the operative field, or improving visibility. AI contributes by interpreting data and providing real-time guidance, although decision-making remains surgeon-led. Applications include orthopedic surgery (e.g., the MAKO robotic arm for joint replacement) and neurosurgery for stabilization and amplification.
- Semiautonomous automated surgery: Robots perform specific tasks autonomously under surgeon supervision. AI calculations direct activities such as suturing, tissue resection, or bone cutting. The example system is ROSA (Robotic Surgical Assistant) for brain and spine surgeries. The applications are spinal combinations and exact tumor resections.
- Fully autonomous robotic surgery: Robots perform surgery with negligible human mediation, depending completely on AI. AI frameworks analyze information, make choices, and execute surgical plans freely. The applications are independent soft-tissue surgery and microsurgical procedures, which are still in experimental stages.
- Image-guided robotic surgery: It uses AI-driven imaging technologies. It consolidates computer vision and AR for upgraded visualization. The applications are

minimally invasive surgeries (e.g., laparoscopy) and cancer surgeries with precise tumor margin identification.

- Orthopedic robotic surgery: It is centered on joint and bone methods, utilizing AI for exact arrangement and situation. Preoperative planning integrates AI-based models for implant customization. The case framework is the NAVIO Surgical Framework. The applications are knee and hip substitutions and spinal surgery.
- Microsurgical robotic frameworks: It is outlined for intricate surgeries requiring extraordinary accuracy at a minuscule level. AI compensates for micro-movements, improving dexterity and stability. The applications are ophthalmic surgery (e.g., retinal procedures) and reconstructive surgery, including nerves and vessels.
- Endoluminal robotic systems: It is specialized for minimally invasive surgeries performed on interior tubular organs (e.g., intestines, blood vessels). AI-based routes guarantee secure maneuvering in restricted spaces. The applications are endoscopic surgeries, gastrointestinal, and vascular strategies.
- AI-assisted diagnostic robots: AI-assisted diagnostic robots are equipped with imaging and ML capabilities that allow them to detect abnormalities during surgery. They can guide biopsies with robotic precision and facilitate early identification of irregularities during exploratory procedures.

15.6 Case study

The United States has been at the cutting edge of creating and actualizing AI-based automated surgery frameworks. The da Vinci System allows surgeons to operate from a console equipped with 3D high-definition vision and intuitive controls. Its robotic arms translate the surgeon’s hand movements into precise actions inside the patient’s body, assisted by a high-definition endoscope that provides magnified visualization.

- Surgeon's support: The specialist sits in a comfortable environment prepared with 3D high-definition vision and intuitive controls.
- Robotic arms: These highly articulated arms mirror the surgeon's hand developments, deciphering them into exact and scaled developments inside the patient's body.
- Endoscope: A high-definition 3D camera gives specialists an amplified and three-dimensional view of the surgical field.
- Image analysis: AI calculations are being created to analyze real-time surgical video bolsters. These calculations can recognize basic anatomical structures, such as nerves and blood vessels, giving specialists with profitable real-time input.
- Predictive analytics: AI can analyze endless sums of surgical information to foresee potential complications, such as dying or nerve harm. This data can assist specialists in altering their approach during the strategy and make strides in understanding results.

- Surgical planning: AI can help in preoperative arranging by analyzing persistent information (such as CT scans and MRIs) to make 3D models of the surgical location. This allows surgeons to better visualize the life systems and arrange the ideal surgical approach.
- Haptic feedback: Inquiries are underway to consolidate haptic feedback into the da Vinci framework, giving specialists a more reasonable sense of touch and feel amid the strategy.
- Prostatectomy: Robotic-assisted laparoscopic prostatectomy has become the gold standard for the treatment of prostate cancer within the United States.
- Nephrectomy: Robotic-assisted laparoscopic nephrectomy is progressively utilized for the evacuation of kidney tumors.
- Hysterectomy: Robotic-assisted hysterectomy may be a common, negligibly intrusive approach for the evacuation of the uterus.
- Myomectomy: Automated surgery is utilized to expel uterine fibroids while protecting richness.
- Cardiac valve repair: Robotic-assisted mitral valve repair is progressively performed, offering a less obtrusive alternative to conventional open-heart surgery.
- Lung surgery: Robotic-assisted lobectomy is utilized for the expulsion of lung tumors.
- Colectomy: Robotic-assisted colectomy is utilized for the expulsion of parts of the colon.
- Hernia repair: Robotic-assisted hernia repair offers a negligibly obtrusive approach to this common surgical method.

AI-based robotic surgery is revolutionizing healthcare in the USA, offering enhanced precision, efficiency, and patient outcomes. A notable example is Virtua Health's introduction of advanced spine surgery using AI-assisted robotics, marking a significant milestone in Southern New Jersey. Studies have shown that AI-assisted robotic surgeries lead to better patient outcomes across various specialties. For instance, spinal surgeries using AI-assisted robotics reported a 25% reduction in operative time and a 30% decrease in intraoperative complications compared to manual methods. Surgical precision improved by 40%, with enhanced targeting accuracy during tumor resections and implant placements. AI-driven robotic systems reduce surgical trauma and inadvertent damage to surrounding tissues. Patients experience shorter hospital stays, lower postoperative pain scores, and faster functional restoration. Reduced complications, shorter hospitalizations, and fewer revision surgeries lead to significant cost savings. Diligent Robotics' Moxi robot, acquired by Serve Robotics, has completed over 1.25 million deliveries in 25 US hospitals, enhancing clinical workflow efficiency. AI-assisted robotic systems are being used in urologic oncology, spinal surgery, and pediatric cases, demonstrating improved surgical margins and accelerated recovery. As

AI technology continues to evolve, we can expect even more innovative applications in robotic surgery, further transforming healthcare in the USA.

15.6.1 The use of the da Vinci framework within the US has illustrated noteworthy benefits for patients

- Reduced pain: Minimally invasive strategies regularly result in less postoperative torment and distress.
- Faster recovery: Patients regularly encounter shorter clinic stays and quicker recuperation times compared to conventional open surgery.
- Diminished blood loss: Robotic surgery regularly results in less blood misfortune, lessening the need for blood transfusions.
- Improved cosmesis: Smaller incisions lead to less obvious scarring.

15.6.2 In spite of its victories, the broad appropriation of artificial intelligence-based automated surgery within the US, moreover, faces challenges

- Cost: The high cost of robotic systems can limit access for certain patients and healthcare institutions.
- Expertise improvement: Specialists require specialized preparation to function mechanical frameworks successfully, which can be time-consuming and exorbitant.
- Ethical considerations: The expanding independence of AI in surgery raises moral questions almost understanding security, specialist responsibility, and the potential for work relocation.
- Information protection and security: Guaranteeing the protection and security of understanding information utilized to prepare and work with AI algorithms is crucial.

15.6.3 Future research and development are aimed at improving autonomy, advancing artificial intelligence algorithms for tissue identification and predictive analytics, integrating with augmented reality/virtual reality technologies, and establishing ethical and regulatory frameworks to ensure safe adoption. Progressing inquiring about and advancement are centered on

- Improved independence: Creating more independent mechanical frameworks that can perform progressively complex errands with negligible human intercession.
- Progressed AI calculations: Creating more modern AI calculations for assignments such as real-time tissue recognizable proof, prescient analytics, and haptic input.

- Integration with other advances: Coordination of AI-based automated surgery with other rising advances, such as increased reality and VR, to advance and improve surgical exactness and proficiency.
- Addressing ethical considerations: Creating ethical rules and administrative systems to guarantee the secure and capable utilization of AI in surgery.

The da Vinci Surgical Framework, with its expanding integration of AI, speaks to a point of interest and accomplishment within the field of mechanical surgery within the United States. This innovation has revolutionized the way numerous surgical methods are performed, advertising noteworthy benefits for patients and surgeons alike. As AI proceeds to advance, it is able to anticipate seeing indeed more modern and imaginative mechanical frameworks changing the future of surgery within the United States.

Whereas Russia has a solid establishment in logically inquiring about and innovative advancement, the selection and advancement of AI-based mechanical surgery inside the nation display a one-of-a-kind set of challenges and openings.

- Restricted access: The accessibility of progressed mechanical surgical frameworks, such as the da Vinci Surgical Framework, is right now restricted inside Russia. These frameworks are costly, and their dissemination is concentrated in major urban centers, fundamentally in Moscow and St. Petersburg.
- Reliance on conventional surgical methods: Whereas mechanical surgery is picking up footing, numerous healthcare teachers in Russia still depend intensely on conventional open surgery and negligibly obtrusive laparoscopic procedures.
- Government bolsters: The Russian government recognizes the potential of automated surgery and has taken steps to advance its advancement and usage. Activities incorporate:
- Financing investigation and advancement: Supporting inquiries about regions such as mechanical framework plans, AI calculation advancement, and surgical recreation.
- Investing in training programs: Providing preparation programs for specialists and other healthcare experts within the utilization of automated surgical frameworks.
- Progressing innovation: Encouraging the procurement and dissemination of mechanical surgical frameworks to territorial healing centers.
- Financial imperatives: The high cost of mechanical surgical frameworks and related advances presents a critical financial boundary for numerous healthcare institutions in Russia.
- Need for skilled personnel: A deficiency of prepared specialists and specialists with mastery in automated surgery can prevent the broad appropriation of this innovation.
- Technological dependence: Dependence on foreign-manufactured automated frameworks can make a reliance on universal providers and constrain mechanical independence.

- Integration with healthcare framework: Coordination of automated surgery into the existing healthcare framework can be challenging, requiring critical ventures in foundation, preparation, and information administration.
- Creating indigenous advances: Russia has the potential to create its own imaginative automated surgical frameworks and AI-powered advances, lessening dependence on outside imports.
- Center on telemedicine: Leveraging telemedicine innovations to associate inaccessible districts with specialized mechanical surgery centers in major cities.
- AI-driven inquiry about: Conducting inquiries about AI-driven surgical strategies, such as image-guided surgery, prescient analytics, and personalized surgical arranging.
- Collaboration with international accomplices: Collaborating with worldwide investigators, teachers, and companies to share information, abilities, and assets within the field of automated surgery.
- Russian Academy of Sciences: Conducting investigations and improvements in areas such as robotics, AI, and medical imaging.
- Service of Well-being of the Russian Federation: Directing the advancement and execution of healthcare arrangements, including the appropriation of modern innovations such as mechanical surgery.
- Driving Medical Institutions: Major clinics and inquiry centers in Moscow and St. Petersburg are effectively included within the appropriation and progression of mechanical surgery.

The Russian government has recognized the potential of robotic surgery, investing in research, training programs, and procurement of systems for regional hospitals. While challenges stay, the future of AI-based automated surgery in Russia holds significant guarantees. By contributing to investigation and advancement, preparing programs, and framework, Russia can overcome these challenges and use this transformative innovation to move forward with persistent care and development of its position within the global healthcare scene. AI-based robotic surgery is transforming healthcare in Russia, offering enhanced precision, efficiency, and patient outcomes. Russia has been actively adopting robotic surgery technology, with numerous hospitals integrating systems like the Da Vinci and Senhance robots. The Pirogov National Medical and Surgical Center in Moscow has been a pioneer in robotic surgery since 2018, successfully performing various procedures in urology, gynecology, abdominal and thoracic surgery, neurosurgery, and orthopedics. The center has reported improved outcomes, reduced complications, and faster recovery times. In Moscow, hospitals have performed over 2000 high-tech operations with robotic complexes between January and July 2025, matching the total number performed in 2024. The city is seeing a surge in robotic surgery, with systems such as Da Vinci and Senhance being used in leading oncology centers. Robotic surgery is being applied in various specialties, including urology for prostatectomy and kidney stone removal, gynecology for hysterectomy and endometriosis treatment, and orthopedics for joint replacement. The technology is also being used in pediatric surgery, with successful procedures performed at the Irkutsk Regional Children's Clinical Hospital. The Russian government is supporting the development of robotic surgery, with investments in equipment and training programs. Moscow's healthcare department has equipped hospitals with cutting edge robotic systems, and specialists are receiving continuous training and accreditation.

AI-based robotic surgery is revolutionizing healthcare in Germany, offering enhanced precision, efficiency, and patient outcomes. Germany has been at the forefront of adopting robotic surgery technology, with numerous hospitals integrating systems like the Da Vinci and Dexter robots. The country's healthcare landscape is witnessing a significant transformation, driven by the growing adoption of AI-powered robotic systems. In urology, robotic-assisted prostatectomy has become a standard procedure, with German centers utilizing multiparametric MRI to guide surgery and improve outcomes. The technology is also being applied in visceral surgery for colorectal resections and esophageal, gastric, pancreatic, and liver surgeries. Orthopedic surgeons are leveraging innovative robot-assisted procedures for joint replacement surgery, eliminating the need for bone pins and external cameras. The benefits of AI-based robotic surgery are evident in Germany's healthcare sector. Improved surgical precision and accuracy are leading to reduced complications and faster recovery times. Patients are experiencing shorter hospital stays, lower postoperative pain scores, and accelerated functional restoration. The economic benefits are also significant, with reduced healthcare costs and fewer revision surgeries. Despite the growing adoption, challenges persist. High costs, lack of reimbursement, and limited cooperation between disciplines are hindering widespread implementation. German surgeons prioritize diagnostic accuracy, patient selection, and transparent counseling to ensure successful outcomes. Germany's robotic surgery landscape is expected to evolve further, driven by advancements in AI and ML. Integration of these technologies will enhance decision-making and precision, expanding applications in outpatient and ambulatory surgical centers. Training and education will remain crucial, ensuring surgeons are equipped to harness the full potential of AI-based robotic surgery. As Germany continues to invest in robotic surgery, the country's healthcare system is poised to reap significant benefits. With improved patient outcomes, reduced costs, and enhanced surgical capabilities, AI-based robotic surgery is set to transform Germany's healthcare landscape. The future of surgery in Germany looks increasingly robotic, promising better lives for patients across the country.

AI-based robotic surgery is transforming healthcare in New Zealand, with several hospitals adopting advanced robotic systems like the da Vinci Surgical System. This technology enables surgeons to perform complex procedures with greater precision, accuracy, and control, resulting in improved patient outcomes. In New Zealand, robotic surgery is being used in various specialties, including urology, general surgery, gynecology, and orthopedics. For instance, robotic-assisted prostatectomy has become a standard procedure, with patients experiencing shorter hospital stays, reduced blood loss, and fewer complications. The technology is also being applied in knee and hip replacements, providing real-time data for precise implant positioning. The benefits of robotic surgery in New Zealand are evident. Patients are experiencing faster recovery times, less pain, and reduced scarring. The technology is also improving surgical training and education, with surgeons able to practice and refine their skills using simulation-based learning. Several hospitals in New Zealand are leading the way in robotic surgery, including Braemar Hospital in Waikato, which has invested in a multimillion-dollar da Vinci Xi robot. This investment is expected to improve access to high-quality surgical care for patients in the region. The New Zealand government is also supporting the adoption of robotic surgery, recognizing its potential to improve healthcare outcomes and reduce costs. As the technology continues to evolve, it is likely to play an increasingly important role in New Zealand's healthcare system.

AI-based robotic surgery is transforming healthcare in South Africa, with several hospitals adopting advanced robotic systems, such as the da Vinci Surgical System. This technology enables surgeons to perform complex procedures with greater precision, accuracy, and control, resulting in improved patient outcomes. In South Africa, robotic surgery is being used in various specialties, including urology, general surgery, gynecology, and orthopedics. For instance, robotic-assisted prostatectomy has become a standard procedure, with patients experiencing shorter hospital stays, reduced blood loss, and fewer complications. The technology is also being applied in knee and hip replacements, providing real-time data for precise implant positioning. The benefits of robotic surgery in South Africa are evident. Patients are experiencing faster recovery times, less pain, and reduced scarring. The technology is also improving surgical training and education, with surgeons able to practice and refine their skills using simulation-based learning. Several hospitals in South Africa are leading the way in robotic surgery, including Groote Schuur Hospital and Tygerberg Hospital, which have invested in da Vinci Xi Surgical Systems. These hospitals have performed over 500 procedures, including robotic-assisted laparoscopic surgery, partial nephrectomy, and complex reconstructive procedures. Despite the growing adoption, challenges persist. High costs, lack of reimbursement, and limited cooperation between disciplines are hindering widespread implementation. However, the South African government is supporting the adoption of robotic surgery, recognizing its potential to improve healthcare outcomes and reduce costs. As the technology continues to evolve, it is likely to play an increasingly important role in South Africa's healthcare system. With improved patient outcomes, reduced costs, and enhanced surgical capabilities, AI-based robotic surgery is set to transform South Africa's healthcare landscape.

15.7 Challenges

As AI-based robotic surgery advances, several challenges must be addressed to realize its full potential, that is, high costs, limited availability outside major urban centers, and dependence on foreign-manufactured systems hinder widespread adoption. It is essential for ensuring broad adoption, effectiveness, and ethical integration into healthcare.

- Information quality and amount: AI frameworks require broad, high-quality datasets for preparation. The need for standardized, annotated surgical datasets hampers AI improvement. Variability in data across regions and institutions limits generalizability.
- Real-time handling: AI-driven robots need to prepare huge volumes of information in genuine time. High computational demands may cause delays or errors during surgery.
- Interoperability: Seamless integration of AI frameworks with existing healthcare frameworks and gadgets is complex. Standardizing communication protocols across diverse robotic systems is critical.
- Responsibility and risk: Determining liability for adverse outcomes caused by AI decisions remains unclear. Who is accountable—the specialist, the robot producer, or the AI designer?
- Bias in AI calculations: Training datasets may reflect demographic or systemic biases, leading to disparities in surgical outcomes.
- Patient consent and belief: Patients may hesitate to trust AI in critical decisions. Transparent communication about AI's role in surgery is crucial to building confidence.
- Regulatory approval: Strict approval is required to guarantee that AI frameworks are secure and solid. Long and complex approval processes may delay the selection of imaginative innovations.
- High costs: The improvement, procurement, and upkeep of AI-based automated frameworks are costly. Smaller hospitals and clinics in low-resource settings often struggle to afford such technologies.
- Training requirements: Specialists and staff require broad training to function in progressively automated frameworks viably. The shortage of trained professionals in underserved regions limits accessibility.
- Financial disparities: Wealthier healthcare systems adopt AI-based robotic surgery more readily, potentially widening global healthcare disparities.
- Surgeon–AI collaboration: Ensuring surgeons trust and effectively collaborate with AI systems is essential. Resistance to change among clinicians may slow adoption.
- Error mitigation: Even minor errors in AI algorithms can pose significant risks. Rigorous testing and validation are critical to prevent failures.

- Cybersecurity risks: AI-driven systems are vulnerable to cyberattacks and data breaches. Ensuring data security and safe operation is essential in healthcare settings.
- Job displacement concerns: Concerns about automation replacing surgeons may create resistance among professionals. It is important to emphasize AI as a tool to augment rather than replace human expertise.
- Public perception: Educating the public about the benefits and safety of AI-based robotic systems is vital.
- Technological advancements: Develop robust, adaptable AI systems capable of real-time decision-making. Cultivate collaborations to make expansive, different datasets for preparation.
- Policy and regulation: Establish clear guidelines for accountability in AI-driven surgery and streamline approval processes while maintaining strict safety standards.
- Financial and preparing support: Provide funding mechanisms to make robotic systems affordable in low-resource settings and create global training programs for surgeons and healthcare staff.
- Open awareness and education: Promote awareness of AI's role as a supportive tool, not a replacement for human judgment. Build public trust through transparency and successful case studies.
- Ethical AI development: Address algorithmic biases to ensure equitable care, and involve diverse stakeholders in building inclusive AI systems.

15.8 Overcoming barriers and possible solutions in artificial intelligence-powered robotic surgery

Collaborative efforts to create larger and more diverse datasets through anonymized data sharing are essential to improving AI systems in robotic surgery. AI-powered robotic surgery holds immense promise for revolutionizing healthcare, but several barriers must be addressed to fully realize its potential. Here's a breakdown of the key challenges and potential arrangements:

Information confinements and security concerns: AI algorithms depend on tremendous amounts of high-quality information to memorize and perform successfully. Limited or biased datasets can lead to inaccurate predictions and potentially compromise patient safety.

- Nonstop checking and updates to AI calculations to adjust them to the most recent therapeutic information.
- Coordination of AI explainability instruments to form decision-making forms more straightforward, improving belief while guaranteeing human oversight amid strategies.

Integration and workflow challenges: Integrating AI-powered robotic systems into existing hospital workflows can be complex.

- Building up clear conventions for information approval and AI calculation confirmation.
- Proactive planning of both framework and workforce, including comprehensive preparing programs.
- Cultivating collaboration among key stakeholders—including innovation engineers, specialists, and administrative offices.

Ethical and regulatory considerations: Ensuring patient privacy and data security is fundamental. Furthermore, exploring administrative obstacles and guaranteeing AI-powered advances meet stringent security and viability standards are crucial.

- engaging stakeholders to address concerns to ensure a delicate understanding of information.
- Locks in all partners, counting innovation engineers, specialists, and administrative offices, to address security and administrative concerns.

Cost and availability: The starting venture in AI-powered mechanical frameworks can be significant, constraining get to for numerous healthcare offices.

- Investigating imaginative financing models, such as renting or shared-risk assentions.
- Contributing to inquiries about and improvements to diminish the taken a toll of these advances.

Trust and acceptance: Transparent communication about AI's capabilities and limitations, clear emphasis on the surgeon's role as the ultimate decision-maker, and presentation of successful case studies are essential to building trust among both patients and healthcare professionals.

- Straightforward communication around the capabilities and impediments of AI in surgery.
- Emphasizing the role of specialists as the extreme decision-makers, with AI serving as a steady instrument.
- Displaying effective use cases and illustrating the benefits of AI-powered mechanical surgery.

15.9 Future perspective

The future of AI-powered robotic surgery holds great promise as technology continues to advance. Here are a few key angles that will shape the end of the AI-powered robotic surgery. Fig. 15.2 illustrates key future perspectives of AI-powered robotic surgery:

- Progressed AI calculations for accuracy and personalization: AI frameworks will likely become more advanced, using deep learning and reinforcement learning to not only improve accuracy but also adapt to individual patient anatomy and surgical inclinations. These algorithms could anticipate and suggest ideal surgical ways,

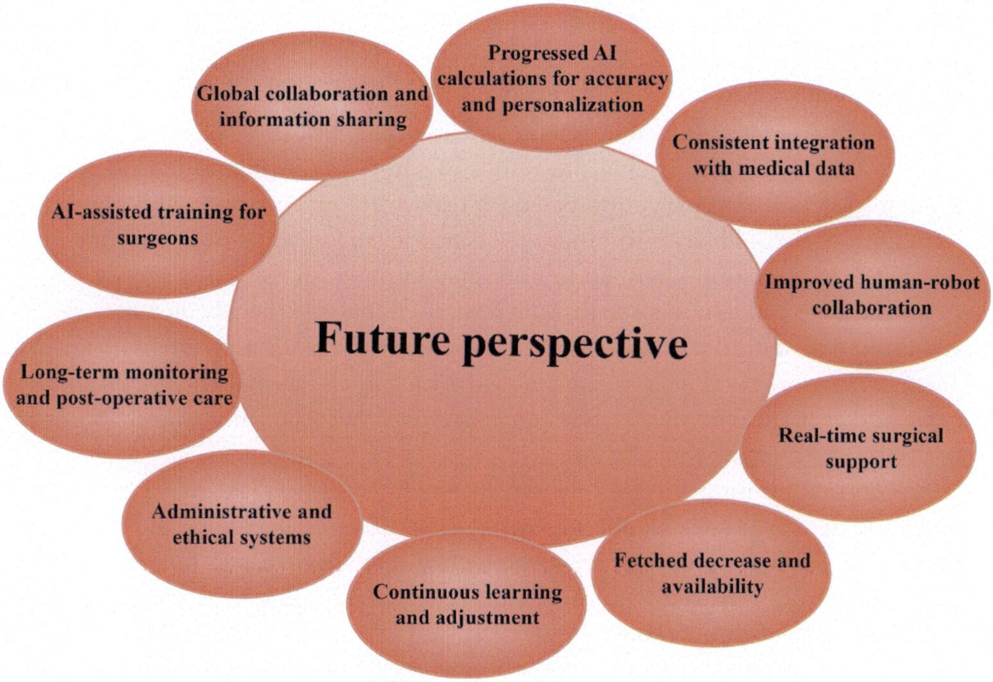

Figure 15.2 Future perspective.

diminishing human blunder and making strides in surgical results on a patient-by-patient basis.

- Consistent integration with medical data: In the future, AI-powered robots will integrate more effectively with diverse sources of medical data, including real-time imaging, electronic health records, hereditary data, and patient medical history. This integration will enable specialists to make better decisions, anticipate complications, and optimize treatment plans before, during, and after surgery.
- Improved human–robot collaboration: As robotic frameworks become more intuitive, human specialists will collaborate consistently with AI, utilizing progressed haptic input, expanded reality (AR), and real-time direction frameworks to upgrade their decision-making and execution. Instead of supplanting human surgeons, AI will become a powerful partner that expands human ability and judgment.
- Real-time surgical support: With advancements in AI's capacity to handle vast amounts of data quickly, real-time observing and help will be accessible, cautioning specialists to any deviations from the arranged strategy or potential dangers. This dynamic support system will help reduce complications and move forward with quiet security amid surgery.

- Cost reduction and accessibility: As AI-driven robotic systems become more widespread and affordable, surgical costs may decrease, making advanced care more accessible globally.
- Continuous learning and adjustment: AI-powered frameworks may persistently learn and advance by analyzing the results of surgeries, distinguishing designs, and making strides in their execution based on real-world information. This continuous learning cycle could lead to improved surgical strategies and better outcomes, optimized recuperation plans, and diminished complications.
- Administrative and ethical systems: The future will require the advancement of clear administrative and ethical guidelines to administer the utilization of AI in surgery. These frameworks will need to address issues such as responsibility, straightforwardness, and patient consent, while guaranteeing that the innovation is secure, successful, and impartial.
- Long-term monitoring and postoperative care: After surgery, AI may play a part in observing recuperation, identifying early signs of complications, and helping in restoration. Robotic systems may assist with minimally invasive follow-up procedures or give ceaseless criticism to healthcare experts, leading to faster recovery and improved long-term outcomes.
- AI-assisted training for surgeons: Robotic surgery frameworks may advance to back preparation for modern specialists, providing simulated environments to practice and refine skills. AI could monitor the trainee's progress, giving criticism, and guaranteeing they are arranged for real-life surgery with negligible dangers.
- Global collaboration and information sharing: With AI, surgical data could be shared globally, allowing surgeons from different regions to collaborate and learn from one another. Cloud-based systems could aggregate anonymized surgical data, fostering a global community dedicated to advancing robotic surgery and developing best practices.

15.10 Conclusion

AI-based robotic surgery holds immense potential to transform healthcare by improving precision, efficiency, and patient outcomes. However, its full realization is hindered by technological, ethical, regulatory, financial, and societal challenges. Addressing these barriers requires a multidisciplinary approach that includes advancements in AI technologies, robust regulatory frameworks, equitable financial models, and comprehensive education and training initiatives. Building trust among healthcare professionals and patients, ensuring data security, and mitigating algorithmic bias are essential steps toward the responsible integration of AI into surgical practice. Collaboration among governments, industry leaders, and the medical community will be critical to overcoming these obstacles and ensuring that the benefits of AI-based robotic surgery are safe, accessible, and equitable for all.

References

Ali M., Peña R.M.G., Ruiz G.O., Ali S. A comprehensive survey on recent deep learning-based methods applied to surgical data. arXiv, Mexico arXiv; 2022. Available from: https://arxiv.org. https://doi.org/10.48550/arXiv.2209.01435.

Bhatt S., Sharma S., Bhadula S. NeuroAI: emerging artificial intelligence and image processing techniques in neuroscience for enhanced medical diagnosis of brain tumor. In: 1st International conference on pioneering developments in computer science and digital technologies, IC2SDT 2024 – Proceedings; 2024. p. 204–209, Available from: https://doi.org/10.1109/IC2SDT62152.2024.10696499, http://ieeexplore.ieee.org/xpl/mostRecentIssue.jsp?punumber=10695942.

Boal MWE, Anastasiou D, Tesfai F, Ghamrawi W, Mazomenos E, Curtis N, et al. Evaluation of objective tools and artificial intelligence in robotic surgery technical skills assessment: a systematic review. British Journal of Surgery 2024;111(1). Available from https://doi.org/10.1093/bjs/znad331.

Brian R, Murillo A, Gomes C, Alseidi A. Artificial intelligence and robotic surgical education. Global Surgical Education - Journal of the Association for Surgical Education 2024;3(1). Available from https://doi.org/10.1007/s44186-024-00262-5.

Byrd IV TF, Tignanelli CJ. Artificial intelligence in surgery—a narrative review. Journal of Medical Artificial Intelligence 2024;7:29. Available from https://doi.org/10.21037/jmai-24-111.

Iftikhar M, Saqib M, Zareen M, Mumtaz H. Artificial intelligence: revolutionizing robotic surgery: review. Annals of Medicine & Surgery 2024;86(9):5401–9. Available from https://doi.org/10.1097/ms9.0000000000002426.

Knudsen JE, Ghaffar U, Ma R, Hung AJ. Clinical applications of artificial intelligence in robotic surgery. Journal of Robotic Surgery 2024;18(1). Available from https://doi.org/10.1007/s11701-024-01867-0.

Moglia A, Georgiou K, Georgiou E, Satava RM, Cuschieri A. A systematic review on artificial intelligence in robot-assisted surgery. International Journal of Surgery 2021;95:106151. Available from https://doi.org/10.1016/j.ijsu.2021.106151.

Nautiyal N., Agarwal P., Sharma S. Rechain: a secured blockchain-based digital medical health record management system. In: International conference on innovative trends in information technology, ICITIIT; 2023. Available from: https://doi.org/10.1109/ICITIIT57246.2023.10068707, http://ieeexplore.ieee.org/xpl/mostRecentIssue.jsp?punumber=10068516.

Qian C, Ren H. Deep reinforcement learning in surgical robotics: enhancing the automation level. Elsevier BV; 2025. p. 89–102. Available from http://doi.org/10.1016/b978-0-443-13271-1.00055-8.

Sharma, S., Chauhan, Y., & Tyagi, R. (2023). Artificial Intelligence based Applications in Medical Tourism. 2023 14th International Conference on Computing Communication and Networking Technologies (ICCCNT), 1–6. https://doi.org/10.1109/icccnt56998.2023.10308020.

Singh S, Sharma S, Bhadula S, Mohan S. Industry 4.0 internet of medical things enabled cost effective secure smart patient care medicine pouch. Springer Science and Business Media LLC; 2023. p. 149–70. Available from http://doi.org/10.1007/978-3-031-20443-2_7.

Thai MT, Phan PT, Hoang TT, Wong S, Lovell NH, Do TN. Advanced intelligent systems for surgical robotics. Advanced Intelligent Systems 2020;2(8). Available from https://doi.org/10.1002/aisy.201900138.

Vasey B, Lippert KAN, Khan DZ, Ibrahim M, Koh CH, Layard Horsfall H, et al. Intraoperative applications of artificial intelligence in robotic surgery: a scoping review of current development stages and levels of autonomy. Annals of Surgery 2023;278(6):896–903. Available from https://doi.org/10.1097/sla.0000000000005700.

CHAPTER 16

The road ahead: innovations and future trends in artificial intelligence medical robotics

Geetha Manoharan[1] and M. Nivedha[2]
[1]School of Business, SR University, Warangal, Telangana, India,
[2]School of Creative & Cultural Business, Robert Gordon University, Aberdeen, Scotland, United Kingdom

16.1 Healthcare revolution—the impact on artificial intelligence

The upward push of artificial intelligence (AI) has ushered in a brand-new technology of innovation across diverse sectors, with healthcare being an outstanding beneficiary. AI-powered tendencies in clinical robotics hold massive ability for revolutionizing surgical approaches and enhancing general healthcare transport (Shuaib, 2024). AI algorithms, coupled with improvements in sensor technology and mechatronics, are using the improvement of sophisticated robotic systems capable of performing complicated surgical duties with splendid precision and dexterity (Briganti and Le Moine, 2020). The incorporation of AI in clinical robotics additionally facilitates remote surgical procedures, permitting surgeons to perform on patients from a distance (Elendu et al., 2023). This capability has profound implications for providing healthcare and getting right of entry to underserved populations in far-flung or rural areas.

One promising location of development is the use of AI to research surgical videos and offer real-time remarks to surgeons. AI algorithms can pick out critical steps in a surgical treatment, assess the health practitioner's method, and offer guidelines for improvement. Furthermore, AI-powered robots can robotically adjust their actions based on real-time feedback from sensors, optimizing surgical performance and minimizing the risk of complications. The combination of AI into scientific medicinal drugs is creating an unheard-of possibility for improving patient care, optimizing workflow, and facilitating knowledgeable decision-making (Aravazhi et al., 2025). AI structures are being researched widely in healthcare programs, where they are being trained not simply from one data modality but from multivariate records generated throughout more than one clinical activity, such as imaging, genomics, analysis, and remedy undertaking, where institutions between these disparate modalities are not constantly well-understood (Serag et al., 2019).

AI algorithms can analyze big datasets, which include genomic facts, to tailor treatment plans to man or woman sufferers (Elendu et al., 2023). By predicting health

AI-Powered Developments in Medical Robotics
DOI: https://doi.org/10.1016/B978-0-443-27783-2.00002-8

risks and personalizing treatments, AI can decorate affected person effects and optimize healthcare useful resource allocation (Li et al., 2024). The upgrades in surgical accuracy introduced by AI-powered structures might also cause a paradigm alternate in surgical education. Surgical residents can hone their skills in a danger-unfastened, practical setting, thanks to digital truth and augmented fact simulations pushed by AI (Veluru, 2024). These simulations offer individualized remarks, allowing surgeons to boost up their learning and enhance their skill ability.

The potential of AI to automate repetitive and mundane obligations offers a vast possibility to enhance the usage of human assets in a far-wished direction (Akinrinmade et al., 2023). AI can investigate widespread quantities of complicated records far more quickly than people can, declaring details that a human eye may omit and making sure an objective and correct evaluation of the statistics (Diaconu et al., 2023). AI algorithms can also analyze patient facts to perceive people at excessive risk of growing into certain situations, allowing proactive interventions and preventive care strategies (Gala et al., 2024). Using AI in surgical structures has validated big benefits, together with a 15% reduction in operation time and a 10% growth in accuracy as compared to standard processes (Mizna et al., 2025). Another way AI improves affected person results is by way of helping with early contamination detection. By using and analyzing clinical information, AI algorithms can spot minute anomalies that might factor into the early stages of a disease. This allows for a spark-off intervention, which significantly complements remedy results.

The combination of AI algorithms into robot systems permits the development of sensible surgical gear that may adapt to various anatomical structures and tissue residences. This flexibility is precious in minimally invasive surgical operations, in which surgeons operate through small incisions using specialized units (Li et al., 2024). AI is getting used an increasing number of regularly in several scientific specialties to interpret statistics, diagnose illnesses, and assist medical specialists in achieving good patient consequences (Ghaffari et al., 2024). The usage of AI in healthcare has the capacity to lower an affected person's financial duties relating to their remedy and enables the screening and analysis of diseases (Li et al., 2024).

The capability of AI to enhance surgical results, grow patient protection, and streamline healthcare approaches is simple. Regardless of the capacity of blessings, there are ethical issues related to the usage of AI in healthcare (Jotterand and Bosco, 2020). Worries about information privacy, sets of rules bias, and the capability for task displacement need to be cautiously addressed to make sure the responsible and equitable implementation of AI in scientific robotics (Jeyaraman et al., 2023; Jha et al., 2023).

16.1.1 Machine learning algorithms for surgical planning

The improvement of AI-powered remedy-making plans structures leverages machine learning techniques to investigate large amounts of patient facts, considering factors like clinical history, anatomical versions, and remedy success rates (Dhopte and Bagde, 2023). Those systems can then generate customized surgical plans that optimize effects and reduce risks (Joshi et al., 2021). AI algorithms' capability to understand styles, flag outliers, and anticipate risks can greatly resource scientific planning. Gadgets getting to know algorithms have the capability to discover intricate relationships among variables, which will cause the development of greater accurate predictive fashions that enhance surgical making plans and decision-making (Maleki Varnosfaderani and Forouzanfar, 2024).

AI algorithms are critical for preoperative making plans, allowing surgeons to simulate tactics, anticipate capability-demanding situations, and optimize surgical strategies. AI algorithms facilitate the analysis of scientific pics, together with CT scans and MRIs, to create certain 3D models of the affected person's anatomy. AI is especially helpful in minimally invasive remedies because of its ability to modify to new inputs and generalize properly. The integration of AI into surgical making plans permits surgeons to move away from a one-size-fits-all technique and include customized remedy techniques that cater to the desires of every patient (Li et al., 2024).

16.1.2 Computer vision for enhanced surgical navigation

PC vision plays a pivotal role in improving surgical navigation via offering real-time photo guidance and anatomical represntation. AI-powered laptop imaginative and prescient algorithms can analyze surgical videos to become aware of vital anatomical systems, track tool actions, and provide surgeons with real-time remarks (Mascagni et al., 2022). The incorporation of PC imaginative and prescient in surgical robots improves surgical accuracy and protection by means of giving surgeons progressed imaginative and prescient and control. With the usage of six-degree-of-freedom lightweight robot fingers as actuators, infrared stereo cameras as visible sensors, and various surgical systems, image-guided surgical robot structures are getting more and more not unusual (Yuan et al., 2020). By means of employing state-of-the-art photo processing strategies, laptop vision algorithms can extract quantitative statistics from surgical scenes, presenting goal metrics for comparing surgical performance and identifying regions for development.

Surgeons can use photograph-guided technologies to look at anatomy that would otherwise be hidden due to the fact that traits in the imaging era, computers, and operative strategies (Brodke and Randolph, 2003). Surgeons need to work closely with the technical team that allows them to make additional development, as they should use their expertise to assess the generation's viability (Wilson et al., 2024). Overlays of pre and intraoperative imaging with the surgical console can deliver the doctor insightful information that can help with decision-making at some point with tough tactics (Pakhomov et al., 2019). It is viable to mix preoperative imaging facts with real-time intraoperative video streams thanks to the combination of laptop imaginative and prescient and augmented reality (AR). This allows surgeons to peer at virtual models of anatomical structures overlaid onto the surgical discipline, giving them a more thorough draw close of the surgical surroundings (Ruijters, 2021).

16.1.3 Artificial intelligence-driven robotic systems for enhanced surgical precision

AI-driven robot structures are revolutionizing surgical exercise by means of imparting more desirable precision, dexterity, and manipulation. Those structures combine the talents of surgical robots with the intelligence of AI algorithms to perform complicated surgical duties with extra accuracy and performance. AI-powered robot systems can research from vast quantities of surgical records and adapt their movements to optimize surgical results (Pugalenthi et al., 2025). Robot surgical treatment combines scientific technology, robotics, and engineering to decorate the precision of surgeons' moves during complicated techniques (Rivero-Moreno et al., 2023).

Robotic surgical treatment has been broadly adopted across surgical and clinical specialties, demonstrating its capability to improve hospital therapy globally as PC processing, optics, mechanics, and haptic comments technologies boost speed (Probst, 2023). AI algorithms are hired for numerous critical elements of the system, along with target localization, image fusion, needle trajectory planning, and real-time feedback. With the use of state-of-the-art sensors and actuators, AI-driven robot systems can perform minimally invasive surgeries with smaller incisions, resulting in much less ache, scarring, and healing instances for patients.

16.1.4 Computer vision in robotic surgery

Using AI-driven picture processing for the identification of the prostate in preoperative and intraoperative snapshots, with the purpose of automating the picture fusion method, in addition to the identification and localization of prostate lesions, is being explored (Maris, 2024). AI algorithms may be used to gain information concerning the suitable positioning of orthopedic implants and associated headaches (Gitto et al., 2024). AI strategies are expected to automate clinical workflow, integrate facts from a couple of domain names to beautify the decision-making manner, and generate diagnostic, prognostic, and predictive biomarkers. The development of novel sensors and actuators will facilitate the introduction of AI-driven robot systems capable of performing an increasing number of complicated surgical responsibilities with greater precision and autonomy.

16.1.5 The impact of artificial intelligence on medical education

The upward push of AI-powered scientific robotics has implications for clinical schooling and training. Clinical colleges ought to include management abilities in their curricula for the effective use of AI, further to communication and empathy talents essential for shelling out the substantial expertise attained with the assistance of AI, due to the fact AI will deeply affect the existing health system (Tokuç and Varol, 2023). The curriculum must be notably restructured to consist of AI equipment even while maintaining the crucial human components of medical education, such as patient interaction and ethical decision-making (Pohn et al., 2025). AI-incorporated clinical training opens new potentialities for improved teaching and learning studies and improved outcomes (Narayanan et al., 2023). New techniques of training and evaluation will be required as AI-driven surgical robots grow to be extra full-size in operating rooms to guarantee that surgeons are equipped with the use of those advanced technologies.

Simulation-based educational platforms can provide scientific college students and surgeons with possibilities to exercise surgical methods in digital surroundings, presenting insightful feedback and recommendations (Hamilton, 2024). Educators can create custom-designed getting-to-know-you stories that fit the wishes of each student with the aid of utilizing AI-stronger coaching gear like digital simulations and real-time feedback structures (Pohn et al., 2025). Those simulations can mirror more than a few surgical eventualities, enabling trainees to hone their technical skills and decision-making procedures without endangering patients (Narayanan et al., 2023). Future clinical specialists must be organized to include AI and machine-mastering breakthroughs, and clinical training ought to emphasize studies, skills beyond memorization, and a fee-based technique to medical exercise (Mir et al., 2023). The integration of AI into clinical training is poised to transform the manner in which physicians are trained, equipping them with the knowledge and skills essential to navigate the complexities of current healthcare (Xu et al., 2024).

16.1.6 Natural language processing for human–robot interaction

Additionally, NLP algorithms are enabling greater natural and intuitive conversation among surgeons and robot structures. Surgeons can use voice commands or natural language commands to control the robotics' actions, streamlining the surgical workflow and minimizing cognitive burden. NLP allows scientific robots to comprehend and interpret human language, facilitating seamless communication and cooperation between surgeons and robots. Because the capabilities of medical robots keep adapting, NLP will play an increasingly critical position in permitting extra intuitive and green human–robotic interplay. AI is being explored as a tool to aid scientific professionals in making defensible, rational selections about affected person care.

The improvement and use of AI technologies have the capability to dramatically exchange the transport of healthcare, enhance patient outcomes, and improve the effectiveness and efficiency of surgical operations (Mir et al., 2023). The creation of AI-pushed robots with state-of-the-art sensors and actuators will make it possible to carry out an increasing number of intricate surgical operations with higher degrees of accuracy and autonomy (Mir et al., 2023). The incorporation of AI into scientific training is poised to transform the way medical doctors are taught, giving them the information and skills essential to barter the intricacies of current healthcare (Glauberman et al., 2023).

In the end, to effectively leverage the blessings of AI in medicinal drugs, it's far more imperative that clinical experts acquire a thorough education on the thoughts and applications of AI. The intention ought to be to strike a collectively beneficial stability between the effective use of AI and automation and the human strengths and judgment of skilled physicians (Amisha et al., 2019). However, this innovative technology continues to be in its infancy, given the ethical and criminal problems surrounding "information" (Akinrinmade et al., 2023).

16.2 Data-driven techniques for enhanced surgical efficiency

AI-powered tendencies in clinical robotics have brought about the mixing of information-driven techniques that substantially decorate surgical performance. Gadgets getting to know algorithms are used to investigate widespread amounts of surgical records, which include preoperative photos, intraoperative video feeds, and postoperative results, to perceive styles and correlations that can enhance surgical planning, execution, and consequences (Faiyazuddin et al., 2025). Predictive analytics, powered by way of AI, enable surgeons to count on potential complications and make proactive choices to mitigate risks throughout the surgical procedure. Statistical analytics equipment that can procedure and extract beneficial data from big datasets has become vital as the quantity of affected person information produced in healthcare grows (Faiyazuddin et al., 2025). Those strategies also guide real-time selection-making, enabling surgeons to conform their method based totally on the patient's unique anatomy and physiological reaction (Chan and Zary, 2019).

The usage of AI in medical robotics holds substantial promise for improving surgical precision, minimizing invasiveness, and accelerating restoration instances for sufferers (Jarvis et al., 2020). AI algorithms can examine scientific photos at a faster pace and with accuracy than human radiologists, assisting in picking out diffused anomalies and manual surgical interventions (Chang, 2019). Surgical robots geared up with AI-powered navigation structures can help surgeons in correctly focusing on tumors or different lesions, minimizing harm to surrounding wholesome tissue. AI algorithms can also process widespread quantities of surgical records to pick out styles and correlations that could improve surgical strategies and consequences (Topol, 2019). The usage of AI in surgical treatment makes it feasible to gain unparalleled tiers of precision, personalization, and efficiency, which ultimately results in higher affected person outcomes (Han et al., 2024).

The fulfillment of AI-driven strategies in clinical robotics hinges on the availability of facts, nicely annotated datasets that may be used to educate gadgets getting to know models. Moreover, the development of robust and dependable AI algorithms calls for close collaboration among surgeons, statistical scientists, and engineers to ensure that the era is aligned with clinical needs and moral considerations. Hospitals are using records analytics to maximize using their operating rooms, body of workers, and equipment, doubtlessly decreasing wait instances and elevating patient satisfaction (Moreland et al., 2025). Affected person protection may be improved with AI algorithms that detect problems, accumulate new and existing facts, and assist quality development initiatives (Bates et al., 2021). The capability to gather and examine statistics successfully enables AI to provide insights into patient care and protection, resulting in real-time (Abukhadijah and Nashwan, 2024).

AI can look at giant volumes of medical facts and find styles and correlations, which could result in greener and more effective healthcare systems and better patient effects. In the scientific field, in which data is often fragmented, incomplete, unlabeled, or unavailable, acquiring large quantities of data categorized statistics might be hard due to the fact that AI algorithms want it to feature efficaciously (Al-Antari, 2023). To be able to create AI models that can be reliable and secure for usage in clinical settings, it is far more vital to handle these troubles via growing techniques for facts augmentation, transfer getting to know, and federated getting to know, in addition to guaranteeing statistics privacy and security. AI systems can help clinical experts in making sound, rational decisions approximately affected person care, offering another perspective to consider while making diagnoses and treatment plans.

Clinical records are at risk of issues that includes mistakes, omissions, a lack of standardized metadata standards, fusion-demanding situations, insufficient control techniques, nonstandardized cleansing approaches, and storage in unstructured formats like text and photographs, which makes fact control and integration extra hard (Al-Antari, 2023; Zhang and Zhang, 2023). It's miles crucial to clear up these issues with a view to completely realizing AI's potential to enhance healthcare. Standardizing data infrastructure, digitizing clinical statistics, and growing sturdy structures to shield patient records' confidentiality are vital for knowing the full potential of AI in healthcare (Basu et al., 2020). AI equipment ought to be carefully monitored for overall performance adjustments, as they may be utilized in unique environments, with numerous patient populations, and over time, as health facts can also trade. Continuous tracking makes certain AI systems live sincerely and dependably in actual international medical settings with the aid of permitting the detection and mitigation of biases, drift, and faults (Morone et al., 2025).

16.2.1 Predictive analytics for surgical outcomes

The combination of predictive analytics into surgical workflows represents a good-sized development in AI-powered scientific robotics, allowing surgeons to make more knowledgeable decisions and optimize patient consequences. By means of studying preoperative affected person information, together with demographics, medical records, and imaging results, AI algorithms can predict the probability of specific complications or adverse occasions throughout the surgical operation. Predictive fashions can also be used to customize surgical planning, tailoring the approach to the individual affected person's anatomy and physiological traits. The combination of AI into healthcare necessitates cautious attention to data privacy, safety, and moral implications (Yu et al., 2018). It's essential to place safeguards in the vicinity that defend patient confidentiality, guarantee compliance with regulations like GDPR and HIPAA (Yekaterina, 2024), and promote openness in AI decision-making tactics.

With the aid of anticipating capacity risks and headaches, surgeons can take proactive measures to mitigate those risks, together with adjusting surgical strategies, optimizing anesthesia protocols, or getting ready for ability interventions (Bin Akhtar, 2025). Predictive analytics also can be used to identify sufferers who are at excessive threat of readmission or different submit-operative complications, bearing in mind centered interventions to enhance affected person effects and decrease healthcare charges. (Yekaterina, 2024). AI has the capacity to revolutionize affected person safety via higher tracking, diagnosis, prediction, and management of affected person ailments, and trends in this vicinity have given healthcare vendors new equipment to decrease dangers (Epelde, 2024). Predictive analytics is a huge use of AI in healthcare that helps patients with forecasting, population fitness control, and resource allocation (Faiyazuddin et al., 2025).

16.2.2 Real-time data analysis in the operating room

The capacity to analyze records in real-time throughout all surgical approaches is another key area of innovation in AI clinical robotics. By way of integrating AI algorithms with surgical robots and imaging systems, surgeons can gain advantageous access to a wealth of statistics that can inform their decision-making and improve surgical precision. Real-time information evaluation can also be used to screen the affected person's physiological popularity during a surgical procedure, detecting subtle adjustments that can imply the onset of headaches. AI allows the evaluation of real-time information from diverse resources, consisting of imaging, robotics, and sensors, thereby equipping surgeons with complete insights for knowledgeable decision-making.

By continuously monitoring surgical overall performance and patient results, AI algorithms can perceive regions for improvement and offer comments to surgeons, fostering a tradition of continuous studying and improvement. In surgical procedures,

AI-powered robots can enhance accuracy, lower invasiveness, and shorten healing times for sufferers, for that reason redefining surgical precision. As AI will become an increasing number of incorporated into healthcare, it's far more vital that healthcare professionals receive the training and schooling vital to apply this technology in an ethical and accountable way.

16.2.3 Big data and surgical robotics

The incorporation of huge statistical analytics into surgical robotics is revolutionizing surgical procedures, enabling surgeons to make more informed decisions and optimize patient outcomes. With the aid of aggregating and reading sizeable quantities of surgical data, which includes affected person demographics, scientific records, surgical techniques, and consequence data, AI algorithms can identify patterns and correlations that would not be possible for human beings to detect (Faiyazuddin et al., 2025). This perception permits surgeons to customize surgical processes for stepped-forward results and to identify viable troubles earlier than or during operations, which in the long run improves affected person care. The usage of virtual, augmented, and mixed fact in clinical training and procedures, in addition to synthetic intelligence and robotics in healthcare, will all be blanketed (Yamin, 2018). This additionally consists of using huge statistics.

By way of identifying fine practices and standardizing surgical techniques, large-scale information analytics can reduce variability in surgical consequences and improve overall satisfactory of care. AI algorithms can use massive amounts of facts to predict affected persons' consequences after surgical treatment, permitting clinical personnel to personalize treatment regimens. Facts on surgical procedures, patient demographics, scientific history, and outcomes may be accrued and evaluated to locate tendencies and insights that could beautify surgical selection-making and patient effects (Mayol, 2023). To fully realize the ability of AI in healthcare, a robust infrastructure for information garage, processing, and analysis is important.

16.2.4 Enhanced surgical precision with artificial intelligence

AI algorithms are being developed to decorate the precision and accuracy of surgical robots, permitting surgeons to perform complicated tactics with more dexterity and manipulation (Yip and Das, 2018). AI-powered picture steering systems can offer surgeons with real-time comments on the region of essential anatomical structures, which include nerves and blood vessels, assisting them to avoid damage in the course of the surgical procedure. AI-powered robots can probably enhance surgical precision, lower invasiveness, and enhance affected person effects through real-time feedback and superior dexterity. AI is being integrated into a ramification of surgical specialties, including plastic surgical treatment, wherein it is being used for responsibilities like facial reconstruction, breast augmentation, and pore and skin most cancer excision (Kiwan et al., 2024; Duong et al., 2024; Farid et al., 2024). The use of AI in plastic surgical operations can result in greater unique and predictable outcomes, in addition to shorter recovery times for sufferers (Novotny et al., 2024).

By way of automating repetitive tasks and providing real-time comments, AI can lessen the threat of human errors and improve surgical effects. AI-powered robots can carry out intricate responsibilities with extra precision and consistency than human surgeons, decreasing the threat of complications and improving affected person effects (Sharma et al., 2024). These algorithms are capable of researching tricky records and simulating human cognition, which has caused a good-sized development in diagnosis, surgical planning, and remedy choices (Qin and Gu, 2023).

16.2.5 Artificial intelligence-enabled surgical navigation systems

AI-enabled surgical navigation systems are transforming the manner in which surgeons plan and execute complicated processes, imparting them with real-time steerage and visualization tools to improve surgical accuracy and protection. By way of integrating AI algorithms with medical imaging information, inclusive of CT scans and MRIs, these systems can create distinctive 3D fashions of the affected person's anatomy, allowing surgeons to visualize the surgical area in exceptional element. AI algorithms are used to research medical images and generate customized surgical plans that do not forget the affected person's precise anatomy and scientific records.

16.2.6 Autonomous surgical robots

The development of self-reliant surgical robots is one of the most interesting and swiftly evolving areas of AI medical robotics (Probst, 2023). Independent surgical robots have the capacity to revolutionize surgery by performing unique obligations or maybe whole tactics without human intervention. Using AI algorithms and superior sensor technology, those robots can adapt to converting conditions and making real-time decisions, for that reason paving the way for extra specific and green surgical interventions. But the development of truly independent surgical robots raises several moral and regulatory issues that should be addressed before those structures can be broadly followed.

Those structures use real-time records from sensors and cameras to navigate the surgical field, pick out anatomical structures, and keep away from important tissues. AI algorithms are educated to recognize and respond to unexpected occasions, along with bleeding or tissue damage, and to take corrective action without human intervention. The usage of AI to interpret clinical pics is turning into extra not unusual; however, its use in interventional images, like those used for the duration of surgical procedures, continues to be primarily untapped due to the fact current surgical AI structures are skilled in the use of facts collected after stay surgeries, which have boundaries customized to surgical methods.

16.2.7 Personalized surgical approaches

AI algorithms can examine patient information, together with medical records, genetic data, and imaging outcomes, to discover the gold standard surgical method for every individual. This personalized technique to surgical treatment can result in higher consequences and reduced headaches (Sardar et al., 2019). Via studying patient-particular records, AI algorithms can pick out the handiest surgical techniques and predict capability risks and headaches (Shan et al., 2020). Surgeons and laptop scientists are running collectively in a new field referred to as surgical data science with the purpose of using computational strategies primarily based on AI to investigate large amounts of surgical information if we want to highlight the advantages and downsides of surgical procedures and to develop clever approaches.

16.2.8 Integration of augmented reality and virtual reality

The combination of AR and virtual reality technologies with AI clinical robotics can revolutionize surgical education and intraoperative guidance. Augmented truth can overlay digital photos and records onto the real-world surgical subject, imparting surgeons with real-time steerage and visualization tools. Digital fact can create sensible simulations of surgical approaches, allowing surgeons to practice and refine their competencies in safe and managed surroundings.

16.2.9 Artificial intelligence-driven surgical skill assessment and training

AI algorithms are being evolved to evaluate surgical talents and provide personalized training hints (Mascagni and Padoy, 2021). Those systems can analyze surgical performance data, which includes device movements and tissue manipulation, to identify areas where surgeons can improve their technique. AI-powered surgical schooling systems can offer surgeons with real-time remarks and steering, assisting them to broaden the skills and understanding they want to perform complex procedures safely and effectively. Surgeons who have in no way used AI say that they could be open to the use in their surgical education and schooling.

Cloud computing and remote surgical treatment are emerging tendencies that have the potential to convert the delivery of healthcare, particularly in underserved areas. Cloud computing allows surgeons to access and percentage facts and resources from anywhere in the world, facilitating collaboration and far-flung sessions. Far-off surgical operations allow surgeons to carry out techniques on patients who are located some distance away, using robot systems managed over an at-ease network. AI-enabled robotics and telehealth are increasing the attainability of nursing care, which makes healthcare offerings and remote tracking of sufferers' fitness conditions more available. The worldwide marketplace for AI in healthcare is expected to attain $188 billion by 2030, which highlights the increasing effect of AI on the healthcare industry (Burrell, 2023).

16.3 Ethical and regulatory considerations

As AI medical robotics become more universal, it's far more important to address the ethical and regulatory issues associated with this technology. The usage of AI in healthcare is revolutionizing clinical practices and diagnostics, and to ensure that AI systems are used adequately and ethically, it's miles more critical to set up clear recommendations and guidelines (Elendu et al., 2023). The usage of AI in healthcare raises some moral worries, along with statistical privacy and security, algorithmic bias, and the ability for job displacement (Elendu et al., 2023; Mennella et al., 2024). Addressing demanding situations of moral issues, the shortage of standardization, and doubtful criminal liability are vital for the widespread use of AI in healthcare (Lambert et al., 2023). It's far more important to make sure that AI systems are transparent, responsible, and truthful, and that they are utilized in a manner that blessings all contributors of society (Ochasi et al., 2024). To ensure the moral use of AI in healthcare, it's critical to cope with informed consent, record privacy, algorithmic equity, and transparency (Jeyaraman et al., 2023; Naik et al., 2022; Pham, 2025; Gerke et al., 2020). Addressing those ethical dimensions is critical for fostering belief and ensuring the responsible implementation of AI and robotics in healthcare (Elendu et al., 2023; Farhud and Zokaei, 2021). Existing ethical frameworks need to be evaluated, gaps identified, and, in addition, research performed to establish ethical practices and guidelines for the accountable integration of AI and robotics in healthcare (Elendu et al., 2023).

Regulatory frameworks for AI in healthcare are still evolving, necessitating the improvement of clear pointers and standards to ensure patient safety and facts privacy (Jha et al., 2023). Corporations need to set up new requirements, an authentic policy, and safety guidelines concerning AI rollout in healthcare (Esmaeilzadeh, 2020). It is also vital to shield the affected person's information and privacy, which includes implementing sturdy security features and adhering to ethical concepts, inclusive of knowledgeable consent and facts minimization. It is essential to remember the viable results of AI at the healthcare personnel, ensuring that AI is used to reinforce human capabilities as opposed to replace healthcare professionals (Ahmed, 2024). AI structures are designed to offer a clinical body of workers with higher know-how and to assist them in making higher decisions. The purpose of AI is to address several of the activities that are now accomplished by using humans, liberating them as much as cognizance on better-degree duties and allowing them to characteristic at their great.

16.3.1 Cloud robotics for remote surgery

Cloud robotics, which integrates cloud computing with robotics, allows real-time fact processing and far-off acess to robotic structures, greatly increasing the possibilities for far-flung surgical procedures (Tang et al., 2023). Cloud robotics structures can shop and technique enormous quantities of records, allowing AI algorithms to analyze and adapt in real-time (Elendu et al., 2023). Cloud robotics allows surgeons to remotely manipulate robot systems from everywhere in the world, offering right of entry to specialized surgical information in underserved regions. This approach not only handiest complements surgical precision but also allows training and mentorship opportunities on an international scale.

16.3.2 Ethical and legal frameworks for artificial intelligence in surgery

The sizeable adoption of AI in surgical operations necessitates the development of complete ethical and legal frameworks to address problems inclusive of liability, statistics privacy, and algorithmic bias (Weiner et al., 2025; Mennella et al., 2024). It's miles crucial to set up clear pointers for the use of AI in surgical decision-making and to make sure that patients are fully knowledgeable approximately the risks and benefits of AI-assisted techniques. The protection and reliability of AI gear are crucial, which calls for the proper functioning of the technology and the absence of algorithmic bias (Hildt, 2025). Those frameworks need to additionally cope with troubles, which include data possession, highbrow belongings rights, and the capacity for misuse of AI-powered surgical gear.

Clinical ethics, including affected person autonomy, beneficence, nonmaleficence, and justice, is the muse of moral AI and robotics practices in healthcare, emphasizing the significance of patient welfare and equitable care (Elendu et al., 2023).

Moreover, patient facts privacy has to be protected through sturdy security measures and compliance with healthcare rules, in addition to duty mechanisms. Transparency is accomplished through explainable AI strategies that provide insights into AI selection-making procedures, which promotes acceptance as true and facilitates human oversight. Collaboration between ethicists, legal experts, healthcare experts, and policymakers is essential to create comprehensive recommendations that make sure AI and robotics are implemented responsibly in healthcare (Abujaber and Nashwan, 2024; Elendu et al., 2023).

The collaboration between surgeons and AI structures is important to enhance accuracy, reduce variability, and enhance affected person effects, but the integration must be carefully managed to ensure safety. Algorithmic bias wishes to be addressed by educating AI models on diverse datasets and constantly monitoring overall performance throughout exclusive patient subgroups (Bongurala et al., 2025).

Safeguarding the affected person's privacy will become vital to preserve agreement within medicinal drug and healthcare systems (Zhang et al., 2025). Most of the moral issues mentioned above have legal answers or ramifications; at the same time, there may be nothing sacrosanct in our division between the 2, we now shift to demanding situations we associate extra at once with the prison device (Gerke et al., 2020).

16.4 Ethical considerations and challenges

AI's use in healthcare raises numerous moral worries, which necessitate cautious attention and resolution. Those challenges include problems associated with privacy, information safety, algorithmic bias, and the ability for process displacement (Jeyaraman et al., 2023). AI algorithms should be obvious and explainable to healthcare professionals and sufferers, which allows them to recognize how selections are made and construct considerations inside the era (Tang et al., 2023). Algorithmic bias, which could lead to unfair or discriminatory outcomes, must be identified and mitigated by using schooling AI fashions on numerous datasets and constantly monitoring their overall performance across specific patient subgroups (Corfmat et al., 2025). The established order of accountability mechanisms is likewise important to decide who is answerable for the movements and selections made by using AI structures, especially in conditions where errors arise.

16.4.1 Challenges and opportunities

Notwithstanding the superb ability of AI-powered scientific robotics, numerous challenges need to be addressed to make sure of their secure and powerful deployment. Those challenges encompass the want for big, best datasets to train AI algorithms, the development of robust and reliable AI fashions, and the combination of AI systems into present medical workflows. The dangers of incorporating AI into healthcare, which encompass obligation attribution, ought to be cautiously analyzed. These dangers additionally embody problems inclusive of the need for AI validation, transparency, explainability, and the involvement of sufferers and clinicians in the improvement of AI (Vo et al., 2023). The legal and regulatory frameworks of healthcare AI in numerous areas are explored, emphasizing the need for coordinated, multidisciplinary projects to oversee the mixing of AI units in healthcare (Pesapane et al., 2021). Addressing those demanding situations will require collaboration among researchers, clinicians, and policymakers to set up clear guidelines and standards for the improvement and deployment of AI-powered scientific robotics.

16.4.2 Technical challenges

The mixing of AI into clinical robotics faces technical hurdles that need to be triumphed over to ensure reliability and effectiveness (Wei et al., 2025; Shandhi and Dunn, 2022). One of the predominant challenges consists of the necessity to explain AI selections to advantage the self-belief of the physicians. AI fashions are often created by using human beings without medical schooling; a result, cease customers do not have any impact over how the outcomes are acquired (Mudgal et al., 2022). Further studies are needed to discover subject matters of algorithmic bias and unfairness, even as growing mitigations to address those, to lessen brittleness and improve generalizability, and to develop techniques for stepped-forward interpretability of system gaining knowledge of predictions (Kelly et al., 2019). This venture includes growing AI algorithms that can take care of noisy and incomplete statistics, in addition to adapting to adjustments inside the surgical surroundings.

16.5 Opportunities for future research

Several opportunities exist for future studies within the discipline of AI-powered scientific robotics. This consists of the development of new AI algorithms for surgical making plans, navigation, and execution, in addition to the creation of more state-of-the-art human–robot interfaces that permit surgeons to interact with robots in a natural and intuitive manner. Research is required to demonstrate the capacity of AI-based totally technology to provide insights into precision remedy. Moreover, there is a need for more research on the moral and societal implications of AI in healthcare, in addition to the development of techniques to address those demanding situations.

16.6 Discussion and conclusion

AI-powered medical robotics has the capability to revolutionize surgical techniques, improve affected person results, and decorate healthcare transport. But a successful integration of AI in clinical robotics requires careful consideration of regulatory, ethical, and practical challenges (Han et al., 2024). With the aid of addressing those demanding situations proactively and embracing a collaborative technique, we can liberate the total capacity of AI in scientific robotics and pave the way for a destiny of improved surgical performance and advanced affected person care (Pohn et al., 2025). The issues of autonomy and obligation are brought up by way of the sizeable use of AI in the medical subject (Chen et al., 2023). The idea of knowledgeable consent is, in addition, complicated via using AI in figuring out affected person treatment or tactics (Saripan et al., 2021). Who bears duty, as an instance, if an AI system makes a wrong diagnosis or suggests a volatile path of motion? (Farhud and Zokaei, 2021). Using AI raises problems of duty, transparency, permission, and privateness due to the fact healthcare selections have traditionally been made nearly completely by people (Davenport and Kalakota,

2019). It's miles vital to have agree with in AI structures; accordingly, clinicians must be sure that AI systems can be trusted to provide the nice care of every patient. AI structures must be comprehensible and transparent, as well as have methods for managing errors, which will hold consideration (Tang et al., 2023).

AI-powered traits in scientific robotics keep big promise for reworking surgical processes and enhancing affected person results. Via leveraging information-driven strategies, AI can decorate surgical precision, efficiency, and safety, which leads to higher consequences for patients (Aravazhi et al., 2025). To maximize AI's capacity in healthcare, interdisciplinary collaboration, moral recommendations, and the safety of affected persons' rights are critical (Yelne et al., 2023). Destiny research will focus on developing explainable AI algorithms, incorporating AI with other technology, and investigating the long-time results of AI-pushed interventions. The development of AI in medication relies on researchers, doctors, and regulators running collectively to solve difficulties and guarantee moral and accountable implementation (Maleki Varnosfaderani and Forouzanfar, 2024; Li et al., 2024; Alowais et al., 2023).

As AI maintains to adapt, it has the potential to revolutionize scientific robotics and improve the lives of patients round the world (Akinrinmade et al., 2023; Davenport and Kalakota, 2019; Jotterand and Bosco, 2020; Wubineh et al., 2023). The use of AI in clinical remedy is unexpectedly increasing, but there may be a lack of oversight, law, and guidance on key factors of affected person-targeted care (Witkowski et al., 2024). Medical specialists want to emerge as acquainted with these breakthroughs so that it will enhance healthcare shipping to most people (Amisha et al., 2019). It's miles important to address those issues and foster consideration amongst stakeholders for you to completely realize the promise of AI in healthcare (Akinrinmade et al., 2023; Briganti and Le Moine, 2020; Elendu et al., 2023; Morone et al., 2025). AI can revolutionize healthcare by enhancing decision-making, boosting productivity, and cutting prices (Faiyazuddin et al., 2025). For AI to be used efficiently in medicinal drugs, healthcare data ought to be standardized, interoperable, and of excessive satisfactory quality (Goktas and Grzybowski, 2025). Addressing those demanding situations will pave the way for a future in which AI and clinical robotics work together to deliver more powerful, efficient, and customized healthcare for all (Gao et al., 2024; London, 2022; Murdoch, 2021; Kiseleva et al., 2022).

References

Abujaber AA, Nashwan AJ. Ethical framework for artificial intelligence in healthcare research: a path to integrity. World Journal of Methodology 2024;14(3). Available from https://doi.org/10.5662/wjm.v14.i3.94071.

Abukhadijah HJ, Nashwan AJ. Transforming hospital quality improvement through harnessing the power of artificial intelligence. Global Journal on Quality and Safety in Healthcare 2024;7(3):132–9. Available from https://doi.org/10.36401/JQSH-24-4, https://meridian.allenpress.com/

innovationsjournals-JQSH/article/7/3/132/501393/Transforming-Hospital-Quality-Improvement-Through.

Ahmed SK. Artificial intelligence in nursing: current trends, possibilities and pitfalls. Journal of Medicine, Surgery, and Public Health 2024;3:100072. Available from https://doi.org/10.1016/j.glmedi.2024.100072.

Akinrinmade AO, Adebile TM, Ezuma-Ebong C, Bolaji K, Ajufo A, Adigun AO, et al. Artificial intelligence in healthcare: perception and reality. Cureus 2023;15(9). Available from https://doi.org/10.7759/cureus.45594.

Al-Antari MA. Artificial intelligence for medical diagnostics—existing and future AI technology!. Diagnostics 2023;13(4):688. Available from https://doi.org/10.3390/diagnostics13040688.

Alowais SA, Alghamdi SS, Alsuhebany N, Alqahtani T, Alshaya AI, Almohareb SN, et al. Revolutionizing healthcare: the role of artificial intelligence in clinical practice. BMC Medical Education 2023;23(1) Available from https://doi.org/10.1186/s12909-023-04698-z, https://bmcmededuc.biomedcentral.com/.

Amisha, Malik P, Pathania M, Rathaur VK. Overview of artificial intelligence in medicine. Journal of Family Medicine and Primary Care 2019;8(7):2328. Available from https://doi.org/10.4103/jfmpc.jfmpc_440_19.

Aravazhi PS, Gunasekaran P, Benjamin NZY, Thai A, Chandrasekar KK, Kolanu ND, et al. The integration of artificial intelligence into clinical medicine: trends, challenges, and future directions. Disease-a-Month. 2025;71(6) Available from https://doi.org/10.1016/j.disamonth.2025.101882, https://www.sciencedirect.com/science/journal/00115029.

Basu K, Sinha R, Ong A, Basu T. Artificial intelligence: how is it changing medical sciences and its future? Indian Journal of Dermatology 2020;65(5):365. Available from https://doi.org/10.4103/ijd.ijd_421_20.

Bates DW, Levine D, Syrowatka A, Kuznetsova M, Craig KJT, Rui A, et al. The potential of artificial intelligence to improve patient safety: a scoping review. npj Digital Medicine 2021;4(1) Available from https://doi.org/10.1038/s41746-021-00423-6, https://www.nature.com/npjdigitalmed/.

Bin Akhtar Z. Artificial intelligence within medical diagnostics: a multi-disease perspective. Artificial Intelligence in Health 2025;0(0):5173. Available from https://doi.org/10.36922/aih.5173.

Bongurala AR, Save D, Virmani A. Progressive role of artificial intelligence in treatment decision-making in the field of medical oncology. Frontiers in Medicine 2025;12 Available from https://doi.org/10.3389/fmed.2025.1533910, journal.frontiersin.org/journal/medicine.

Briganti G, Le Moine O. Artificial intelligence in medicine: today and tomorrow. Frontiers in Medicine 2020;7 Available from https://doi.org/10.3389/fmed.2020.00027, journal.frontiersin.org/journal/medicine.

Brodke D, Randolph GB. Image guidance for spinal surgery. Operative Techniques in Orthopaedics 2003;13(3):152–8. Available from https://doi.org/10.1016/S1048-6666(03)00020-X.

Burrell DN. Dynamic evaluation approaches to telehealth technologies and artificial intelligence (AI) telemedicine applications in healthcare and biotechnology organizations. Merits 2023;3(4):700–21. Available from https://doi.org/10.3390/merits3040042.

Chan KS, Zary N. Applications and challenges of implementing artificial intelligence in medical education: integrative review. JMIR Medical Education 2019;5(1):e13930. Available from https://doi.org/10.2196/13930.

Chang A. The role of artificial intelligence in digital health. Springer Science and Business Media LLC; 2019. p. 71–81. Available from http://doi.org/10.1007/978-3-030-12719-0_7.

Chen C, Chen Z, Luo W, Xu Y, Yang S, Yang G, et al. Ethical perspective on AI hazards to humans: a review. Medicine 2023;102(48):e36163. Available from https://doi.org/10.1097/md.0000000000036163.

Corfmat M, Martineau JT, Régis C. High-reward, high-risk technologies? An ethical and legal account of AI development in healthcare. BMC Medical Ethics 2025;26(1) Available from https://doi.org/10.1186/s12910-024-01158-1, https://bmcmedethics.biomedcentral.com/.

Davenport T, Kalakota R. The potential for artificial intelligence in healthcare. Future Healthcare Journal 2019;6(2):94–8. Available from https://doi.org/10.7861/futurehosp.6-2-94.

Dhopte A, Bagde H. Smart smile: revolutionizing dentistry with artificial intelligence. Cureus 2023;15(6). Available from https://doi.org/10.7759/cureus.41227.

Diaconu C, State M, Birligea M, Ifrim M, Bajdechi G, Georgescu T, et al. The role of artificial intelligence in monitoring inflammatory bowel disease—the future is now. Diagnostics 2023;13(4) Available from https://doi.org/10.3390/diagnostics13040735, http://www.mdpi.com/journal/diagnostics/.

Duong TV, Vy VPT, Hung TNK. Artificial intelligence in plastic surgery: advancements, applications, and future. Cosmetics 2024;11(4) Available from https://doi.org/10.3390/cosmetics11040109, www.mdpi.com/journal/cosmetics.

Elendu C, Amaechi DC, Elendu TC, Jingwa KA, Okoye OK, John Okah M, et al. Ethical implications of AI and robotics in healthcare: a review. Medicine (United States) 2023;102(50):E36671. Available from https://doi.org/10.1097/MD.0000000000036671, https://journals.lww.com/md-journal/pages/default.aspx.

Epelde F. Revolutionizing patient safety: the economic and clinical impact of artificial intelligence in hospitals. Hospitals 2024;1(2):185–94. Available from https://doi.org/10.3390/hospitals1020015.

Esmaeilzadeh P. Use of AI-based tools for healthcare purposes: a survey study from consumers' perspectives. BMC Medical Informatics and Decision Making 2020;20(1). Available from https://doi.org/10.1186/s12911-020-01191-1.

Faiyazuddin M, Rahman SJQ, Anand G, Siddiqui RK, Mehta R, Khatib MN, et al. The impact of artificial intelligence on healthcare: a comprehensive review of advancements in diagnostics, treatment, and operational efficiency. Health Science Reports 2025;8(1) Available from https://doi.org/10.1002/hsr2.70312, onlinelibrary.wiley.com/journal/23988835.

Farhud DD, Zokaei S. Ethical issues of artificial intelligence in medicine and healthcare. Iranian Journal of Public Health 2021;50(11). Available from https://doi.org/10.18502/ijph.v50i11.7600.

Gala D, Behl H, Shah M, Makaryus AN. The role of artificial intelligence in improving patient outcomes and future of healthcare delivery in cardiology: a narrative review of the literature. Healthcare (Switzerland) 2024;12(4) Available from https://doi.org/10.3390/healthcare12040481, www.mdpi.com/journal/healthcare/.

Gao W, Liu Y, Yu Z, Cui D, Liu W, Liang X, et al. AI.vs.clinician: unveiling intricate interactions between AI and clinicians through an open-access database. arXiv, China arXiv; 2024. Available from: https://arxiv.org. https://doi.org/10.48550/arXiv.2406.07362.

Gerke S, Minssen T, Cohen G. Ethical and legal challenges of artificial intelligence-driven healthcare. Elsevier BV; 2020. p. 295–336. Available from http://doi.org/10.1016/b978-0-12-818438-7.00012-5.

Ghaffari M, Zhu Y, Shrestha A. A review of advancements of artificial intelligence in dentistry. Dentistry Review 2024;4(2):100081. Available from https://doi.org/10.1016/j.dentre.2024.100081.

Gitto S, Serpi F, Albano D, Risoleo G, Fusco S, Messina C, et al. AI applications in musculoskeletal imaging: a narrative review. European Radiology Experimental 2024;8(1) Available from https://doi.org/10.1186/s41747-024-00422-8, https://www.springer.com/journal/41747.

Glauberman G, Ito-Fujita A, Katz S, Callahan J. Artificial intelligence in nursing education: opportunities and challenges. Hawaii Journal of Health and Social Welfare 2023;82(12):302–5. Available from https://www.ncbi.nlm.nih.gov/pmc/articles/PMC10713739/pdf/hjhsw8212_0302.pdf.

Goktas P, Grzybowski A. Shaping the future of healthcare: ethical clinical challenges and pathways to trustworthy AI. Journal of Clinical Medicine 2025;14(5):1605. Available from https://doi.org/10.3390/jcm14051605.

Hamilton A. Artificial intelligence and healthcare simulation: the shifting landscape of medical education. Cureus 2024;16(5). Available from https://doi.org/10.7759/cureus.59747.

Han H, Li R, Fu D, Zhou H, Zhan Z, Wu Y, et al. Revolutionizing spinal interventions: a systematic review of artificial intelligence technology applications in contemporary surgery. BMC Surgery 2024;24(1) Available from https://doi.org/10.1186/s12893-024-02646-2, https://bmcsurg.biomedcentral.com/.

Hildt E. What is the role of explainability in medical artificial intelligence? A case-based approach. Bioengineering 2025;12(4):375. Available from https://doi.org/10.3390/bioengineering12040375.

Jarvis T, Thornburg D, Rebecca AM, Teven CM. Artificial intelligence in plastic surgery: current applications, future directions, and ethical implications. Plastic and Reconstructive Surgery - Global Open 2020;8(10) E3200, Available from https://doi.org/10.1097/GOX.0000000000003200, http://journals.lww.com/prsgo/pages/default.aspx.

Jeyaraman M, Balaji S, Jeyaraman N, Yadav S. Unraveling the ethical enigma: artificial intelligence in healthcare. Cureus 2023;15(8). Available from https://doi.org/10.7759/cureus.43262.

Jha D, Rauniyar A, Srivastava A, Hagos D.H, Tomar N.K, Sharma V, et al. Ensuring trustworthy medical artificial intelligence through ethical and philosophical principles; 2023. https://doi.org/10.48550/arxiv.2304.11530.

Joshi RS, Lau D, Ames CP. Machine learning in spine surgery: predictive analytics, imaging applications and next steps. Seminars in Spine Surgery 2021;33(2) Available from https://doi.org/10.1016/j.semss.2021.100878, http://www.elsevier.com/inca/publications/store/6/2/3/1/8/9/index.htt.

Jotterand F, Bosco C. Keeping the "Human in the loop" in the age of artificial intelligence. Science and Engineering Ethics 2020;26(5):2455–60. Available from https://doi.org/10.1007/s11948-020-00241-1.

Kelly CJ, Karthikesalingam A, Suleyman M, Corrado G, King D. Key challenges for delivering clinical impact with artificial intelligence. BMC Medicine 2019;17(1) Available from https://doi.org/10.1186/s12916-019-1426-2, http://www.biomedcentral.com/bmcmed/.

Kiseleva A, Kotzinos D, De Hert P. Transparency of AI in healthcare as a multilayered system of accountabilities: between legal requirements and technical limitations. Frontiers in Artificial Intelligence 2022;5 Available from https://doi.org/10.3389/frai.2022.879603, www.frontiersin.org/journals/artificial-intelligence#.

Kiwan O, Al-Kalbani M, Rafie A, Hijazi Y. Artificial intelligence in plastic surgery, where do we stand? JPRAS Open 2024;42:234–43. Available from https://doi.org/10.1016/j.jpra.2024.09.003.

Lambert SI, Madi M, Sopka S, Lenes A, Stange H, Buszello CP, et al. An integrative review on the acceptance of artificial intelligence among healthcare professionals in hospitals. npj Digital Medicine 2023;6(1) Available from https://doi.org/10.1038/s41746-023-00852-5, https://www.nature.com/npjdigitalmed/.

Li F, Wang S, Gao Z, Qing M, Pan S, Liu Y, et al. Harnessing artificial intelligence in sepsis care: advances in early detection, personalized treatment, and real-time monitoring. Frontiers in Medicine 2024;11 Available from https://doi.org/10.3389/fmed.2024.1510792, journal.frontiersin.org/journal/medicine.

London AJ. Artificial intelligence in medicine: overcoming or recapitulating structural challenges to improving patient care? Cell Reports Medicine 2022;3(5):100622. Available from https://doi.org/10.1016/j.xcrm.2022.100622.

Maleki Varnosfaderani S, Forouzanfar M. The role of AI in hospitals and clinics: transforming healthcare in the 21st century. Bioengineering 2024;11(4):337. Available from https://doi.org/10.3390/bioengineering11040337.

Maris B. Advancing robotic prostate biopsy through artificial intelligence. Medical Robotics 2024;2. Available from https://doi.org/10.54844/mr.2023.0488.

Mascagni P, Alapatt D, Sestini L, Altieri MS, Madani A, Watanabe Y, et al. Computer vision in surgery: from potential to clinical value. npj Digital Medicine 2022;5(1) Available from https://doi.org/10.1038/s41746-022-00707-5, https://www.nature.com/npjdigitalmed/.

Mascagni P, Padoy N. OR black box and surgical control tower: recording and streaming data and analytics to improve surgical care. Journal of Visceral Surgery 2021;158(3):S18. Available from https://doi.org/10.1016/j.jviscsurg.2021.01.004.

Mayol J. Transforming abdominal wall surgery with generative artificial intelligence. Journal of Abdominal Wall Surgery 2023;2. Available from https://doi.org/10.3389/jaws.2023.12419.

Mennella C, Maniscalco U, De Pietro G, Esposito M. Ethical and regulatory challenges of AI technologies in healthcare: a narrative review. Heliyon 2024;10(4):e26297. Available from https://doi.org/10.1016/j.heliyon.2024.e26297.

Mir MM, Mir GM, Raina NT, Mir SM, Mir SM, Miskeen E, et al. Application of artificial intelligence in medical education: current scenario and future perspectives. Journal of Advances in Medical Education and Professionalism 2023;11(3):133–40. Available from https://doi.org/10.30476/jamp.2023.98655.1803, https://jamp.sums.ac.ir/article_49364_14dacde02ba60a89c88c0918815e4b15.pdf.

Mizna S, Arora S, Saluja P, Das G, Alanesi WA. An analytic research and review of the literature on practice of artificial intelligence in healthcare. European Journal of Medical Research 2025;30(1). Available from https://doi.org/10.1186/s40001-025-02603-6.

Moreland R, Lyons C, Chandra S. AI-driven remote ECG monitoring and predictive analytics: transforming cardiac care from arrhythmia detection to perioperative optimization. American Journal of Healthcare Strategy 2025;1(3):32–41. Available from https://doi.org/10.61449/ajhcs.2025.8.

Morone G, De Angelis L, Martino Cinnera A, Carbonetti R, Bisirri A, Ciancarelli I, et al. Artificial intelligence in clinical medicine: a state-of-the-art overview of systematic reviews with methodological recommendations for improved reporting. Frontiers in Digital Health 2025;7 Available from https://doi.org/10.3389/fdgth.2025.1550731, https://www.frontiersin.org/journals/digital-health#.

Mudgal SK, Agarwal R, Chaturvedi J, Gaur R, Ranjan N. Real-world application, challenges and implication of artificial intelligence in healthcare: an essay. Pan African Medical Journal 2022;43 Available from https://doi.org/10.11604/pamj.2022.43.3.33384, https://www.panafrican-med-journal.com/content/article/43/3/pdf/3.pdf.

Murdoch B. Privacy and artificial intelligence: challenges for protecting health information in a new era. BMC Medical Ethics 2021;22(1). Available from https://doi.org/10.1186/s12910-021-00687-3.

Naik N, Hameed BMZ, Shetty DK, Swain D, Shah M, Paul R, et al. Legal and ethical consideration in artificial intelligence in healthcare: who takes responsibility? Frontiers in Surgery 2022;9 Available from https://doi.org/10.3389/fsurg.2022.862322, journal.frontiersin.org/journal/surgery.

Narayanan S, Ramakrishnan R, Durairaj E, Das A. Artificial intelligence revolutionizing the field of medical education. Cureus 2023;15(11). Available from https://doi.org/10.7759/cureus.49604.

Novotny M, Fast A, Radtke C. Transformative role of artificial intelligence in plastic and reconstructive surgery: innovations, applications and future directions. Artificial Intelligence Surgery 2024;4(4):376–86. Available from https://doi.org/10.20517/ais.2024.52.

Ochasi A, Mahamadou A.J.D, Altman R.B. Justice in healthcare artificial intelligence in Africa. arXiv, United States arXiv; 2024. Available from: https://arxiv.org. https://doi.org/10.48550/arXiv.2406.10653.

Pakhomov D, Premachandran V, Allan M, Azizian M, Navab N. Deep residual learning for instrument segmentation in robotic surgery. Lecture Notes in Computer Science 2019;11861:566–73. including subseries Lecture Notes in Artificial Intelligence and Lecture Notes in Bioinformatics https://doi.org/10.1007/978-3-030-32692-0_65, https://www.springer.com/series/558.

Pesapane F, Bracchi DA, Mulligan JF, Linnikov A, Maslennikov O, Lanzavecchia MB, et al. Legal and regulatory framework for AI Solutions in healthcare in EU, US, China, and Russia: new scenarios after a pandemic. Radiation 2021;1(4):261–76. Available from https://doi.org/10.3390/radiation1040022.

Pham T. Ethical and legal considerations in healthcare AI: innovation and policy for safe and fair use. Royal Society Open Science 2025;12(5) Available from https://doi.org/10.1098/rsos.241873, rsos.royalsocietypublishing.org/.

Pohn B, Mehnen L, Fitzek S, Choi KE, Braun RJ, Hatamikia S. Integrating artificial intelligence into pre-clinical medical education: challenges, opportunities, and recommendations. Frontiers in Education 2025;10 Available from https://doi.org/10.3389/feduc.2025.1570389, www.frontiersin.org/journals/education.

Probst P. A review of the role of robotics in surgery: to DaVinci and beyond!. Missouri Medicine 2023;120(5):389–96.

Pugalenthi LS, Garapati C, Maddukuri S, Kanwal F, Kumar J, Asadimanesh N, et al. From data to decisions: AI in varicose veins—predicting, diagnosing, and guiding effective management. Journal of Vascular Diseases 2025;4(2) Available from https://doi.org/10.3390/jvd4020019, https://www.mdpi.com/journal/jvd.

Qin F, Gu J. Artificial intelligence in plastic surgery: current developments and future perspectives. Plastic and Aesthetic Research 2023;10(3). Available from https://doi.org/10.20517/2347-9264.2022.72, http://doi.org/10.20517/2347-9264.2022.72 Available from.

Rivero-Moreno Y, Echevarria S, Vidal-Valderrama C, Stefano-Pianetti L, Cordova-Guilarte J, Navarro-Gonzalez J, et al. Robotic surgery: a comprehensive review of the literature and current trends. Cureus 2023;15(7). Available from https://doi.org/10.7759/cureus.42370.

Ruijters D. Artificial intelligence in minimally invasive interventional treatment. arXiv, Netherlands arXiv; 2021. Available from: https://arxiv.org. https://doi.org/10.48550/arxiv.2106.15306.

Sardar P, Abbott JD, Kundu A, Aronow HD, Granada JF, Giri J. Impact of artificial intelligence on interventional cardiology. JACC: Cardiovascular Interventions 2019;12(14):1293–303. Available from https://doi.org/10.1016/j.jcin.2019.04.048.

Saripan H, Mohd Shith Putera NSF, Abu Hassan R, Abdullah SM. Artificial intelligence and medical negligence in malaysia: confronting the informed consent dilemma. International Journal of Academic Research in Business and Social Sciences 2021;11(11). Available from https://doi.org/10.6007/ijarbss/v11-i11/11254.

Serag A, Ion-Margineanu A, Qureshi H, McMillan R, Saint Martin MJ, Diamond J, et al. Translational AI and deep learning in diagnostic pathology. Frontiers in Medicine 2019;6 Available from https://doi.org/10.3389/fmed.2019.00185, journal.frontiersin.org/journal/medicine.

Shandhi MMH, Dunn JP. AI in medicine: where are we now and where are we going? Cell Reports Medicine 2022;3(12) Available from https://doi.org/10.1016/j.xcrm.2022.100861, www.cell.com/cell-reports-medicine.

Sharma S, Kumari P, Sabira K, Parihar AS, Rani DP, Roy A, et al. Revolutionizing dentistry: the applications of artificial intelligence in dental health care. Journal of Pharmacy and Bioallied Sciences 2024;16(3):S1910. Available from https://doi.org/10.4103/jpbs.jpbs_1290_23, http://www.jpbsonline.org.

Shuaib A. Transforming healthcare with AI: promises, pitfalls, and pathways forward. International Journal of General Medicine 2024;17:1765–71. Available from https://doi.org/10.2147/ijgm.s449598.

Tang L, Li J, Fantus S. Medical artificial intelligence ethics: a systematic review of empirical studies. Digital Health 2023;9 Available from https://doi.org/10.1177/20552076231186064, https://journals.sagepub.com/home/DHJ.

Tokuç B, Varol G. Medical education in the era of advancing technology. Balkan Medical Journal 2023;40(6):395–9. Available from https://doi.org/10.4274/balkanmedj.galenos.2023.2023-7-79.

Topol EJ. High-performance medicine: the convergence of human and artificial intelligence. Nature Medicine 2019;25(1):44–56. Available from https://doi.org/10.1038/s41591-018-0300-7, http://www.nature.com/nm/index.html.

Veluru CS. Impact of artificial intelligence and generative AI on healthcare: security, privacy concerns and mitigations. Journal of Artificial Intelligence & Cloud Computing 2024;3(1):1–6. Available from https://doi.org/10.47363/jaicc/2024(3)347.

Vo V, Chen G, Aquino YSJ, Carter SM, Do QN, Woode ME. Multi-stakeholder preferences for the use of artificial intelligence in healthcare: a systematic review and thematic analysis. Social Science & Medicine 2023;338:116357. Available from https://doi.org/10.1016/j.socscimed.2023.116357.

Wei Q, Pan S, Liu X, Hong M, Nong C, Zhang W. The integration of AI in nursing: addressing current applications, challenges, and future directions. Frontiers in Medicine 2025;12 Available from https://doi.org/10.3389/fmed.2025.1545420, journal.frontiersin.org/journal/medicine.

Weiner EB, Dankwa-Mullan I, Nelson WA, Hassanpour S. Ethical challenges and evolving strategies in the integration of artificial intelligence into clinical practice. PLOS Digital Health 2025;4(4) Available from https://doi.org/10.1371/journal.pdig.0000810, https://journals.plos.org/digitalhealth/article?id=10.1371/journal.pdig.0000794.

Wilson JP, Fontenot L, Stewart C, Kumbhare D, Guthikonda B, Hoang S. Image-guided navigation in spine surgery: from historical developments to future perspectives. Journal of Clinical Medicine 2024;13(7):2036. Available from https://doi.org/10.3390/jcm13072036.

Witkowski K, Okhai R, Neely SR. Public perceptions of artificial intelligence in healthcare: ethical concerns and opportunities for patient-centered care. BMC Medical Ethics 2024;25(1) Available from https://doi.org/10.1186/s12910-024-01066-4, https://bmcmedethics.biomedcentral.com/.

Xu Y, Jiang Z, Ting DSW, Kow AWC, Bello F, Car J, et al. Medical education and physician training in the era of artificial intelligence. Singapore Medical Journal 2024;65(3):159–66. Available from https://doi.org/10.4103/singaporemedj.SMJ-2023-203, https://journals.lww.com/SMJ/pages/default.aspx.

Yamin M. IT applications in healthcare management: a survey. International Journal of Information Technology 2018;10(4):503–9. Available from https://doi.org/10.1007/s41870-018-0203-3.

Yekaterina K. Challenges and opportunities for AI in healthcare. International Journal of Law and Policy 2024;2(7):11–5. Available from https://doi.org/10.59022/ijlp.203.

Yelne S, Chaudhary M, Dod K, Sayyad A, Sharma R. Harnessing the power of AI: a comprehensive review of its impact and challenges in nursing science and healthcare. Cureus 2023;15(11). Available from https://doi.org/10.7759/cureus.49252.

Yip M, Das N. Robot autonomy for surgery. World Scientific Pub Co Pte Lt; 2018. p. 281–313. Available from http://doi.org/10.1142/9789813232266_0010.

Yu KH, Beam AL, Kohane IS. Artificial intelligence in healthcare. Nature Biomedical Engineering 2018;2(10):719–31. Available from https://doi.org/10.1038/s41551-018-0305-z, www.nature.com/natbiomedeng/.

Yuan HX, Li QQ, Zhao H, Wang LJ, Song R. An automatic calibration method for puncture surgical robot. IOP Conference Series: Materials Science and Engineering 2020;717(1):012008. Available from https://doi.org/10.1088/1757-899x/717/1/012008.

Zhang J, Zhang Z-M. Ethics and governance of trustworthy medical artificial intelligence. BMC Medical Informatics and Decision Making 2023;23(1). Available from https://doi.org/10.1186/s12911-023-02103-9.

Zhang F, Zhai D, Bai G, Jiang J, Ye Q, Ji X, et al. Towards fairness-aware and privacy-preserving enhanced collaborative learning for healthcare. Nature Communications 2025;16(1) Available from https://doi.org/10.1038/s41467-025-58055-3, https://www.nature.com/ncomms/.

Index

Note: Page numbers followed by "*f*" indicate figures and "*t*" indicate tables.

B

H

I

J

L

M

Q

R

S

T

Printed and bound by CPI Group (UK) Ltd, Croydon, CR0 4YY
21/09/2026
02219915-0011